Chemicals without Harm

Urban and Industrial Environments

Series editor: Robert Gottlieb, Henry R. Luce Professor of Urban and Environmental Policy, Occidental College

For a complete list of books published in this series, please see the back of the book.

Chemicals without Harm

Policies for a Sustainable World

Ken Geiser

The MIT Press
Cambridge, Massachusetts
London, England

© 2015 Massachusetts Institute of Technology

All rights reserved. No part of this book may be reproduced in any form by any electronic or mechanical means (including photocopying, recording, or information storage and retrieval) without permission in writing from the publisher.

MIT Press books may be purchased at special quantity discounts for business or sales promotional use. For information, please email special_sales@mitpress.mit.edu

This book was set in Sabon by Toppan Best-set Premedia Limited. Printed and bound in the United States of America.

Library of Congress Cataloging-in-Publication Data

Geiser, Ken.
Chemicals without harm : policies for a sustainable world / Ken Geiser.
   pages   cm
Includes bibliographical references and index.
ISBN 978-0-262-01252-2 (hardcover : alk. paper)—ISBN 978-0-262-51206-0 (pbk. : alk. paper)
1. Green chemistry.   2. Chemicals—Safety measures.   3. Chemical industry—Waste minimization.   I. Title.
TP155.2.E58G324   2015
660—dc23
                                                                          2014045725

10 9 8 7 6 5 4 3 2 1

To Dillon, Lindsey, and Maeve and to the future of our children

Contents

Preface

This book is about chemicals and chemical policy. It starts from the premise that there is a problem with chemicals—the economy that supports our lifestyle and has raised the prospects of millions of people around the world is based on hazardous chemicals. Chemicals are used in the production of the vast reservoir of products and services that lie at the heart of national economies. These chemicals have enriched the lives of people throughout the world and improved their productivity and enjoyment of life. However, many of these same chemicals can pose significant risks to human health and too often disrupt and compromise the careful balances of natural ecosystems. Over the last 50 years we have invested heavily in regulations, barriers, and protections to control hazardous chemical exposures, emissions, and effluents. But we have been like parents carefully draping children to protect them from the rain, and we have little noticed that the rain showers have turned into torrents. A control strategy for chemical use is costly and never perfect; it leaves open all the points in the life cycle of chemicals where chemicals leak and flow into air sheds, water bodies, and soils and where they persist, accumulate, transform, and reappear in different guises. Ten years of this, maybe twenty, but fifty or a hundred years of this strategy and the environment is so altered that there no longer exists anything that could be called a background level.

The problem with synthetic chemicals is similar to the problem with many other technologies—we rush to develop and enjoy the fruits of novel technologies, long before we fully understand their consequences and create the systems needed to address their costs. The hazardous chemical control strategy might have been a good one—if there had only been a few hundred really dangerous chemicals. But that did not turn out to be the case. Many chemicals present hazardous properties. There truly are really dangerous chemicals, but there are many others that present varied types and degrees of hazard, and there are, also, many that are relatively benign. We could spend the next fifty years trying to build better and better controls around more and more chemicals, or we could reconsider this strategy. If we are going to make big efforts to address the chemicals problem, it would be better if we worked to develop safer chemicals and moved thoughtfully and progressively to convert our economy to safer and more sustainable chemicals.

There is no assumption here that this is an easy path or that we can simply synthesize loads of safe chemicals and substitute them here and there in a myriad of applications. Transforming our economy to safer chemicals is a grand mission, and it will take more than a generation. But if we are going to ever create an economy that is safer and more sustainable, we should get moving, pull together the many forces that are working to build that economy, better integrate and support them, and develop the goals, plans, and policies that will guide us and keep us on track.

There are solutions to the chemicals problem. We can have safer products, safer workplaces, and safer communities. We can have a vibrant, innovative, and rewarding economy, and it can be founded solidly on highly effective synthetic chemicals—just new and different ones. We need a sustainable chemical industry and robust product manufacturing industries; they just need to be more thoughtfully directed to delivering safer chemicals and products. We need chemists, toxicologists, and environmental health scientists who are knowledgeable, creative, and committed to assessing and making chemicals that are safer and more sustainable.

During the 1980s, I helped to draft a series of bills that would become the Massachusetts Toxics Use Reduction Act, and for more than a decade, I had the opportunity to lead the Massachusetts Toxics Use Reduction Institute at the University of Massachusetts Lowell. For nearly twenty-five years I taught in an interdisciplinary academic department dedicated to occupational and environmental health and served in a university devoted to science and technical innovation. Working with others, I helped to establish the Lowell Center for Sustainable Production that has championed the development of safer chemical policy. During these years I joined many others in developing and managing several safer chemical state, interstate, and non-governmental organizations. Through all of this I have learned how to think about chemicals and industrial production from chemists, industrial product managers, business leaders, government agents, health scientists, environmental advocates, students, staff, and academic colleagues. I hope that I have served these people and institutions well, but I know that I have learned much from them and that all that experience has provided me a clearer understanding of what has worked and not worked in chemical policy and what might be more effective. This book is built on that learning.

This is not a study. It offers no new research. It is more of a proposal—or a set of proposals, some quite simple and easy to adopt and others well beyond what is currently likely. I am aware that the argument here suggests changes of such magnitude that it would take enormous forces and decades to achieve. However, I hope the ambition of these thoughts does not reduce the value of the recommendations. We need to think big and audacious ideas or we will never have a chance to realize them.

I use the concept of a mental frame or framework here. The idea of problem framing has long been of interest to me. In my graduate work, I was influenced by the work of Don Schon, Marty Rein, William Gamson, and Lisa Peattie, each of whom was interested in public controversies and the way in which action and practice are shaped by underlying stories that explain a problem and determine the way in which solutions are framed. I use that perspective here.

In this writing I have wandered far from my original training in environmental law and policy. That framework held a special place for federal government policy and government regulation, in particular. However, the government that I studied and engaged with for some forty years has changed dramatically. There are no big visions and grand gestures now. We are quite far from the Great Society Program, the Apollo Project, and the building of the National Interstate and Defense Highway System. Indeed, we have not seen a major new federal environmental statute in thirty years. I am aware of how difficult it would be in the foreseeable future to achieve major statutory changes in federal chemical policy. That is why much of what I describe here could be accomplished either through federal initiative or without it. So it is not surprising that I have looked elsewhere for the forces of change. What I have found is hundreds of local, state, corporate, trade union, nonprofit, and international initiatives that are struggling to advance a safer system of chemicals. So my challenge became more like an auto mechanic picking up components here and there and trying to conceive and assemble a new and different vehicle. Certainly, some of these ideas are more developed than others. However, even if the solutions that I suggest here do not seem practical or achievable, I hope that my effort to reframe the chemicals problem is useful for others who might see different solutions.

Like any writing that arises out of the lived history of the writer, the subject has been broadly shaped by colleagues. I am grateful for all of the ideas and support that I have received over these past several years. Many people have offered ideas and comments and provided time for interviews and substantive discussions. I thank Paul Anastas, Ingela Andersson, Monica Becker, Bjorn Beeler, Ann Blake, Bill Carroll, Richard Clapp, Cathy Crumbly, Buzz Cue, Clive Davies, Richard Dennison, Joe DiGangi, Mike Ellenbecker, Art Fong, John Frazier, Terri Goldberg, Elizabeth Harriman, Lauren Heine, Helen Holder, Tom Lent, Annie Leonard, Richard Liroff, Kaj Madsen, Tim Malloy, Rachel Massey, Gina McCarthy, Roger McFadden, Greg Morose, Marty Mulvihill, Kevin Munn, Dara O'Rouke, Pierre Quiblier, Margaret Quinn, Debbie Raphael, Meg Schwartzman, Alex Stone, Beverley Thorpe, Yve Torre, Howard Williams, Mike Wilson, and Martin Wolf. I am grateful for all the information, ideas, and comments; however the proposals developed here are my own.

Many other people took time to read chapters and review even quite crude text for which I am immensely grateful. For these reviews, I thank Mike Belliveau, Charlotte Brody, Ryan Bouldin, Amy Cannon, Gary Cohen, Sally Edwards, Joel Garrett, Jim Geiser, David Kriebel, Joanie Parker, Mark Rossi, Ted Smith, Joel Tickner, Bill Walsh, John Warner, and Rand Wilson.

By focusing on chemicals it should be clear that we are not just addressing a human health problem or an environmental problem. This is an industrial problem, a technology problem, a science problem, and sadly, in our country, a political problem. However, for every problem, there is an opportunity. We can do more and better. We need new and better policies, programs, and practices, and we need safer chemicals—effective, appropriate, and highly functional chemicals, if not without harm then at least with a lot less.

1

The Problem with Chemicals

Dr. John Warner sits in a light-filled office in a nondescript suburban office park north of Boston. He will tell you that 60 percent of the chemicals on the market today are dangerous and have no safer alternative as a substitute.

John is a chemist and one of the guiding lights of a dynamic international movement among chemists called green chemistry. He and his colleagues are challenging the fundamental way that chemists are trained and conduct their research. For John, it is not enough for a chemist to know organic and inorganic chemistry, chemical synthesis, and chemical process control. Green chemistry requires that chemists also understand toxicology, pharmacology, and environmental science as well as the public health and environmental laws and regulations that affect chemicals. However, even though he established the nation's first doctoral degree program in green chemistry at the University of Massachusetts's Boston campus, John grew frustrated with the academic approach to chemistry. Ten years ago, he left the university world to set up the Warner-Babcock Institute for Green Chemistry, a chemical research and development laboratory for promoting green chemistry in private industry.

Across the country in an upscale office building in downtown San Francisco, Dara O'Rourke describes a new vision for the commercial product market. Dara sees a day when shoppers around the world snap little cell phone photos of the bar codes of products that they are planning to buy and instantly read out the product ingredients and the color-coded scores that tell them how safe and socially acceptable the products are. These are not just abstract thoughts. Dara, a faculty member at the University of California at Berkeley, is the founder of *GoodGuide*, a new innovative, Internet-based consumer education tool. *GoodGuide* provides an "app" for mobile smart phones that allows consumers to see color-coded ratings on the health, environmental, and social effects of products before they purchase them.

Further down the San Francisco peninsula, Helen Holder sits in the corporate offices of Hewlett Packard (HP), one of the world's leading manufacturers of consumer electronics, and reviews hazard information on hundreds of chemicals that could become constituents in laptops and printers. Helen is the Corporate Materials Selection Manager for HP, and her job is to evaluate and recommend the safest and highest performing materials for the company's products. She has helped the firm's hundreds of component suppliers find alternatives to the lead, mercury, cadmium, and brominated flame retardants that are now banned in electronic products sold in Europe. Her work has made HP a leader in its industry and earned the company high marks on environmental activists' scorecards. She understands the powerful role that a corporation like HP can play in shifting suppliers to safer chemicals, and she uses that role to push for higher standards for the entire computer and electronics industry.

Charlotte Brody is a nurse by training and one of the founders of Health Care without Harm, an international network of health care practitioners and advocates engaged in making hospitals and clinics safer and more sustainable. Today, she directs the Health Initiatives at BlueGreen Alliance, a coalition of fourteen trade union and environmental organizations working to build a safer, cleaner, and more competitive economy. "Lots of people are getting more information about hazardous chemicals," Charlotte claims, "but workers were being left out." Working with several trade unions, she devised a new online chemical hazard communication tool that provides workers with the health and safety information they need to protect themselves and to negotiate for changes in workplace technologies that would reduce the use of hazardous chemicals.[1]

Roger McFadden is the Director of Sustainability for Staples, the world's largest office supply retailer. For Roger, the issue of hazardous chemicals in products is central to his business. "My company is committed to selling products that its customers know are safe and environmentally friendly. It's a matter of customer trust and social responsibility," Roger says with a big, gregarious smile. From Roger's office outside Denver he can see the majestic foothills of the Rocky Mountains. "Protecting the environment is the new icon of competitive advantage," he says glibly, not revealing how hard he is working to pressure Staples' hundreds of vendors to reveal the chemical constituents of the products that they offer Staples to sell to the firm's worldwide customer base.[2]

What links these committed individuals together is a vision of a more sustainable world that is free of many of the risks associated with the manufacture and use of toxic and hazardous chemicals. Each is a contributor to a transformation that is taking place in the way we conventionally think about chemicals and chemical policy. They are not alone. There are hundreds of scientists, corporate managers, environmental activists, and government leaders across the country who are fashioning a new approach to chemicals—a new, twenty-first-century approach to chemical design and management.

For years our conventional approach to chemicals has been wildly entrepreneurial and opportunistic. Today, there are literally thousands of synthetic chemicals used to make our clothing, cosmetics, personal care products, vehicles, electronic gadgets, household products, recreation equipment, and toys. Many of those who live in highly industrialized societies are able to live rich, long, and comfortable lives because of the products of the modern chemical industry. But these chemicals—the chemicals of the synthetic chemical revolution—also harbor a darker side. Many of these highly useful chemicals are also persistent, toxic, and dangerous to our health and the environment that supports us. As C. P. Snow once noted, "Technology…is a queer thing. It brings you great gifts with one hand, and it stabs you in the back with the other."[3]

This is a basic irony of technological development, but it is not destiny. Now, in the twenty-first century, we could change the course of the synthetic chemical revolution that we have inherited. We could be designing molecules, synthesizing compounds, mixing chemicals, and manufacturing products that are safe and compatible with human health and ecological systems. However, to do so will require a dramatic shift in how we think about chemistry, the chemical industry, and the government we expect to protect us.

1.1 Chemicals in the Environment and Us

The academic fields of chemistry and chemical engineering largely emerged from the scientific developments of the eighteenth and nineteenth centuries. Scientific knowledge and conceptual understanding of molecules and chemical reactions came from the early British work on gases and dyestuffs and the German work on coal-based chemical derivatives. During the nineteenth century, leading scientists such as Hilaire de Chardonnet, Alfred Nobel, Ernest and Alfred Solvey, John Wesley Hyatt,

Charles Martin Hall, and Herbert Dow pioneered new chemical synthesis and process innovations. By the early 1900s, the new chemical industries were producing a host of novel synthetic chemicals ranging from dyes and inks, to kerosene, aluminum, dynamite, sodium nitrate, ammonium cyanide, and chlorinated hydrocarbons, which were rapidly transforming everything from foods and drugs to domestic products and building materials.[4]

Industrialized nations began the twentieth century with a wealth of minerals that could be mined to make inorganic chemicals and coal and petroleum that could be processed to make organic chemicals. Two major World Wars later, and these economies have been transformed into gigantic commercial engines for converting synthetic chemicals into thousands of inexpensive, useful, and appealing commercial products. Corporations and government agencies poured large investments into chemical research and huge chemical manufacturing facilities that could pump out millions of pounds of polymers, solvents, fibers, paints, inks, mastics, resins, pharmaceuticals, and pesticides.

However, this impressive wave of chemical innovation brought problems. Some of these chemicals brought injury and disability to workers, and once they dribbled out into the environment, they contaminated the air and polluted the rivers. When Rachel Carson began *Silent Spring* during the late 1950s, she intended to examine the chemical threats to wildlife, but when she connected those hazards to human disease, she ignited a firestorm of public concern. Over the decades since then, public awareness and fear have focused on pesticides on foods, mercury in fish, dioxins in waste incineration, lead in paints, halogens in plastics, heavy metals in packaging, phthalates in cosmetics, and the buildup of persistent and bioaccumulative chemicals in soils, sediments, and our own human bodies. The newest research shows that people are exposed daily to low levels of a constantly changing, complex mixture of synthetic chemicals. The result is that most of us—those of us living in highly industrialized societies and those living in much less developed areas—carry within us a rich mixture of synthetic chemicals, many of which did not exist in our grandparents' time.[5]

It is not difficult to find the sources of these chemicals. Workers often experience chemical exposures at work. Although some of the most dangerous chemicals are reasonably controlled at U.S. worksites, there are many jobs in this country where workers still work with significant exposures to hazardous chemicals. Chemical processing, metal working, paper making, chemical product formulation, construction, mining, janitorial cleaning, and hairdressing expose workers to recognized carcinogens, reproductive toxins, neurotoxins, sensitizers, and many other chemicals of concern.

Consumers experience exposure to hazardous chemicals as they appear in household products. Formaldehyde, a recognized carcinogen, shows up in air fresheners, toilet bowl cleaners, and many household cleaning products. Toluene, a neurotoxin, is used in various adhesives, paint thinners, sealants, disinfectants, and nail polishes. Shampoos, cosmetics, soaps, and lotions may contain diethanolamine, a compound that can degrade into a nitrosodiethanolamine, a probable carcinogen. Garments may be dyed with azo dyes, treated with disinfectants, and coated with perfluorinated compounds, leaving small residues of potential carcinogens and endocrine disruptors. Percholoethylene, a probable carcinogen, is used to dry clean garments. Oil- and enamel-based paints, paint strippers, and paint thinners contain volatile compounds such as acetone, methylene chloride, and petroleum distillates. Biocides are added as preservatives in many formulated products. Upholstered furniture, sleepwear, and mattresses may be treated with brominated flame retardants. Phthalates, which are linked to birth defects and endocrine disruption, are used as plasticizers in some shower curtains, raincoats, shampoo bottles, furniture, and children's toys. Indeed, children's products are little safer than common domestic products. More than 17 million toys were recalled in 2007 because they violated federal lead paint standards. In 2010 the Consumer Products Safety Commission recalled 55,000 units of children's costume jewelry that contained high levels of cadmium, and 12 million promotional drinking glasses sold at McDonald's were recalled because of cadmium in the painted coating.[6]

As products enter our homes, the constituent chemicals can migrate from the product into the air we breathe, the food we eat, or the things we touch. The U.S. Environmental Protection Agency (EPA) has found levels of a dozen common organic compounds to be two to five times greater inside homes than outside. Studies taken of household air show that the dust that floats about in our houses is littered with persistent chemicals degraded from household furnishings and products. According to the Environmental Working Group, an environmental advocacy organization, the average American is exposed daily to more than 100 chemicals of concern in cosmetics and personal care products applied directly to the skin.[7]

In 2008 the federal Centers for Disease Control (CDC) tested blood and urine from a broad, national sample of Americans for 212 substances ranging from heavy metals to polycyclic aromatic hydrocarbons, dioxins, phthalates, and various pesticides and found traces of these chemicals widely dispersed throughout the nation's population. While brominated flame retardants were found in nearly all participants, 90 percent of the sample showed traces of bisphenol A, a production chemical used to make polycarbonate and epoxy resins. Perfluoroctanoic acid, a byproduct in the production of stain resistant and non-slip surface coatings, was found in a majority of participants.[8]

Once in us, some synthetic chemicals can cause acute and severe injury. In 2011 the American Association of Poison Control Centers reported that more than 2,334,000 people in the United States, including 1,145,000 children, sought help for harmful exposures to products containing hazardous chemicals, including personal care products, pesticides, hobby supplies, and paints.[9]

The chronic, longer term health effects of chemical exposures are more difficult to determine. However, the weight of evidence suggests that a concerning number of synthetic chemicals found in workplaces and products is associated with cancer, reproductive dysfunctions, developmental disorders, and immunological damage. Many pesticides and agricultural chemicals in common use contain known or suspected carcinogens and endocrine disrupters, and chronic, low-level exposure to some of these is linked with neurological, developmental, and other effects in children. Isocyanates, amines, formaldehyde, and aldehydes used in building products (paints, caulks, adhesives, foams, insulation) are known chemical sensitizers. Polybrominated diphenyl ethers, pentachlorophenol, and bisphenol A can interfere with adult hormonal activity, potentially resulting in developmental disorders among fetuses and children. Research on the health and environmental effects of the new generation of nano-scaled and synthetic biochemicals is just now emerging but already suggests a reasonable basis for concern.[10]

The National Research Council estimates that chemical exposures play a role in at least one in four cases of developmental disorders. Asthma has been linked to long-term exposure to urban air pollution and chemicals such as formaldehyde and phthalates. The U.S. President's Cancer Panel released a report in 2010 identifying many cancer-causing agents that are common in industrial, agricultural, medical, and military workplaces. Epidemiological studies of occupational exposures show strong correlations between exposures to chemicals such as benzene, asbestos, and arsenic and specific cancers.[11]

The harms from hazardous chemical exposure rest heaviest on vulnerable populations. Communities of color, indigenous peoples, and low-income communities bear a disproportionate burden of hazardous chemical exposure and the potential adverse health effects. Children are particularly susceptible to the effects of chemical exposures because their organs and immune systems are still developing.[12]

Chemical contamination also threatens the health of eco-systems and wildlife. A recent assessment of western national parks by the EPA found widespread chemical contamination with persistent, bioaccumulative, and toxic chemicals and endocrine-disrupting chemicals among several indicator species. A study of chemical contaminants in wild bird eggs in Maine found mercury and several chlorinated, brominated, and perfluorinated compounds, all known to cause adverse health effects in animals.[13]

This is only what we know. Although we have good science on some chemicals, for most substances, we do not have enough studies to know the potential hazards, and we certainly do not know the consequences of the multiple and continuous exposures to the broad mixes of chemicals that we get every day. Thus, we are faced with a complex chemicals problem. We rely on a host of chemicals that are central to our economy but could be dangerous to our health. However, instead of throwing our government and industrial talents to developing industrial chemicals that are safe and environmentally compatible, we have erected complex legal and physical infrastructures to ensure that people are not exposed to dangerous chemicals—at least not "unreasonably" exposed—and that the environment is not jeopardized—at least not "significantly."

1.2 Toxic Chemical Policies

The development of synthetic, organic chemicals has never been put to a democratic vote. No government has turned to its citizens and asked what chemicals should be part of their economy and what chemicals should not. Instead, governments rely largely on the private market to make decisions about the chemicals that go into products and make up the structural and functional materials of modern society. Each year the chemical manufacturing industries pump out thousands of synthetic chemicals, and thousands of large and small product formulators and manufacturers turn those chemicals into commercial, industrial, and agricultural products. The chemicals used to make these products are determined within the multiple transactions of the vast chemical market that links chemical manufacturers and product manufacturers.

Large and powerful interests guard that chemical market. With billions of dollars of chemical sales each year, those who own and invest in chemical manufacturing corporations have aggressively fought to limit government efforts to regulate the chemicals market. While government agencies are empowered by laws to regulate chemicals used in workplaces and formed into products and chemicals that show up in wastes and emissions, few chemicals are actually prohibited from manufacture and use. Instead, chemical manufacturing firms and their "downstream" manufacturing customers work within a loose legal environment of chemical control laws that offers exposure limits for some chemicals, restricted uses for others, testing and registration requirements, product labeling standards, and liability penalties for chemical damages.

With an estimated 30,000 chemical substances in use in industry, it would take a monumental effort by governments to identify, test, and regulate each of the chemicals used in workplaces and products. Indeed, for many chemicals there simply is not enough information to determine whether they pose risks substantial enough to warrant regulation. Until recently many of the largest volume production chemicals—those manufactured in volumes over a million pounds per year—had little or no health or environmental effects data publically available, and even today there is insufficient information on most of the substances manufactured at lower volumes.

Where adequate data do exist, current government policies often require extensive and costly risk assessments to determine human health threats, and when government officials do decide that regulations are necessary, they are often required to balance their risk management options against economic consequences to ensure that current industries are not overburdened financially. Thus, the government laws that have been passed to regulate chemical manufacture and use provide far less than comprehensive and protective vehicles for ensuring the safety of industrial and agricultural chemicals and the chemicals that show up as constituents in commercial products.

1.3   New Initiatives in Chemicals Policy

In 2006, the European Union enacted a new, far-reaching policy overhauling most of Europe's conventional laws for managing industrial chemicals. This broad policy shift marks a significant historical development in the procedures by which governments seek to manage the risks of chemicals. Commonly referred to as REACH (Regulation, Evaluation, and Authorization of Chemicals), this new regulation establishes a new European chemicals agency and sets out a comprehensive program to register all chemicals in commerce, test and evaluate several thousand chemicals of concern, and require special government authorization for the use of those chemicals of highest concern. It requires governments of the twenty-seven member nations of the European Union to rewrite and harmonize their differing national chemical management laws, and it creates a broad array of new responsibilities for firms manufacturing and using chemicals in Europe.

The fact that the European Union initiated such a major overhaul and saw it through to parliamentary approval is fairly remarkable considering the complexity of the policy and the significant opposition that it generated. Given that the European chemical market is the largest in the world, the procedures put forward under REACH have a significant impact on the global chemical market and, by default, have become an international standard.[14]

Today, there is a growing interest in the United States in going beyond a singular focus on toxic and hazardous chemicals and developing broader policies for managing chemicals. These chemical policies are comprehensive policies that address a broad range of chemicals and place a high priority on replacing higher hazard substances with lower hazard substances. Such policies transcend the chemical-by-chemical control policies of the past, which promoted efforts to phase out the most dangerous chemicals, but failed to consider their substitutes or the many other chemicals of concern. This focus on both hazardous and safer chemicals makes these chemical policies more proactive and more market changing than the narrowly conceived chemical control policies of the past.

Converting to aqueous cleaners and degreasers, switching to low-volatile coatings, transitioning to biobased mastics and adhesives, selecting non-chlorinated polymers, and converting to environmentally friendly inks and dyes are procedures encouraged by these new, safer chemicals policies. Indeed, rather than "chemical bans," the terms "chemical conversions" and "chemical substitutions" best characterize the active component of these policies.

Brand name manufacturers in the apparel, furniture, household products, automotive, and electronic industries are drawing up lists of hazardous chemicals to avoid and substituting safer alternatives where they find opportunities. Leading retailers in clothing, housewares, groceries, and office supplies are requiring that product manufacturers limit the use of chemicals of broad public concern in the products that they retail. The health care industry has begun to conduct professional reviews of the hazardous chemicals used in hospitals and clinics.

These safer chemical policies are putting pressure on the manufacture of highly hazardous chemicals, but they are also encouraging chemists to develop newer, safer chemicals such as lactic acid, citrates, methyl esters, cellulose, glycerol, sorbitol, and polysaccharides. There is a growing demand for new chemicals, new chemical processes, and nonchemical technologies that can substitute for the toxic legacy of the twentieth-century chemicals. Here lies a great opportunity for science and engineering. "Green chemistry" and "green engineering" are terms used to describe the development of these new, safer chemistries and chemical processes. Green chemistry offers a new perspective to conventional chemistry by seeking chemical synthesis that reduces or avoids hazardous molecules or processes. Green engineering encourages technical and process procedures that encourage recycling, avoid wastes, reduce resource consumption, and promote inherently safer technologies. Many university centers and corporate programs around the world are now promoting green chemistry and engineering.

Today, well into this new century, we can see initiatives that are pushing forward to correct the unresolved problems of the synthetic chemical revolution, and the results promise a healthy, new frontier. The transformation that is now just emerging is taking the lessons of some two hundred years of chemistry and chemical production and marrying them with a commitment to public health and environmental protection. The chemicals of the future are being fashioned today around concepts such as green, clean, benign, zero-waste, atom-efficient, safe, and sustainable chemistry. These changes are just beginning. However, there is no guarantee that this transition will succeed. Massive barriers and openly hostile resistances exist. But if we can reframe our approach to chemicals, reconsider the needs of our economy, and redesign the policies of our corporations and governments, we could be fashioning a new, twenty-first-century chemical industry and a safer, sounder, and more sustainable economy.

1.4   The Objectives of This Book

Can we create a truly safer chemicals market? Can we rid our economies of chemicals that endanger us? Can we develop and adopt new, safer, and more sustainable chemicals? Can we make chemicals without harm? This book offers broad outlines for a strategy and identifies and examines many initiatives that are currently under way to provide answers to these questions.

There is no need to argue that we are better off because of the advances generated by the synthetic chemical revolution—we are. Nothing in this book should be interpreted to suggest that we do not need high-performing and cost-effective chemicals; we just need different ones. Nor will we argue here that there are too many dangerous chemicals on the market. That case has been made by many others.[15] Instead of focusing on the problems that hazardous chemicals cause, this book seeks to identify potential solutions. The central question is how to preserve and extend the tremendous gains made possible by synthetic chemicals while better protecting human health and the environment. If we are serious about building a sustainable economy, then we need to become more thoughtful, more responsible, and more creative about what chemicals we make and how we use them.

This book is about a broad, international quest for a safer, sounder, and more sustainable approach to chemicals management. The book offers a new narrative about the chemicals problem—a story focused on the systems of chemical production and consumption—and it identifies an emerging movement determined to more effectively address the problems raised by dangerous chemicals. There is a central and positive argument to the book:

We can develop and use safer alternatives to the chemicals that threaten our health and environment; however, this will require a new chemical strategy focused on broad changes in science, the chemical economy, and government policy.

This book is about policy—chemical policy. However, it is also about science, politics, and the economy. We will need government and corporate policies that not only seek to reduce the use of toxic and hazardous chemicals but also to promote the development and adoption of safer, cleaner, and "greener" chemicals. We will need to transform the chemical manufacturing industries and move from a dependence on finite supplies of petroleum to more biocompatible chemicals secured from renewable resources. We will need to restructure the downstream product industries that manufacture, distribute, and sell products and manage those products once they become wastes. We will need new approaches to science, both the environmental health sciences that provide information about chemical hazards and the molecular sciences that can develop new chemicals and chemical processes that take into account environment and health effects.

The book considers all the chemicals on the chemicals market.[16] It makes no distinction among industrial chemicals, chemicals in commercial products, agricultural chemicals, pesticides, and food additives—these are all synthetic chemicals manufactured by the chemical industry. It makes sense to consider chemicals as a whole because they are all bound together in complex and inter-related chemical production and consumption systems.

The book largely focuses on the United States. This is quite arbitrary because the chemical market and the chemical industry are global. However, there needed to be some boundaries, and the United States has fallen so far behind international expectations that finding new directions for the United States offers a much needed contribution to international chemical policies.

The book argues for a broad and comprehensive strategy for converting the chemical economy by focusing on the chemical market, the chemical industry, and the transformation of chemistry. The argument begins with an examination of past government policies to control chemicals and argues that these policies were compromised by economic, procedural, political, and implementation issues. However, this history provides useful lessons on what worked and what failed. The text moves on to present an analysis of emerging government initiatives outside the United States that represent a new wave of chemical policies built on a new set of assumptions and capacities. Arguing that the solution to the chemicals problem lies in reframing the problem, a systems view of chemical production and consumption is used to identify general principles for a more comprehensive, hazard-based, and transformative policy framework. Following a brief overview of the chemical industry and its markets, the text moves on to examine three strategic fronts that are advancing the conversion of the current systems of chemical production and consumption toward safer chemicals. The chapters that follow develop a safer chemical policy framework that include processes for characterizing, classifying, and prioritizing chemicals; generating and using new chemical information; promoting transitions to safer chemicals; and developing safer chemicals. The book closes by considering the potential role for government in advancing these policies.

To convert the current outdated chemical control policies into safer chemical policies will require a significant transformation of laws, chemical, and product manufacturing and science. We can be respectful of the generations of scientists, industrialists, and government officials who devoted their careers to creating the rich and diverse array of chemicals that today provide the material foundation of modern economies. However, the task ahead for this century is not more commercially viable chemicals but safer and more sustainable chemicals. Indeed, the directions ahead for developing safer chemicals through basic chemistry, biochemistry, biotechnology, and nanotechnology offer many exciting challenges and innovative opportunities.

If John Warner, Dara O'Rourke, Helen Holder, Charlotte Brody, Roger McFadden, and hundreds of other nonconventional thinkers have their way, the chemistry of the twenty-first century will be far safer, cleaner, and more sustainable than the chemistry of the past. We can have chemicals with less harm, but it will take a substantial transformation of our economy and our society, and that is what this book is about.

I

Chemical Control Policies

[W]e need no longer remain in a purely reactive posture with respect to chemical hazards. We need no longer be limited to repairing damage after it has been done; nor should we allow the general population to be used as a laboratory for discovering adverse health effects. There is no longer a valid reason for continued failure to develop and exercise reasonable controls over toxic substances in the environment.

—U.S. Office of the President, Council of Environmental Quality, *Toxic Substances* (1971)

2

Regulating Hazardous Chemicals

The U.S. approach to managing toxic and hazardous chemicals relies heavily on government regulations and the judicial imposition of liability. The regulatory response to chemical management ranges from laws protecting the air, water, and ecosystems from the release of hazardous chemicals to laws protecting public health, such as those that regulate chemical exposures at work and in the marketplace. These are subject protection policies—they are intended to protect against the risks of chemical hazards. However, some laws focus more directly on chemicals and regulate their manufacture and use. These are the so-called chemical control policies.

Subject protection policies typically involve the setting of safety standards, the issuing of permits, the promotion of protective technologies, and the monitoring and inspecting of air (indoor or outdoor), water (water bodies or drinking water), or soils and sediment. Chemical control policies involve collecting chemical information, requiring chemical testing where information is insufficient, determining levels of potential harm, and placing restrictions on the manufacture or use of chemicals that may result in significant danger. This may involve the outright prohibition of the use of highly hazardous chemicals; however, in most cases, it involves restricting or setting conditions on the production, use, and disposal of those substances.

The chemical control laws differ from the subject protection laws because they are intended to address chemicals directly as substances, not as production pollutants, workplace hazards, emissions, or wastes. As such, they are seen as the first line of a chemical protection strategy—if truly dangerous chemicals are not manufactured or their use is highly restricted, then exposure to them and their emissions and wastes can be eliminated or effectively minimized. When these statutes were first enacted, they were seen as innovative and groundbreaking; today, however, these laws are perceived as antiquated and limited in their capacity to meet the public's rising expectations about chemical safety.

2.1   Chemical Regulation in the United States

The history of environmental protection laws in the United States begins well back in the nineteenth century. As the nation's economy industrialized, factory discharges polluted rivers, emissions from smokestacks fouled the air, workplaces exposed workers to hazardous chemicals, and misbranded and dangerous chemicals filled the commercial markets. States with heavy industries such as New York, Massachusetts, Pennsylvania, and New Jersey passed smoke and effluent abatement laws and set regulations outlawing the use of chemicals such as mercury, arsenic, and other well-recognized poisons in foods and medicines.[1]

However, the great advances in chemistry during the nineteenth and early twentieth centuries unleashed a flood of novel synthetic chemicals that dramatically transformed everything from agriculture to health care, transportation, communications, construction, and warfare. The early U.S. chemicals industry was largely an inorganic chemicals industry producing sodium carbonate, sodium hydroxide, and potash; however, by 1850, there were nearly 170 chemical production facilities in the country. Following the Civil War, the U.S. industry began to develop rapidly, with annual growth rates in excess of 6 percent. In 1900 there were just over 500 chemical establishments employing some 24,000 workers, but forty years later there were 2,000 chemical production establishments with some 90,000 employees. The First World War propelled the development of the nation's organic chemical production with significant developments in resins, fibers, coatings, inks, dyes, and medicines.[2]

The first widespread dispersive use of synthetic chemicals involved pesticides and fertilizers. Early commercial pesticides were based on lead, mercury, sulfur, copper, and arsenic. By the close of the century, new broad-spectrum pesticides appeared based on cyanide, sodium, calcium chlorate, and fluorine and bromine compounds. Chemically synthesized phosphorus, potassium, and nitrate fertilizers became widely available after World War I. During the 1940s, the broad-spectrum pesticides were replaced with more selective organoochloride- and organophosphate-based bactericides, fungicides, herbicides, and insecticides. These included dichloro-diphenyl-trichloroethane (DDT), 2,4,5-trichlorophenoxyacetic acid (2,4,5-T), chlorodane, endrin, malathion, parathion, and diazinon.

Between 1947 and 1960, the annual use of synthetic pesticides in the United States increased from 124 million pounds to 627 million pounds.[3]

Synthetic chemicals widened the market for conventional household and personal care products. The first synthetic toothpaste, Odol, was put on the market in 1903, and a synthetic detergent, Persil, appeared in 1907, while new hydrogenation processes reduced the costs of fats and oils used in the production of soaps, shampoos, cosmetics, and margarine. Synthetic paints and coatings, some laced with lead and cadmium, came onto the market for household use during the 1930s.[4]

Petrochemicals based on petroleum derivatives transformed the chemicals industry. The sale of automobiles and the fuel demands of the two World Wars drove the market for gasoline and aviation fuel, which in turn propelled rapid advances in petroleum drilling and refining. New catalytic techniques for fractioning oil led to large volumes of inexpensive by-products such as ethylene and propylene, the building blocks of petrochemicals. By the 1930s plastics were being used in toys, kitchenware, containers, furniture, and tools. Polyvinyl chloride (PVC) came onto the market in the 1930s, and during the next decade, polyester and polyethylene terephthalate (PET) fibers became available. Between 1939 and 1946, the annual production of synthetic resins and cellulosic plastics rose from 247 million pounds to 1.2 billion pounds, while polystyrene production went from 1 million pounds in 1939 to 150 million pounds in 1947.[5] By the 1940s Union Carbide was the nation's largest producer of petroleum-based chemicals, with leading products in plastics, pesticides, and agricultural chemicals. DuPont launched a series of commercially successful chemical products such as Teflon, Kevlar, Tyvek, Saran Wrap, Formica, and Naugahyde and synthetic fibers such as Dacron and Orlon. Dow Chemical transformed its chlorine production lines to become a world leader in styrene, polystyrene, ethylene dibromide, ethylene glycol, and magnesium chloride.[6]

The low cost and high performance of these compounds created materials that revolutionized the design and manufacture of industrial, agricultural, and commercial products. Quite rapidly these synthetic chemicals came to substitute for and replace conventional mineral-based and plant- and animal-derived chemicals in the product manufacturing markets. The result was an economy increasingly dependent on the synthetic chemical manufacturing industries and an ever-increasing chemical intensification of the economy.

Initially, the public was pleased with the vast array of new synthetic chemical-based products. "Better Things for Better Living...through

Chemistry," the DuPont slogan, was easily accepted by a growing consumer sector. However, slowly and tragically, the public began sensing dangers. Studies of workers were the first to document adverse effects. By 1910 Alice Hamilton had published her first studies on lead poisoning in the pottery and paint industries, and John Andrews had released findings of phosphorus poisoning in the match-making industry. But these studies implicated the chemical hazards of industrial workplaces; it would take more time for domestic exposures to be recognized. In 1933, Isadore Kallett and F. I. Schink published *100,000,000 Guinea Pigs* documenting the hazards of lead and arsenic and, three years later, Ruth deForest Lamb wrote *American Chamber of Horrors*, a chilling indictment of the food, drug, and cosmetic industries. By the 1940s, Elizabeth and James Miller had identified azo dyes as carcinogenic; during the 1960s, cyclamates, commonly used as artificial sweeteners, were discovered to induce cancer in rats; and in 1970, the first paper on the mutagenic effects on vinyl chloride monomer was presented at the International Cancer Congress.[7]

Increasingly, the public responded. Broadly based community opposition to the fluorination of drinking water supplies appeared during the 1950s, and a consumer movement arose during the 1960s over diethylstilbestrol (DES), a drug given to pregnant women to prevent miscarriages that was linked to vaginal cancer in the daughters of the women who had been exposed. In 1959, just days before Thanksgiving, the market for cranberries collapsed when traces of the herbicide, aminotriazole, a possible carcinogen, were found on the berries. During the following decade, activist lawyer Ralph Nader organized people around pesticide use, nuclear energy plants, and dangerous food additives. However, it was the publication of Rachel Carson's *Silent Spring* in 1962 that drove the chemical contamination issue from an episodic, marginal concern to a fully inflamed, national (indeed, international) public issue.[8]

Throughout much of the nation's history, environmental protection and occupational safety were viewed as responsibilities of the states, industries, and professional associations. Respecting its Constitutional limits, the federal government largely abstained from adopting national laws regarding environmental and occupational health. This changed dramatically during the 1970s. There emerged a strong critique of the inadequacy of the patchwork of state initiatives and a well-articulated desire for uniform national laws. The public concern and social activism of the 1960s and 1970s finally pushed Congress to pass bold, new national laws to regulate hazardous chemical emissions to the air and water, protect industrial workers, and manage industrial and domestic wastes. The federal Environmental Protection Agency (EPA) was created in 1970. That same year, the Clean Air Act and the Occupational Safety and Health Act (OSHAct) were enacted, and two years later the Clean Water Act was passed. In 1974, the Safe Drinking Water Act was passed, and two years after that Congress enacted the Resource Conservation and Recovery Act to manage municipal and industrial wastes.

These ambitious environmental protection laws provided the new EPA with authority to set ambient and emission standards and require permits for the release of hazardous chemicals to the environmental media. The OSHAct focused on protecting the work environment from the dangers of toxic chemicals and authorized a new Occupational Health and Safety Administration (OSHA) to set and enforce workplace chemical exposure standards.

Over this same period, the federal government enacted and amended laws focused directly on the manufacture and use of toxic and hazardous chemicals. Between 1972 and 1976, Congress amended and strengthened earlier laws intended to regulate chemical ingredients used in foods, drugs, and pesticides and passed the Toxic Substances Control Act and the Consumer Product Safety Act. Today these chemical control laws create the structural framework for the nation's chemical control policies.[9]

2.2 Regulating Chemicals in Foods, Drugs, and Cosmetics

Serious federal attention to chemicals used in foods arose when the Department of Agriculture's Division of Chemistry commenced early studies of food adulteration. Dr. Harvey Wiley, the Division's chief chemist after 1883, pioneered early studies of the hazards of food preservatives and lobbied for federal authority to inspect food processing. However, the revelations of filthy meatpacking plant practices by novelist Upton Sinclair and news accounts about the dangers of patent medicines by journalist Samuel Hopkins Adams finally drove Congress to pass the Pure Food and Drug Act in 1906. Although this law required the labeling of chemical constituents of some foods and drugs, it did not provide federal authority to regulate uses of those chemicals. Continued dissatisfaction with the provisions led to proposals for major revisions during the mid-1930s. Early drafts of the revisions were hotly opposed by the food and drug industries. However, public fears aroused by the deaths of some one hundred patients from a sulfanilamide drug formulated with untested diethylene glycol finally led to enactment of a new Federal Food, Drug, and Cosmetic Act (FFDCA) in 1938.[10] This law, plus additional amendments in 1962, 1976, and 1997, set the basic framework for the testing, use, and labeling of dangerous substances in foods and medicines. The amendments of 1962, which followed the discovery that thalidomide, a drug given to pregnant women, caused birth defects, expanded government authority to regulate the efficacy as well as the risks of pharmaceuticals.[11]

Today, toxic chemicals in food, drugs, medical devices, cosmetics, and personal care products are regulated under several federal laws administrated by the federal Food and Drug Administration (FDA). The FFDCA requires premarket approval of new drugs, food additives, and coloring agents and authorizes standards for levels of pesticides, naturally occurring poisons, and toxic additives in or on food products. In addition, the law establishes standards for chemical content in various products and defines departures from those standards as adulteration or misbranding. Chemical additives to foods, drugs, or cosmetics are considered adulterations unless the FDA specifically approves their use based on the evidence of safety submitted by the manufacturer. In making such approvals, the agency may specify conditions of use, the amount of a chemical substance in a product, and any required product labeling.

Chemical ingredients such as sweeteners, artificial flavors, colorants, and preservatives used in processed foods are regulated under the FDA's Center for Food Safety and Applied Nutrition. Food additives were addressed by the 1958 Food Additive Act requiring that they must be certified as "generally recognized as safe" (GRAS) or approved for specific uses. This statute also included what became known as the Delaney Clause after its Congressional sponsor, which prohibited the inclusion in any processed food of any chemical known to cause cancer.[12] Determinations by the FDA on the safety of chemicals in foods are based on test data submitted by manufacturers but may also be validated by independent laboratory analyses performed for the FDA. While the EPA sets tolerances for pesticides appearing in or on foods, the FDA is responsible for monitoring foods in interstate commerce and enforcing compliance with tolerances through its food inspection program.

The FFDCA requires premarket approval of all new or newly imported pharmaceutical products. While the FDA sets the standards for such tests, each drug manufacturer must conduct tests and present test data on both the effectiveness of the drug and its safety. The FDA's Center for Drug Evaluation and Research is responsible for evaluating both prescription and over-the-counter drugs and monitoring drug use for adverse effects. The approval process for a new drug application (NDA) requires a preclinical (animal testing) investigation, followed by a new drug application that includes a general investigation plan and information on pharmacology, toxicology, and manufacturing processes, before an applicant can initiate clinical trials. Clinical trials use human subjects to determine whether a drug is effective and what side effects it might pose. Following the completion of the trials, the agency sets the conditions determining production processes, product labeling, product advertising, and managing special uses. The use of such drugs is further managed through "prescriptions" written by trained physicians. Over-the-counter (OTC) drugs are regulated through conformance with specialized "OTC Monographs" that set out standards for acceptable ingredients, doses, formulations, and labeling.

Unlike the strict regulations applied to food additives and drugs under FFDCA, cosmetics are not subject to FDA regulations. Cosmetic manufacturers may use any ingredient unless the FDA proves it may be harmful; however, cosmetic suppliers are encouraged to participate in an industry-sponsored program for testing and reporting.

2.3   Regulating Pesticides

Passage of the Federal Insecticide Act of 1910, the nation's first federal pesticide law, was driven by farmers concerned about mislabeled and ineffective pest control products, and the law focused largely on setting standards for product quality. In 1947, the law was overhauled to form the Federal Insecticide, Fungicide, and Rodenticide Act (FIFRA), and its mission was extended to address the protection of public health and the environment. This broader law included herbicides, fungicides, and moldicides and required that all pesticides be appropriately labeled and registered with the Department of Agriculture. Coverage of the law was expanded in 1954 to address residues on food and expanded again in 1958 to cover pesticides as food additives. In a 1970 amendment, authority to regulate pesticides was transferred to the new EPA. In 1972, the Federal Environmental Pesticide Control Act completely reformulated the original FIFRA and provided more comprehensive authority to regulate the testing, registering, labeling, sales, use, and disposal of pesticides, and it is this set of authorities that provides the framework for pesticide regulation as it exists today.[13]

FIFRA authorizes the EPA to regulate a wide range of dangerous and lethal substances, repellants, and control agents ranging from insecticides to herbicides, defoliants, fungicides, and disinfectants in order to prevent "unreasonable adverse effects on human health or the environment." The law requires that pesticide manufacturers, formulators, or importers register the active ingredients of each pesticide with the EPA prior to shipment or sales of the product. In considering an application for registration, the agency requires evidence that the product will perform its intended function without adverse effects; will not cause unreasonable harm to non-target organisms, including humans, crops, livestock, and wildlife, or to the environment; and will not result in harmful residues on food or feed. To assess the hazards of pesticides, the EPA requires many tests, including specific acute, subchronic, and chronic carcinogenicity tests and tests to assess mutagenicity and pesticide metabolism.

FIFRA prohibits the sale or use of any pesticide unless it is registered and labeled to indicate approved uses and restrictions. In granting a registration, the EPA may set requirements on the marketing, distribution, use, and disposal of the pesticide, certify pesticides for specific uses, and restrict or prohibit those substances that pose a significant risk. Registrations are either for "general" or "restricted" use and are intended to be renewed every five years. The law authorizes the cancellation or suspension of registrations where registered pesticides are found to pose an unreasonable adverse effect on the environment or human health. If a pesticide is intended to be used on a food crop, the EPA must set a "tolerance" and condition the use of the pesticide such that no residue on foods exceeds the tolerance. In establishing each tolerance, the agency must set a level that provides a "reasonable certainty of no harm" taking into consideration the concentration of the pesticide and all potential dietary and non-food exposures.

The use and disposal of registered pesticides is regulated through product labeling requirements. Product labels defining appropriate practices must conform to EPA standards, and it is a violation of the law to use a pesticide "in a manner inconsistent with its labeling." While many pesticides are permitted for the open market, most substance registrations limit pesticide uses to trained application specialists. Enforcement of proper pesticide use by applicators is implemented through cooperative agreements with state agencies that manage training, record keeping, and inspection programs.

In 1993, the National Research Council released a study showing that pesticide residues on foods have disproportionate effects on children that were not accounted for by EPA's tolerances. This report, in addition to recognition of inconsistencies between FIFRA and FFDCA over pesticide standards, led Congress in 1996 to pass the Federal Food Quality

Protection Act (FQPA). The FQPA creates a single, health-based standard for all pesticide residues on food, with explicit attention to risks to children and infants. Because the standards cover the combined residues from different pesticides and mixtures, the EPA was required to reassess all previously established tolerances. In addition, the FQPA created incentives for nonchemical crop protection tools and set an expedited approval process for lower hazard pesticides.[14]

2.4   Regulating Industrial Chemicals

Chemicals used in industrial production were not directly addressed by the federal government until passage of the Toxic Substances Control Act (TSCA) in 1976.[15] TSCA was intended to regulate those chemicals and chemical mixtures that may present "unreasonable risks of injury to health or the environment." When the Act was enacted, it was heralded as a "major step forward in providing urgently needed authority to protect human health and the environment from dangerous chemicals"[16] It provided the EPA with new authorities to inventory existing industrial chemicals, manage the introduction of new chemicals to the market, require health and environmental testing of substances of concern, and restrict the manufacture and use of the most hazardous chemicals.

The passage of TSCA came at the close of the Congress's ambitious years of environmental protections laws and was hailed by Russell Train, then the administrator of the EPA, as "one of the most important pieces of 'preventive medicine' legislation" ever passed by Congress.[17] The basic concepts of the law were taken from a white paper drafted by the White House Council on Environmental Quality (CEQ) in 1971. In this paper, the CEQ argued that the most effective form of environmental protection would focus "on the pollutant rather than on the particular medium being polluted" and that regulating chemicals "before people are exposed" offers the best means to protect public health.[18] The CEQ envisioned a new comprehensive statute that would fill "gaps" left by the subject protection statutes and encourage the testing of chemicals, identify the most hazardous, and restrict their manufacture, use, and disposal.

The law covers all industrial chemicals with exceptions for chemicals covered by other statutes such as pesticides and drugs; however, "new chemicals" proposed for the market are defined separately from "existing chemicals" already on the market. To separate new and existing chemicals, chemical manufacturers and importers were required to add the identities of their existing chemicals to a special "Inventory of Chemical

Substances" by December 1979. Any chemical not on the inventory by 1980 was considered to be a new chemical and is required to undergo a special pre-manufacture review by the agency prior to its introduction into commerce, whereas existing chemicals need no such review.

Any manufacturer or importer proposing a new chemical must provide a "pre-manufacture notice" (PMN) to the EPA that includes available information on the manufacturing process, disposal methods, and potential health and environmental effects of the substance. The EPA's New Chemicals Program then has up to ninety days to determine whether the chemical might present an unreasonable risk by balancing the potential environmental and health effects against the socioeconomic benefits of the commercialization of the chemical. In making its approval decision, the EPA may ask for additional information and may extend the review period by another ninety days.

TSCA was to promote chemical testing, and the law permits the EPA to ask chemical manufacturers to conduct tests designed to reveal potential environmental and health effects if there is insufficient evidence to determine whether the substance poses an unreasonable risk. Where substances are found to present serious or widespread harm from cancer, gene mutations, or birth defects, the EPA is required to take "appropriate regulatory action." To set priorities for testing, TSCA established a special Interagency Testing Committee to review existing chemicals and to nominate up to fifty substances per year to a priority testing list.

TSCA was to encourage the generation of chemical information, and the law requires that chemical manufacturers or importers maintain records and prepare reports on chemicals and mixtures. This includes information on significant adverse reactions among employees and any other "information which reasonably supports the conclusion that such substance or mixture presents a substantial risk of injury to health or the environment." Under TSCA's "Inventory Update Rule" (now retitled as the "Chemical Data Reporting Rule"), any chemical manufacturer or importer of chemicals manufactured or imported in excess of 25,000 pounds per year must report on the quantities in a special national survey that is conducted every four years.

TSCA was to restrict or prohibit the manufacture, use, distribution, or disposal of a chemical substance or mixture that "presents or will present an unreasonable risk of injury to health or the environment." To do so, the agency must first evaluate the health and environmental effects of the chemical, the extent of potential exposure, the benefits of the chemical, the availability of substitutes, and the potential economic impacts of any such restriction. Furthermore, the EPA is required to use the "least burdensome" remedy to protect public health or the environment and to refer actions to other agencies if the risk may be prevented or reduced to a sufficient extent under other statutes.

2.5 Regulating Chemicals in Consumer Products

Toxic chemicals that appear in consumer products are regulated under several laws administered by the Consumer Product Safety Commission (CPSC).[19] Congress passed the Consumer Product Safety Act (CPSA) in 1972 amid a broadly based consumer protection movement. Since the 1950s, the Congress and the states had been passing various product-specific safety laws covering products such as flammable fabrics and poison prevention packaging. Concerned over the absence of a coordinated approach, Congress established the National Commission on Product Safety in 1967 to propose a comprehensive product safety law. Drawing on language from the Commission's final report, the law was designed to protect the public against unreasonable risks associated with consumer products, develop uniform safety standards for products, and study and prevent product-related illnesses and injuries. In order to carry out these functions, the CPSA established a five-member Consumer Product Safety Commission (CPSC) appointed by the president, to consolidate federal regulatory provisions regarding consumer products. The law and the Commission preempted existing state product safety laws with the intention of creating standardized national product safety regulations, which would balance the cost of meeting those standards with the intended gains in safety.

Under provisions of an earlier statute, the Federal Hazardous Substances Act, the CPSC is authorized to require cautionary labeling of hazardous household products that might cause personal injury or illness as a result of reasonably foreseeable handling, use, or ingestion. In addition, the law allows the commission to ban or regulate hazardous substances such as fireworks, charcoal, antifreeze, turpentine, cleaning fluids, alcohols, and fire extinguishers where reasonable labels may not provide sufficient protections. Any toy or product intended for use by children can also be banned if it contains a hazardous chemical accessible by a child.

The CPSC carries out its mandate by developing voluntary standards, issuing mandatory standards, requiring product labeling, requiring the recall of products or arranging for their repair, conducting research on potential product hazards, and providing consumer product safety information. The CPSC selects a product hazard for review based on public petitions, referrals from other agencies, congressional requests, or staff initiatives. Once a priority is set for a particular product hazard, the commission staff goes through a fairly complex process of rule-making before the proposed standard is submitted to the commissioners for a vote.

Over the years the CPSC has drafted a series of product safety standards, including standards pertaining to hazardous chemicals in products. Wherever possible the commission tries to work with industry in setting these standards and relies heavily on standards adopted by various private, voluntary standard setting organizations such as the American National Standards Institute, the American Society for Testing Materials, and Underwriters Laboratory.[20]

For years the CPSC's rather tepid approach to product safety has led to public concern, but stories of contaminated household products and dangerous toys, many from foreign manufactures, mushroomed after 2000 and led to demands for reforms of the CPSA. In 2008, Congress passed the Consumer Product Safety Improvement Act that amends the CPSA and imposes new testing and documentation requirements for a specified range of products. This amendment severely restricts the use of lead and phthalates in many children's products, calls for independent testing of all children's products, and requires supplier certifications of conformity on all products required to meet CPSC standards.

2.6   A Chemical Control Policy Framework

By the close of the 1970s, the federal government had a broad mix of policies for addressing hazardous chemicals. Collectively, these policies created a general policy framework based on the concept of control. Although each of the laws emerged from a unique history, they all started with the premise that some substances present unreasonable risks and need to be controlled, primarily through the use of government regulations. The concept of control is a central theme. The laws vary on how the risks are to be determined and what criteria are to be used to determine a reasonable (acceptable) risk form and unreasonable (unacceptable) risk. However, where unreasonable risks are identified, the laws authorize government controls ranging from restrictions on marketing and use, conditions for special handling and applicator training to requirements for product labeling and outright prohibition on chemical or product marketing and importation.

While there are differences among these laws, there is a policy framework here, and it is built on three common building blocks. These components are displayed in figure 2.1.

Characterize and Prioritize Chemicals

Three of the chemical control laws require government notification by firms that manufacture or import chemical substances. The information required in these notifications typically includes chemical identification by chemical name, code and trade name, and some chemical and physical characterization information. In addition, the laws may include the identification of the manufacturer or importer, the volume manufactured or imported, and various types of hazard and exposure information. The laws differ considerably on the definitions of chemical substances, mixtures of chemicals, chemical preparations, and chemicals in articles and how much information is required in each instance.

Under the terminology of these laws, the terms "hazard" and "risk" are differentiated. The hazard of a chemical refers to the inherent characteristics of the substance (e.g., flammable, reactive, corrosive, toxic) that are

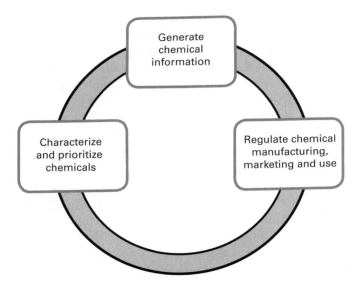

Figure 2.1

Building Blocks of the Chemical Control Policy Framework

likely to cause adverse effects under the conditions in which it is produced or used. Risk, in contrast, is a statistical concept that denotes the expected frequency of undesirable effects arising from exposure to a hazardous chemical. In determining a risk, both the inherent hazards of a chemical and the likely exposures are considered. Toxicity is one of the characteristics that leads to the hazardous properties of a chemical, but nowhere in the chemical control laws has Congress specifically defined what is meant by the term "toxic."

The laws differ in the way they characterize hazards. For instance, the Federal Hazardous Substances Act provides a definition of a hazardous substance that includes the inherent characteristics but only if the substance "may cause substantial personal injury or substantial illness." TSCA defines an "imminently hazardous chemical substance or mixture" by referring to risk as in "a chemical substance or mixture which presents an imminent and unreasonable risk of serious or widespread injury to health or the environment." Under the FFDCA, all detectable levels of pesticide residues on foods are considered unsafe unless a certain tolerance level has been established. In setting these tolerance levels, the EPA uses the toxicological evidence and the potential for exposure presented in each pesticide registration application to assess the risks (see box 2.1).

While the burden for demonstrating the safety of drugs under FFDCA falls on the drug manufacturer, TSCA, FIFRA, and CPSA place the burden of identifying the hazards and determining risks squarely on the government agencies. Based on public information, information supplied by manufacturers or information derived from their own testing the EPA and CPSC must characterize the hazards of each substance, identify the potential routes and magnitude of exposure, and calculate the risks in order to decide whether and how to regulate the substance.

TSCA establishes an interagency process to prioritize chemicals for testing and a requirement to select a minimum number of chemicals for testing, although there is no guidance on prioritizing chemicals for regulation. The CPSA leaves it up to the public media and public petitions to assist the CPSC in establishing priorities. In contrast, the FFDCA and FIFRA provide no prioritization criteria: all drugs, medical devices, and pesticides must be tested.

Generate and Make Available Chemical Information

Each of the four laws provides for the generation and reporting of chemical information, although they differ on the types of information required. Under TSCA, an industrial chemical manufacturer or importer must

Box 2.1
Risk and Risk Assessment

Neither FIFRA, CPSA, nor TSCA define risk, but the laws clearly intend consideration of a chemical's hazards and exposures in a manner that can be determined by a scientifically based logic. In the absence of a clear definition, it fell to the courts to clarify the mechanisms. In 1980, the Supreme Court did just that. In the *Industrial Union Department, AFL-CIO v. American Petroleum Institute* (1980), a case involving occupational safety standards, the court held that where the government is setting chemical safety regulations and there are "no safe levels," it must demonstrate a "significant risk" based on "substantial evidence" that includes an assessment of the hazards and probable exposures. Thereafter, the EPA and other government agencies charged with managing dangerous chemicals have considered risk in standard setting and relied on formalized risk assessment processes.[21]

The procedures of risk assessment follow from a 1983 National Research Council report on managing risks—the so-called "Red Book" (for the color of the cover). This reference separates the science-based (allegedly objective and value-free) procedures of risk assessment from the policy-oriented (permissively subjective and practical) procedures of risk management. With regard to chemical safety, risk assessment combines information on the hazardous characteristics of chemicals with information on the potency of the hazard and the potentially foreseeable pathways of exposure in such a way as to present estimations on the likelihood of adverse effects. The hazard assessment involves consideration of the chemical, physical, and toxicological properties of a substance. A "dose/response" relationship is used to identify the effects of large chemical doses and to estimate the intensity of effects at small doses. The exposure assessment addresses routes of exposure, metabolic pathways, and the differential susceptibility of potentially exposed populations. Based on these tightly defined protocols, risk assessment can offer a well-documented procedure for ordering and interpreting diverse types of information and generating numerical indicators that represent probabilities of harm.[22]

notify the EPA about the hazards of its existing chemicals, and, thereafter, whenever the firm becomes aware of new information on health or environmental effects. Pre-manufacturing notices for new chemicals need to include available test data, but the EPA cannot require specific tests. The CPSA requires reporting on product-related injuries but has limited authority to require product testing.

In contrast, testing is specified, required, and routine under FIFRA and FFDCA. In order to register a pesticide under FIFRA, the EPA requires the manufacturer to demonstrate that the product will perform its intended function without adverse effects. This requires "full description of the tests made and the results thereof upon which the claims are made." Whenever there is a new use of an already registered product, the registration must be amended.

Regulate the Manufacture, Marketing, and Use of Chemicals of High Concern

TSCA, FIFRA, and FFDCA require chemical, pesticide, drug and food additive manufacturers and importers to notify government agencies before the commercialization of new substances. FIFRA requires firms to present the results of a full battery of toxicity studies that meet rigorous standards in order to apply for a registration. FDCA requires drug makers to submit a new drug application (NDA) that contains scientific data and the results of animal and human tests showing that a drug is safe and effective. In contrast, the pre-manufacture notice (PMN) required under TSCA does not require specific toxicity tests or performance assessments but requires only that firms include in their applications any testing that they have completed. The CPSA sets no specific testing requirements for chemicals in existing or new products. While TSCA sets a short time period for the government to act on the presented information, neither FIFRA nor FDCA set such timetables, and pesticide registrations and new drug approvals can take years.

The range of potential government remedies prescribed by these laws also differs. The CPSA's authority is quite limited. Whereas TSCA is intended to regulate all industrial chemicals posing an unreasonable risk, the CPSA directs the CPSC to rely on voluntary consumer product safety standards, rather than mandatory regulations, if the CPSC determines that industry compliance is likely to be sufficient. The standard for tolerances on pesticides is a "reasonable certainty of no harm," which requires less substantiation than the "unreasonable risk" standard required for government action under TSCA. To regulate the manufacture or use of an existing industrial chemical, the EPA must show that the chemical presents an unreasonable risk based on sound toxicological evidence, the benefits of regulation outweigh the costs, and the proposed regulation is the least burdensome approach to managing the risk.

2.7   Shifting Chemicals Policy

The 1970s witnessed a steady stream of federal environmental and workplace protection laws creating a complex array of new government authorities. However, by the close of the decade, the national business community had had enough of this regulatory exuberance. The new federal administration that took office in 1980 was elected with strong business support because, in part, the campaign had promised to reverse the regulatory initiatives of the previous decade. This new administration and several that followed sought to limit rather than expand chemical regulation, and no major new federal laws were passed to control the production or use of chemicals from 1980 until well into the 1990s.

Thereafter, the focus on chemical policy innovation shifted to the states. The states in the United States have often become active when the federal government is unresponsive to public concerns. The states can function as laboratories for testing policy innovations, and often these policies are eventually adopted at the federal level. During the 1980s, this involved a flurry of diverse workplace chemical "right to know" laws that required employers to inform their employees about the hazardous chemicals they might be exposed to at work and provide periodic training in the safe handling of those chemicals. Eventually, twenty states passed some kind of worker right to know law, but in 1983, the movement was abruptly aborted when the federal Occupational Safety and Health Administration preempted these state laws with a national hazardous chemical communication standard.

In one state, California, this right to know concept shifted from a focus on occupational health to a consumer product focus. In 1986, California voters passed a ballot initiative called "Proposition 65" or "The Safe Drinking Water and Toxics Enforcement Act of 1986." Proposition 65 regulates certain toxic substances identified as carcinogens and reproductive hazards and prohibits their release to drinking water supplies or the knowing exposure of the public without an appropriate warning. At least annually, the California Office of Environmental Health Hazard Assessment (OEHHA) must publish a list of chemicals identified as carcinogens, mutagens, and other reproductive toxins—a list that currently contains some 775 substances. The law requires all products containing any of the listed chemicals to carry a label stating, "Warning: This product contains chemicals known to the State of California to cause cancer and birth defects or other reproductive harm."

Broad interpretation of the law has resulted in a wide range of products and premises to be labeled, including cigarettes, wine bottles, paints, adhesives, tools, vehicles, gas stations, parking garages, restaurants, and even newly constructed houses. Enforcement is promoted through lawsuits filed against any product distributor who sells unlabeled products that may contain a listed substance. The law allows a wide array of potential plaintiffs, including the state Attorney General, the district attorneys, and any private party "acting in the public interest." Penalties for failing to provide appropriate warnings can be as high as $2,500 per violation per day.[23]

In 1984, a tragic toxic gas release at a pesticide production plant in Bhopal, India, killed thousands of people living near the facility. This tragedy spurred Congress to enact the Emergency Planning and Community Right to Know Act (EPCRA).[24] One little noticed section—Section 313—of the EPCRA required industrial facilities to report annually to the EPA on the environmental release of a list of some 300 acutely toxic chemicals. When the first-year report from this Toxics Release Inventory (TRI) was released, it documented such large volumes of chemical releases that some states began passing laws and setting up programs aimed at preventing industrial pollution and reducing the generation of hazardous wastes.

Most of these so-called "pollution prevention" laws established state technical assistance programs for assisting firms in reducing their chemical wastes and emissions. The passage of the federal Pollution Prevention Act in 1990 had little effect on the federal administration but did provide funding and support for the maintenance of the state pollution prevention programs. By 1994, twenty-four states had passed laws encouraging some form of pollution prevention planning by industrial facilities.[25] Most of these programs developed broad, multimedia technical assistance services focused on reductions in chemical emissions, effluents, and hazardous wastes. However, some of these state laws did target toxic chemicals more directly. The most developed and best funded of these is a Massachusetts law passed in 1989 that focused on "toxics use reduction."[26]

The goal of the Massachusetts Toxics Use Reduction Act is to reduce the use of toxic chemicals as a means of reducing workplace risks, chemical emissions, and hazardous waste generation. The statute requires that any state manufacturer that uses any of a list of some 1,200 chemicals over a certain volume threshold must prepare facility-specific plans for reducing or eliminating those chemicals, report annually on the volume of chemical use, and pay an annual fee to fund the program administration. To assist firms in assessing chemical use, evaluating alternatives, and preparing facility plans, the law established an Office of Technical Assistance and the Toxics Use Reduction Institute, a research and policy center located at the University of Massachusetts Lowell.[27]

By 1994, all of the largest Massachusetts firms using toxic or hazardous chemicals over a certain volume threshold had prepared toxic use reduction plans that characterized the uses of listed chemicals and detailed plans on how their use could be reduced or eliminated. These toxic use reduction plans were certified by specially trained and state licensed toxic use reduction planners, and they needed to be updated every other year. Annual data collected from some 550 firms on their chemical use and waste generation are posted on an Internet site such that the public can identify chemicals used and wastes generated at regulated facilities and monitor trends over time. Twenty years later, the law continues to be highly successful, with a reported 41 percent reduction in use of the regulated toxic chemicals and a 75 percent reduction in hazardous waste generation. The combination of annual chemical use reporting, facility toxic use reduction planning, and well-funded technical assistance and training services is given broad credit for sharply reducing the use of chlorinated solvents and several carcinogens and reproductive toxins.[28]

Although these state programs were reasonably effective, they did not quell popular concern. During the 1980s and 1990s, the public became increasingly alarmed over a series of dramatically publicized chemical hazards. Mercury in fish, lead in plastic window shades, phthalates in infant teething rings, and toxic residues on apples led to a growing demand for government action. The lack of initiative on dangerous chemicals at the federal level drove state legislatures to pass laws focused on phasing out various chemicals of public concern. Thirty-two states and twenty-one cities have proposed or enacted legislation to phase out the sale of products containing mercury. Fourteen states and one city have enacted laws that restrict the use of lead in certain products. Nineteen states have enacted legislation restricting the use of various toxic chemicals (lead, cadmium, mercury, and hexavalent chromium) in product packaging. Since 2003, eighteen states have approved more than seventy-one laws to restrict toxic chemicals in everyday consumer products.[29]

2.8 The Architecture of Chemical Control Policies

Together, TSCA, FIFRA, CPSA, and FDCA created the structural framework for federal chemical control policy in the United States. In establishing a new risk-based approach to chemicals management, these laws present an impressive innovation in environmental and public health policy. They were drafted or amended with much enthusiasm during the early 1970s and provided government agencies with broad new authorities to regulate unreasonable risks. However, after 1980, the federal government largely ceased enacting regulatory statutes to protect public or environmental health. The Emergency Planning and Community Right to Know Act of 1986 set a context for state and local initiatives but no new federal requirements on chemicals, whereas the Pollution Prevention Act of 1990, long on principles, created no new regulatory authorities. Only the Food Quality Protection Act (1996) and the Consumer Product Safety Improvement Act (2008) broke this long, passive record.

With little new policy coming out of Washington, states began to push new chemical management policies with legislation on pollution prevention, chemical restrictions, toxic use reduction, product labeling, and bans on single chemicals. Although these policy initiatives were important, they varied widely among the states, creating a patchwork of diverse laws and regulations that made the marketing of chemicals and products difficult across the nation. Mercury switches in automobiles are banned in some states but permitted in others, the flame retardants used in most televisions cannot be used in televisions sold in the State of Washington, and products with appropriate warning labels for California requirements are sold in other states where such warnings are unnecessary.

All of these laws were intended to address the problem posed by dangerous chemicals. Although the federal laws differed in procedures, they all included three core building blocks: the characterization and prioritization of chemicals, chemical information generation and disclosure, and regulation of chemical manufacture, marketing, and use.

The laws were typically promoted by public health and environmental protection advocates and written and passed by sympathetic legislators. Where the business community participated, it was hostilely engaged in opposing the legislation or attempting to limits its reach. The result was legislation narrowly focused on managing the risks of dangerous chemicals with little attention to the broader role those substances play in the chemical market or as products of a large and increasingly international industry. Before considering the implications of a broader view on the problem of chemicals, it is useful to examine how these laws were implemented and what they were or were not able to accomplish.

3

Reassessing Chemical Control Policies

In 2005, the Center for Environmental Health, a California-based environmental organization, found lead in the polyvinyl chloride (PVC) plastic of several school lunch boxes that it had tested. The resulting news stories and public concern led several states—including New York and Connecticut—to issue recalls for the soft, insulated plastic lunch boxes made from PVC. The issue was fairly straightforward. The science on the adverse neurological and reproductive effects of lead is overwhelming. Efforts to regulate lead and prohibit its use are legion. However, the federal government was slow to act because of the complexity of its statutes. Lead as a chemical in the environment is regulated by the Environmental Protection Agency (EPA); however, lead in the interior surface of a lunch box is regulated by the U.S. Food and Drug Administration (FDA) because it touches food, and lead in a lunch box is regulated by the Consumer Product Safety Commission (CPSC) because a lunch box is a consumer product.[1]

While the early federal framework for chemical control was pioneering, the various laws created an irregular patchwork of authorities and responsibilities. The most important of these laws were enacted at different times in history and addressed different chemicals problems with different agencies applying different approaches. Even with well-crafted and harmonized authorities, these laws would have suffered as government agencies struggled to implement them. Unlike the environmental statutes that sought to regulate relatively common chemicals in emissions and wastes, these laws were focused directly on the manufacture and use of specialized chemicals with high economic value, sometimes critical industrial functions, and often broad commercial applications. There were bound to be powerful resistances and legal challenges. Writing regulations would need to be deftly negotiated among many competing interests. The challenges that made the bills difficult to write would similarly confront their implementation.

Implementation of environmental and workplace protection laws has been the subject of scholarly study since the 1980s. That research suggests that government program implementation is heavily affected by shifting political expectations, external pressures, agency competition, management capacities and resource availability.[2] Statutes with well-defined problems, clear terms and goals, appropriate authorities, suitable metrics, and reasonable timelines do better at navigating such turbulent conditions. The chemical control laws were not so blessed. Without new legislation, the government agencies responsible for chemical control—the EPA, FDA, and CPSC—made due with the authorities that their statutes provided. However, in implementing the laws, it became clear that there were among them procedural impediments, inadequately defined terms, many gaps in coverage, and big discrepancies between goals and means.

3.1   Regulating Chemicals in Foods and Drugs and Neglecting Cosmetics

Although passage of the original Pure Food and Drug Act of 1906 was driven by the agitation of populist activists, the language was carefully negotiated with a mostly cooperative industry. Since the establishment of the FDA, the relations between the agency and the regulated industries have waxed and waned. At times the food and pharmaceutical industries have openly fought tighter regulations, but after significant incidents of food or drug poisoning, the industries have cautiously sought protection under agency authorities.[3]

FDA's Center for Drug Evaluation and Research is responsible for overseeing both the premarket tests required for new drug applications and the postmarket safety tests once a drug is on the market. Every year, the FDA approves some twenty to forty new drugs and monitors the testing of some 3,000 pharmaceuticals on more than 20 million people to determine their effects. Setting the right standards for these tests has been contentious. Some have considered the FDA's standards for premarket trials to be the strongest in the world, whereas others have criticized this concern for safety as inhibiting innovation and allowing fewer new drugs to come to the market in the United States than in other countries. During the late 1980s, activists held protests outside the FDA headquarters, accusing the agency of unnecessarily delaying the approval of medications to fight HIV. This criticism led Congress in 1992 to pass legislation to streamline the processing of new drug applications by authorizing the FDA to assess fees to finance application reviews.

However, this revenue structure has been controversial. Today, nearly half of the FDA budget for premarket testing comes from the fees that firms pay to speed up drug approvals. This creates a potentially contradictory division in the agency, with some staff pushing for long, careful drug safety tests and others pushing for rapid approvals. This became particularly apparent when a 2004 review by *The Lancet* charged that the FDA left Vioxx, a nonsteroidal painkiller, on the market for five years despite mounting evidence that it increased the risk of heart attacks and strokes.[4]

Responding to these challenges, the National Institute of Medicine conducted a review of the FDA procedures and, in 2007, issued a broadly scoped report, *The Future of Drug Safety: Promoting and Protecting the Health of the Public.* The report offered twenty-five wide-ranging recommendations that would, "bring the strengths of the preapproval process...to the post-approval phase in order to fulfill a lifecycle approach to the study, regulation, and communication about the risks and benefits of drugs." The recommendations included a longer term appointment for the FDA director, updated drug testing and risk assessment protocols, a two-year moratorium on advertising new drugs, increased funding for the agency, and, in the absence of such funding, a 10 cent tax on all drug prescriptions. Some, but not all, of these recommendations are being adopted.[5]

The FDA has pioneered the use of packaging labels on food products. The philosophy behind food labeling is to balance the consumer's need to know against the producer's need to protect trade secrets. The FDA balances these interests by requiring that all ingredients be labeled, but not requiring quantitative proportions of each ingredient. Instead, the ingredients are listed in descending order of weight (pharmaceutical ingredients are also listed on containers but by alphabetical order), generic terms are used for flavors and spices, and "incidental additives" are excluded if they provide no "technical or functional" value. The Nutrition Labeling and Education Act of 1990 (NLEA) required all packaged foods to present standardized food labels that provide ingredient disclosures and nutritional information determined by serving size. Like other food and drug laws, the NLEA preempts state nutrition standards and labeling laws.[6]

Although the word "cosmetics" is used in the title of the law, the FDA has little authority over the safety of cosmetics. Under the Federal Food, Drug, and Cosmetic Act (FFDCA), the FDA is not authorized to require cosmetic manufacturers to register their products or the chemical ingredients in their products, and it cannot require them to substantiate product safety or performance claims. Instead, a private body, the Cosmetic Ingredient Review, funded by the industry trade association, evaluates the safety of cosmetic ingredients. The FDA does maintain a Voluntary Cosmetic Registration Program for manufacturers to register cosmetics sold to consumers, but only a small number of cosmetic suppliers participate. Fragrances present a particularly sensitive concern. The active ingredients often have neurological effects; however, the "fragrance houses" that manufacture them jealously guard against releasing information on the formulations, even to trusted customers.

The FDA does have the authority to conduct inspections of cosmetic manufacturing facilities, although it cannot require reporting of cosmetic-related injuries or require recalls of products that may present a hazard. The FDA's home webpage makes a clear statement to this effect: "FDA's legal authority over cosmetics is different from other products regulated by the agency, such as drugs, biologics, and medical devices. Cosmetic products and ingredients are not subject to FDA premarket approval authority, with the exception of color additives."[7] However, a separate law, the Fair Packaging and Labeling Act, does require manufacturers to declare some cosmetic product ingredients, "allowing consumers to make informed purchasing decisions."[8]

At least in the areas of food and drug safety, the FFDCA and its many amendments have provided sufficient authority to require testing and regulate hazardous chemicals. The safety of food and drugs is particularly sensitive to the public, and it cannot be easily ascertained in advance of purchase by consumers. Government assurance of chemical safety is beneficial to the industries, as well as the public. The willingness of the pharmaceutical industry to support a strong drug safety testing program that includes the efficacy of the drug does serve to quiet public concern and protect the product market.

3.2   Curbing Pesticides

Pesticides are toxic, and their use is dispersive by design. Therefore, pesticide regulations are comparatively stringent, requiring that all pesticide products be registered and their professional use restricted to trained applicators. This strong registration requirement ensures that there is clear identity for each pesticide product, important health and environmental effects data available, and sufficient authority for regulating and, where appropriate, prohibiting product use. In general, the registration requirement creates a normative presumption that industry must justify the safety of a pesticide in order to maintain continued product use.

However, the clear responsibilities laid out under the Federal Insecticide, Fungicide, and Rodenticide Act (FIFRA) are clouded by the procedural requirements to assess risks and account for the benefits of pesticide uses. The responsibility for both determinations rests with the EPA. To provide solid justification for its regulatory decisions, the agency relies on quantitative risk assessment and cost-benefit analysis.[9]

In conducting risk assessments on pesticides, the EPA considers the toxicity of the pesticide and its decomposition products, the quantities and frequency of pesticide applications, the potential exposures of pesticide applicators and workers, the residues remaining on foods at the time they are marketed, and the populations and vulnerable groups (the elderly and children) that are likely to be exposed during food preparation and consumption. To ensure the quality of the data submitted for maintaining pesticide registrations, the EPA has established good laboratory practice standards and harmonized test guidelines. Depending on the type of pesticide, the agency can require more than 100 different tests to determine the effects on humans, soil and water quality, fish, wildlife, and plants.

However, the methods of risk assessment have raised many concerns. Criticism has focused on the way risk assessments condense large amounts of uncertain and ambiguous data, much of which has been modeled and extrapolated beyond what is actually measured, down to a single numerical indicator. Additional concerns focus on the large gaps in the toxicological data on which the hazards of chemicals are assessed, the broad ranges of human susceptibility, and the limited knowledge about the biological mechanisms that bring about disease. Exposure pathways are difficult to predict over the life cycle of substances, and exposure models are often controversial. Further credibility problems appear in the use of dose-response assessments where data are extrapolated across species from test animals to humans and from high testing doses to the low doses of likely human exposures.[10]

Determining the economic calculus of pesticide use has proved no easier. Cost-benefit analysis is used to determine whether the benefits of a chemical management intervention such as a regulation outweigh or are outweighed by the costs. Because such calculations are often quantitative, they tend to overvalue factors that can be monetized and undervalue factors such as good health and ecological quality that are difficult to put a price on. Various surrogate procedures, such as willingness to pay and contingent valuation, provide some approximations of the monetary value of difficult to price factors but can have broad ranges across those potentially affected. Cost-benefit analyses involve so many assumptions, judgment calls, value determinations, and unquantifiable factors that the analysis is too often unconvincing. Like risk analysis, cost-benefit analysis tends to conflate multidimensional problems into simplistic calculations and, if overvalued, can create a false sense of confidence in those making complex chemical management decisions.[11]

Some 1.1 billion pounds of pesticides were used in the United States in 2007, the last year that the EPA presented an estimate. This involves more than 600 bioactive ingredients used in some 25,000 pesticide products. In terms of use, 80 percent was used in the agriculture sector, where herbicides made up the largest volume of that use, and glyphosate and atrazine made up the largest volume of active ingredients (see box 3.1). Only 8 percent was used in the home and garden market, and there 2,4-D and glyphosate made up the largest volume.[12]

On the roughly 500 active ingredients in pesticides registered for food use, the EPA establishes a risk-based tolerance level based on "a reasonable certainty of no harm." This is a substantially more workable standard of proof than the Toxic Substances Control Act's (TSCA's) "unreasonable risk" standard. However, for the ingredients used in most non-food crop-oriented pesticides, the same "unreasonable risk" standard for determining safety standards prevails, resulting in a relatively low level of registration conditions or cancellations.

Even with substantial requirements for chemical information disclosure, FIFRA has not resulted in sufficient information. Pesticide effects and efficacy data must be generated and submitted by applicants for each

Box 3.1
Top Ten Active Ingredients in Pesticides Used in Agriculture in the United States in 2007 (by volume)

Glyphosate	Herbicide
Atrazine	Herbicide
Metam sodium	Fumigant
Metlachlor-S	Herbicide
Acetochlor	Herbicide
Dichloropropene	Fumigant
2,4-D	Herbicide
Methyl bromide	Fumigant
Chloropicrin	Fumigant
Penimenthalin	Herbicide

pesticide registration. With the exception of health and environmental effects data, the applicant may claim much of these data as trade secrets. Pesticide registrants are required to report any new evidence on adverse effects; however, pesticide manufacturers are slow to provide such "new information," and pesticide use may continue long after new science suggests new risks.

The EPA has used its regulatory authority to restrict some of the most dangerous pesticides. Between 1972 and 2007, the EPA used FIFRA to ban or severely restrict the use of sixty-four active ingredients, including DDT and many of the chlorinated insecticides. Since passage of the Food Quality Protection Act (FQPA), the registration of pesticides has tended toward safer active ingredients. Of new active ingredients registered by 1999, the EPA considered 62 percent safer than the conventional pesticides, and 77 percent of the new uses were considered safer than the conventional uses. However, efforts by the EPA to phase out the use of organophosphate ingredients have been stalled for years.[13]

The EPA has developed a substantial body of information on the bioactive ingredients in pesticides; however, there is much less information on roughly 1,200 so-called "inert ingredients" used to preserve, stabilize, and protect the active, pest-killing ingredients and make them easier to apply. These ingredients, which often make up 80 to 90 percent of a pesticide product by volume, include chemicals such as toluene, phenol, chlorobenzene, and various substances linked to cancer, birth defects, and central nervous system disorders. The FQPA required the EPA to evaluate 870 inert ingredients exempted from tolerances; however, there are no labeling requirements for inert ingredients, and their identity is often protected as trade secrets.

The 1972 amendments to FIFRA required the EPA to "re-register" some 35,000 older registrations with regard to current standards. The re-registration process progressed slowly, and in 1988, Congress amended the law, setting a re-registration schedule but permitting the agency to charge a fee for re-registration and an annual fee for maintaining registration. To date, at least 14,000 registrations have been voluntarily canceled by registrants because of the fees and re-registration testing requirements (see box 3.2). However, progress with re-registration continues to be slow. In 2007, an EPA internal evaluation found that on average it took 54 months to re-register a product. At the close of that year, there were 21,000 pesticide products eligible for re-registration; 2,600 had been re-registered, 600 had registrations amended, 5,000 had registrations canceled, and some 13,000 had actions pending. Even this progress is questionable. A recent analysis of the pesticide registrations revealed that some 65 percent are registered

Box 3.2
Examples of Insecticides with Canceled or Withdrawn Registrations

Alkylmercury fungicides	DDT and DDD
Aldrin and dieldrin	Heptachlor and chlorodane
Mirex	DBCP
Chlordecone (Kepone)	2,4,5–T
Endrin	Nitrofen
Aldicarb	Toxaphene
Dinoseb	Alachlor
Chlordane	Parathion-ethyl

through a "conditional registration," a procedure that was intended only for new pesticides that needed more time for testing.[14]

However, among the most significant limits of FIFRA has been its narrow focus on controlling pesticides rather than promoting safer alternatives to managing pests. The FQPA required the EPA to develop a program to encourage safer pesticides and promote integrated pest management (IPM), a pest management approach that relies on ecological principles and discretely managed pesticide applications. While the agency has established voluntary programs such as the Pesticide Environmental Stewardship Program and PestWise to promote safer pesticides and pesticide use, these programs are not scaled to present serious alternatives to conventional pesticides.

Over the years, the pesticide laws have restricted some of the most egregious hazards, pesticide tolerances for human exposure have been reduced, and pesticide use has evolved toward more targeted and efficacious products. Still the use of pesticides has increased; the authority for managing them is split between agencies, and the procedural requirements for assessing risks and costs have slowed regulatory progress. The continued dissipative use of large volumes of pesticides and the concentrations that show up in homes, worksites, and ecological sinks have left the environment awash in low levels of pesticide residuals. Although the biological and ecological effects remain difficult to assess, they are not insignificant.[15]

3.3   Stalling on Industrial Chemicals

The compromises hammered into TSCA's statutory language restrained the law from meeting its mission, stalled the generation of sufficient chemical information, and resulted in little outright chemical regulation. These compromises created a law that granted broad powers to the government but made them so difficult to use that, in essence, the procedural hurdles trumped the statutory authorities.[16]

The differing procedural treatments for new and existing chemicals tends to reward those manufacturing existing chemicals with few requirements and hinder those offering new chemicals who, while not overly burdened, still need to present their products for evaluation. While TSCA requires the EPA to evaluate new chemicals prior to manufacture, it offers no authority to require testing. The EPA estimates that under TSCA, some 36,000 new chemical pre-manufacture notices (PMNs) were reviewed between 1979 and 2003. By its own records, the EPA notes that 67 percent of PMNs contained no test data and 85 percent contained no human health data. Therefore, the agency makes assessments largely based on models that predict the hazards of a new chemical based on its similarity to other, better-studied chemicals and on negotiations with the applicant for more information. Only about 10 percent of the total number of PMNs reviewed has resulted in restrictions, additional testing, withdrawn submissions, or denials.[17]

TSCA requires industry to present the basic toxicological information for existing chemicals; however, to get the information, the EPA must ask for it. In what is often pegged as TSCA's "Catch 22," the agency needs to have information sufficient to establish an unreasonable risk in order to request the information necessary to determine an unreasonable risk. The result is that for the past 25 years, the agency has asked for very little. Instead, the agency often relies on voluntary agreements with chemical manufacturers on what data should be presented and what tests conducted for each individual chemical.[18]

TSCA does provide the EPA with the authority to require tests on existing chemicals recommended by the Interagency Testing Committee; however, in practice, the agency has used its testing rule authority to require testing of just 200 chemicals. Additional studies are conducted by academic centers and private laboratories, but there is no systematic means for collecting them or ensuring their quality. The continuing absence of this information is particularly problematic because, in practice, the lack of toxicity information for many chemicals is often treated as if it were evidence of their safety.[19]

An even bigger data gap involves the lack of information on where and how chemicals are used in the economy. There is no chemical constituent reporting system for products and no chemical transport tracking system.

The Inventory Update Rule, which for many years required chemical manufacturers to report on the production of chemicals, covers only 6,200 of the 84,000 chemicals on the TSCA Inventory, exempts polymers, chemicals imported in products and various impurities and byproducts, and then only requires reporting on a four-year basis.[20]

TSCA's liberal thresholds for protecting confidential business information (CBI) have permitted the withholding of large amounts of potentially valuable chemical information. The EPA has drafted several schemes to require up-front substantiation of CBI claims and minimize inappropriate claims, but none of these has been formally adopted. A recent analysis found that confidentiality claims protected significant portions of the chemical information submitted to the EPA for approximately 17,000 of the 84,000 chemicals on the TSCA Inventory, and of the 20,400 chemicals added to the Inventory since 1979 as new chemicals, 13,600 had their identity protected as confidential business information.[21]

In 1998, the EPA conducted a study of the chemicals sold over 1 million pounds per year and found only 7 percent had sufficient data to meet minimum screening data requirements and 43 percent of the substances had no data at all. This resulted in a voluntary partnership between the EPA, the American Chemistry Council, and the Environmental Defense Fund to urge the reporting of existing test data. This so-named High Production Volume (HPV) Challenge Program focused on some 2,300 chemicals manufactured or imported into the country in annual volumes of a million pounds or more. The tests required were consistent with international standards set by the Organization for Economic Cooperation and Development (OECD), but these standards only cover screening level data, and they do not include important health end points such as carcinogenicity, neurotoxicity, and endocrine disruption. Although the HPV Program required all of the data to be submitted by 2005, by 2007, final reports on only 51 percent had been received and 267 of the HPV chemicals still had no manufacturer willing or capable of participating.[22]

Like FIFRA, TSCA has come to rely on risk assessment and cost-benefit analysis. However, here too the risk-focused approach has often delayed, rather than spurred, protective action. Major risk assessments for some chemicals have taken years. The EPA's risk assessment for trichloroethylene took more than a decade, and its risk assessment of dioxin has been in progress since the 1980s. The requirement under TSCA to balance environmental benefits and economic costs has favored those factors that can be monetized and, therefore, tended to protect the status quo rather than drive changes that have significant costs.[23]

With the exception of the polychlorinated biphenyls that were restricted by law under TSCA, the EPA has not been able to phase out one chemical of concern. The EPA must effectively prove beyond reasonable doubt that a chemical poses a risk and that a restriction will not cause unnecessary economic costs before it can determine that a substance needs to be regulated. However, since the passage of the Act, the EPA has used its power to restrict the use of a dangerous chemical just five times, and in none of these actions was the chemical banned. By requiring the least burdensome remedy and deference to any other statutory authorities, TSCA generates few protections even to those substances that present the most obvious risks.[24]

Take asbestos as an example. In 1989, the EPA issued a regulation under TSCA restricting the manufacture, distribution, and use of most asbestos-containing products. The agency spent some ten years determining that asbestos was a potential carcinogen posing an unreasonable risk and developing a regulation that included multiple public hearings for identifying safer alternatives that could be used as substitutes. However, soon after promulgation of the standard, the asbestos industry brought suit against the agency, and in 1991, the Fifth Circuit Court of Appeals in *Corrosion Proof Fittings v. EPA* ruled against the EPA's extensively constructed case. The Court found that the EPA had failed to sufficiently demonstrate that the regulation was based on a full analysis of the safety of the alternatives, that the regulation was the least burdensome remedy, and that the health and environmental benefits outweighed the costs to industry.[25]

After such a sharp judicial reprimand, the EPA ceased using its regulatory powers to restrict the manufacture and use of chemicals, and in the years since the asbestos decision, the agency has not again tried to ban an industrial chemical. Instead, the agency has opted for various voluntary approaches, such as consent decrees and negotiated settlements. As examples, a consent decree was negotiated with chemical manufacturers for phasing out penta- and octa-brominated diphenyl ethers, and a voluntary PFOA Stewardship Program was jointly designed by the EPA and the industry to encourage the adoption of substitutes to the use of perfluorinated compounds.

After more than three decades of TSCA implementation, the performance record is underwhelming. The federal Government Accounting Office has issued a string of reports critical of the EPA's implementation of TSCA, noting that the agency has published few rules for standardizing chemical tests, has been slow to require tests, has not made test results readily available, and has, over thirty years, made no more than four referrals of chemicals in need of regulation to other agencies.[26] This dismal record has led Mark Greenwood, a former director of the EPA's Office of Pollution Prevention and Toxics (the office responsible for TSCA implementation), to observe that by the 1990s, TSCA was "widely known as a 'broken' statute," and Lynn Goldman, also a former director of the office, to write, "TSCA currently places too high a bar for the EPA to jump to assure the health of the public and protection of the environment."[27]

3.4   Crippling Product Safety

The Consumer Product Safety Act (CPSA) has also suffered a disappointing implementation. While the CPSC was initially hailed as "the most powerful regulatory agency ever created," time would prove it to be the weakest of the federal health and safety regulatory agencies. As with TSCA, Congress provided the 1972 CPSA with significant data collection and regulatory authorities and then hobbled the CPSC with complex procedural requirements and an overly high unreasonable risk standard. Moreover, in 1982, a Congress deeply sympathetic to business complaints passed amendments to the CPSA that sharply limited the powers of the CPSC.[28]

Whatever authority was left at the CPSC was then stripped down with sharp budget cuts during the 1980s, and for the next several decades, the commission was largely neglected. The commission's 2007 budget was 40 percent less than it had been in 1974, adjusting for inflation, and its 800 employees had decreased to 393. Indeed, during this past decade, the president failed for two years to appoint enough commissioners to maintain the quorum needed for decision making.[29]

Unlike TSCA, the CPSA provides no premarket notification for new products prior to their commercialization. CPSC Commissioner Thomas H. Moore emphasized this directly in his 2007 testimony to a Senate subcommittee: "We do not have the luxury of getting ahead of a problem.... We have to wait until one develops and then try to solve it, usually after it has killed or injured consumers." The hazards of existing products fare no better. While manufacturing firms are required to report any hazards, including chemical hazards that they identify in their products currently on the market, firms often wait years to file such information.[30]

The CPSA gave the CPSC authority to collect information on product-related injuries. With no substantial power to require manufacturers to conduct product testing, the CPSC has relied on post-injury data collection from a sample of hospital emergency rooms to identify product-related injuries worthy of concern. Although these data are frequently criticized for underestimating hazards, they focus largely on acute product-related injuries and are inadequate in estimating chronic hazards from product-related chemical hazards. Although the CPSC is required to collect and disseminate information to the public about dangerous products, the 1982 amendments forbid the CPSC from releasing information about a product's hazards that has not been approved by the manufacturer.[31]

The original language of the CPSA provided broad authority to conduct product safety tests and set protective safety standards. However, the Commission has conducted its testing shrewdly. When the CPSC considered the hazards of phthalates in toys, it did so without addressing the potential hazards of other substances in toys, such as lead, pesticides, or flame retardants, or addressing their aggregate consequences. During the 1970s, the CPSC commenced drafting a limited number of safety standards; however, the complex procedural and risk assessment requirements necessary to demonstrate unreasonable risks slowed the process and resulted in few of the standards being finalized. Thereafter, CPSC promulgated few mandatory regulations in large part because the 1982 amendments prohibit the commission from promulgating mandatory standards if voluntary standards would "adequately reduce the risk," and it was likely that there would be "substantial compliance with the voluntary standard."[32]

The CPSC was given authority to require recalls of dangerous products. Such recalls typically involve negotiated agreements between the CPSC and the relevant manufacturers and retailers that result in public notices urging consumers to return the relevant products. However, the public notices are seldom widely broadcast by the media, and with a limited field inspection staff, the CPSC cannot ensure that businesses comply.

With limited authority, budget, and staff, the CPSC has been further crippled with the appointment of commissioners who have failed to instill public trust. During the mid-2000s, controversy arose over questionable meetings between commissioners and industry representatives that resulted in revisions to the test methods used for lead in the vinyl lunchboxes and reduced the estimates of childhood exposures.[33]

The 2008 amendments to the CPSA, besides focusing on lead and phthalates, increased the CPSC budget and reestablished some of the authorities for product testing lost in the 1982 amendments. However, much controversy has arisen over the new law's significant impacts on small businesses, many of which do not have the resources to test products, and unintended institutions, such as public libraries, which have had to prohibit children's access to older books that may contain small amounts of lead in their bindings.

While the commission claims to have jurisdiction over some 15,000 products, a large number of products are excluded from coverage because they are under the jurisdiction of other federal statutes. Instead, the CPSC collects data on consumer injury and death to assess trends in consumer product hazards and waits to regulate a product until there is substantial evidence of harm. There is not much that is preventive here; the chemical control policies of the CPSC are largely reactive and provide little incentive for a comprehensive search for safer chemicals.

3.5   Reconsidering the Chemical Control Laws

Recognizing the ambitious mission of the chemical control laws, it is not surprising how far they have fallen from guaranteeing a safe and environmentally protective chemical market. Some have performed better than others. The regulation of drugs and pesticides has been more successful than the regulation of industrial chemicals or chemicals in commercial products.

The new drug testing program managed by the FDA, unlike the food additive program, is generally considered effective. Among the chemical control laws, this program is unique in that both industry and the government have strong incentives to ensure drug efficacy and safety (an ineffective drug or drug causing significant harm can rapidly jeopardize physician acceptance and consumer trust). The government sets the standards for the safety and performance tests, but the burden for financing and conducting the clinical trials is on the industry. The testing programs are well established; there is plenty of information available, and the agency has adequate authority to prohibit market entry. Because the FDA oversees the efficacy and safety of the product, the agency and the industry have a common interest in product quality, which results in significant information sharing. In addition, the FDA is focused on one class of chemicals—pharmaceuticals—in one economic sector—health care—and the FDA's authority in that sector covers all drugs in a comprehensive and inclusive manner.

The EPA's record in regulating pesticides is more mixed; however, it does offer some similarities. The agency is provided comprehensive oversight of all chemicals used as pesticides. While pesticides are used in many commercial products and non-farm settings, the EPA and FDA have focused largely on just one economic sector—agriculture. The registration requirement allows for an orderly and comprehensive process for classifying, labeling, and collecting information and the basis for setting a registration maintenance fee to offset program expenditures. Because the agency sets the standards for testing while the industry has responsibility for conducting the tests and in determining "a reasonable certainty of no harm," FIFRA provides an achievable standard for action. There are also differences between the two statutes. Drug use is a narrowly controlled enterprise, whereas pesticides are used ubiquitously. Because the health effects of exposures to such a broad mix of diffuse pesticide use and residues is difficult to link directly to specific consumer harms, the safety and economic incentives of the EPA and the industry do not align, although the industry does use the presence of the registration as evidence of a pesticide's safety.

Even in noting these two exceptions, the implementation of federal chemical control laws offer a disappointing level of performance. The primary burden for ensuring industrial chemical safety or the safety of consumer products is on the government, there is inadequate information available, testing protocols are poorly formalized, and the laws either provide too little authority or too much procedural encumbrance. Although it is easy to recognize the many structural barriers to success, significant problems are also built into the core assumptions of the laws.

Not only are the policies narrow in their focus they are also highly reductionist—a term used by some critics to describe a limited, reduced way of addressing the problem of hazardous chemicals. This reductionist approach focuses on individual substances as unique compounds divorced from the class or group of similar chemicals. The focus is not on changing the intrinsic hazards that chemicals present but on managing the human or environmental exposures. Little attention is directed to the complex and integrated industrial production systems that manufacture chemicals or the functions and technologies that create a demand for them. There are no statutory incentives, and only modest program resources directed at the development of new, safer alternatives to chemicals of high concern. Left unaddressed is the sanctity of the chemical market that externalizes the health, environmental, and social expenses, such that the commercial price of chemicals lies far below their total social costs. Finally, this reductionism focuses so narrowly on single chemical regulation that it presents no incentive for an overarching policy on the manufacture, use, and disposal of all chemicals. With the value of hindsight, much can be seen and now questioned about the assumptions and principles that were built into the chemical control policies of the twentieth century:

Chemical policies need only focus on the most dangerous substances. Industrial chemicals and consumer products long on the market are presumed to be safe. The pesticide law's presumption that a pesticide is dangerous is inverted by TSCA to mean that most industrial chemicals on the market—the existing chemicals—are not candidates for government regulations. This presumption of safety is even more pronounced under the CPSA, where existing commercial products are assumed to be safe until some study or incident proves otherwise.

Chemical policies address chemicals one by one. Locked into the early statutes is a one-by-one orientation toward hazardous chemicals. While TSCA provides authority to restrict the manufacture and use of chemical groups or mixtures, in practice, the EPA maintains a chemical-by-chemical approach that focuses on the testing and managing of individual substances. This same narrow focus pervades each of the other statutes.

Chemical policies are fragmented with diverse requirements for chemicals in wastes, emissions, workplaces, and products. The separate chemical control laws differ in their coverage and authorities from the media-specific environmental protection laws. Rather than draft TSCA as an omnibus framework for supporting and coordinating these statutes, TSCA was designed to defer to other statutes and function only where other laws provided inadequate protection. Thus, authority over chemicals is prescribed by more than twenty statutes and divided among some ten federal agencies, and even within the EPA there are separate sections for chemical emissions for air, water, and wastes.

Chemical policies focus on limiting adverse chemical exposure rather than addressing the inherent hazards of chemicals. Focusing on risks is central to each of the chemical laws. However, a focus on risk in seeking safety often requires lengthy efforts to predict dose/responses and potential human exposures and typically encourages exposure management and engineering controls rather than design changes that reduce or eliminate the need for hazardous chemicals. Real exposures over the lifetime of a chemical are often hard to confidently predict, and a trust in exposure control gives comfort to the idea that chemicals are "properly used."

Chemical policies use "sticks" without "carrots." Because the chemical policies of the 1970s developed out of a need to defend human health and the environment, they were designed to restrict chemical uses, releases, and exposures. Setting restrictions is seldom balanced by parallel functions that reward and encourage safer and more sustainable technologies and practices. By solely focusing on the worst substances, government authorities are provided little authority or capacity to fund research, provide incentives, or otherwise promote safer alternatives.

By reducing the problem of hazardous chemicals to a narrow campaign to regulate a few bad actors, the chemical control policies sought to chip away at the exposed tip of a very large iceberg, comfortably avoiding the weighty complexities of the production and consumption systems that invisibly lay below the surface. This reductionism made the hazardous chemicals problem appear manageable and not fundamentally threatening to large corporate powers of an otherwise directed national economy.

3.6   The Political Context of Chemical Controls

The U.S. economy has long been praised for its innovativeness and dynamic development. Based on a rich underpinning of material and energy resources, the twentieth-century industrial economy lunged forward driven by a seemingly endless fountain of molecular inventions and novel technologies. From the 1930s to the 1980s, the United States reigned as one of the world's greatest industrial engines. Its innovative approach boldly and brazenly embraced new technologies and materials well before their risks to workers, consumers, or the environment could be determined or managed. Risk taking lies at the heart of this form of technological development, and American culture has often been enthusiastically permissive toward new technologies. If negative consequences appear, the American response is typically quick to frame the problems in terms of individual providence and only later to consider the conditions socially—most often in the aftermath of a dramatic incident.[34]

By working within the confines of a free market ideology, those who drafted the chemical control laws formalized an institutional framework that gave license to chemical manufacturers to make what substances they could market and saddled government agencies with the responsibility for ensuring safety. TSCA and CPSA provided a mix of authorities, but the procedures required to exert those authorities strangled the EPA's and CPSC's initiatives. Only the authority granted to the FDA under FFDCA to oversee the testing of pharmaceuticals provided effective means to manage the market introduction of relatively safe and effective chemicals. The other chemical control authorities required government to identify a significant level of endangerment before government agencies could act.

Given the statutorily limited and procedurally compromised authority of the chemical control laws, inadequate resources further doomed their success. If Congress intended for the EPA and CSPC to carry the primary burdens for collecting chemical data, testing chemicals, conducting risk assessments, negotiating standards, and monitoring industry for compliance, it certainly did not provide the funding. The CPSC has long been chronically underfunded. The Commission's testing facility at a former Nike missile site in Gaithersburg, Maryland, is old and outdated. Between 2004 and 2007, the CPSC workforce was cut by 15 percent. Compared with the EPA's air and water regulatory programs, funding for its chemicals control programs has been small. In 2013, the toxic substance program at EPA had a budget of $54 million and the pesticides program had a budget of $129 million compared with a budget 20 times larger for water quality protection.[35]

The success of the chemical control laws has been further compromised by the power asymmetry between government agencies and those that they are intended to regulate. Regulations on air emissions can cost firms significant money; regulations on chemicals can cost firms their market. No agency director can act with principled impunity in regulating chemical manufacture and use when the regulated corporations are among the country's largest and most politically connected institutions. While the public will accept certain punishments for clearly noncompliant firms, seldom can a government agency afford to be seen as constraining production at employment-generating factories. So regulation must be balanced against economic effects even if not required by statute. Add to this the accountability that agency directors have to elected officials who by economic necessity are beholden to corporations for large campaign contributions, and the result has been that even well-intended administrators are strategically cautious.

Nor has there been significant administrative leadership in promoting chemical control laws. Since the 1980s, the federal government has been dominated by presidential administrations or congressional majorities hostile to business regulation. During the 1980s and again in the 2000s, the EPA has been led by administrators unsympathetic or only weakly committed to its regulatory mission. Commissioners appointed to serve on the CPSC have often been poorly selected. Congressional oversight committees have repeatedly been chaired by hostile and partisan representatives. There has been little presidential leadership. Environmental protection is often mentioned in presidential addresses, but tough new regulations are seldom noted. The Presidential Office of Information and Regulatory Affairs (OIRA) and the little known Office of Advocacy in the Small Business Administration have maintained a critical stance toward EPA and FDA regulations guarding against any initiative that looks like "regulatory excess." The Paperwork Reduction Act of 1980, the Regulatory Flexibility Act of 1980, the Government Performance and Results Act of 1993, and the Small Business Regulatory Authority Fairness Act of 1996 have provided further support to OIRA's pension for cost-benefit thinking in limiting and rejecting chemical regulations or management programs that have adverse effects on business—small businesses, for sure—but all businesses by implication.[36]

This anti-regulatory, pro-business philosophy has been driven by an articulate and influential movement hostile to government action in general and public regulation in particular. The "new right, market fundamentalist" movement embodied in policy research and advocacy institutions, such as the Heritage Foundation, Competitive Enterprise Institute, and the Cato Institute, regularly releases reports and testimony critical of proposed regulations, questions the science on which regulations are based, and debates the problems the regulations are intended to address. Business trade associations voicing these same arguments have increasingly become more strident and ideological. The result is a powerful and well-funded network of lobbyists, public relations firms, and trade associations that are often so effective in their opposition to aggressive chemical regulation that government agencies are reluctant to even propose such regulations.[37]

It is easy to find fault with the statutory text, procedural encumbrances, and underlying assumptions of the chemical control laws. However, much of their disappointing performance has been due to the political context in which federal agencies struggle to implement the laws. With limited presidential or congressional support, cautious administrators, insufficient budgets, hostile business interests, powerful anti-governmental lobbyists, and a culture comfortable with a "reasonable" amount of risk, it has been difficult to use the regulatory apparatus provided by the chemical control laws to substantially restrict many highly hazardous chemicals.

3.7   An Unfulfilled Mission

The chemical control policy framework of the last century has left a disappointing legacy. The policies have created a fragmented and ill-coordinated array of regulatory instruments and government programs. Some laws have worked better than others. FIFRA and FDCA have established comprehensive oversight over pesticides and pharmaceuticals. However, the EPA has largely ignored the vast majority of existing industrial chemicals, and in practice, the CPSC only attends to hazardous chemicals in those products that have raised public concern.

Today, there is more knowledge on what chemicals are in commerce than thirty years ago and more knowledge on the environmental and human health effects of some of the most common chemicals. The environmental and toxicological sciences have advanced significantly, and there are many respectable chemical research programs and testing laboratories. The use of some highly hazardous substances, such as carbon tetrachloride, pentachlorophenol, aldrin, dieldrin, endrin, and DDT, has been restricted, and some hazardous substances that might otherwise appear in foods or pharmaceuticals do not.

However, these laudable successes pale in comparison with the scale of the chemicals problem. There remains little or no health and environmental effects data on the large majority of industrial chemicals. Retail stores remain filled with insufficiently tested products. Highly toxic pesticides continue to be used and dispersed. There is no government oversight for cosmetics. Although the active ingredients of pesticides and pharmaceuticals are labeled, most commercial products offer no way for consumers to identify other chemical ingredients.

There are lessons here in this reassessment, and they are worth reviewing before considering future developments:

• The chemical-by-chemical approach focusing only on the most hazardous chemicals is too limited in scope and too long, slow, and costly.

• The absence of sufficient chemical information significantly compromises government policy and limits regulatory effectiveness.

• No matter how ambitious the policy or broad the authority, statutory laws can be severely compromised by complex procedural requirements.

• Both risk assessment and cost-benefit analysis have slowed and increased the costs of regulatory initiatives.

• The political context of government agencies creates nearly insurmountable hurdles to implementing regulations that directly confront powerful business interests.

• Government agencies alone cannot carry the burden of ensuring that chemicals are sufficiently safe.

• The success in addressing pharmaceuticals demonstrates the value of focusing on product efficacy and working within economic sectors.

• Although considerable government effort has been put forth to study and characterize some chemicals, far less effort has been made to develop safer alternatives.

Some thirty years of conservative government in the United States have stalled the development of federal chemical policies and limited progress on the management of hazardous chemicals. The vacuum left by this neglected agenda has inspired a mix of spontaneous state initiatives, but these are largely uncoordinated and not scaled to significantly affect the national chemicals market. However, other nations have moved forward during this period. Europe and some Asian countries have developed innovative new chemical policies, and new international treaties and agreements now set the stage for the global management of chemicals. We turn now to these more recent developments.

II
Reframing Chemical Policies

The choice, after all, is ours to make. If, having endured much, we have at last asserted our "right to know," and if knowing, we have concluded that we are being asked to take senseless and frightening risks, then we should no longer accept the counsel of those who tell us that we must fill our worlds with poisonous chemicals; we should look about and see what other course is open to us.

—Rachel Carson, *Silent Spring* (1962)

4

Considering New Initiatives

During the 1970s, the United States was an international pace setter in environmental policymaking. Environmental leaders throughout the world closely followed the policy initiatives crafted in Washington and drew on those early statutes as precedents for environmental policy laws in their own countries. That inspired leadership in setting environmental policy vanished in the decades that followed.

Today, the European Union and the international treaties of the United Nations are setting the standards for environmental policy innovation. Since 2000, the European Union has embarked on an ambitious series of environmental policies ranging from energy efficiency and climate protection to integrated pollution control, product stewardship, integrated waste management, and chemical regulation. As the world's largest economy, the twenty-seven counties of the European Union have the capacity to set global environmental standards, particularly for worldwide markets such as the global market for chemicals and chemical products.

Although the United Nations does not have the legal power to regulate the global manufacture and use of chemicals, it does have the international legitimacy to convene representatives of nations who can draft and ratify treaties that bind participating countries to international chemical management agreements. Today there are five broad multilateral chemical conventions and several regional treaties and codes of conduct that have direct effects on the use and disposal of chemicals. Recently, China, Korea, and Taiwan have launched new chemical policies largely modeled on the European initiatives. These Asian and European initiatives and the United Nations international conventions and strategies on chemicals and wastes are setting global directions for the sound management of chemicals worldwide.

4.1   Chemical Policy in the European Union

During the 1970s, Northern European countries took the lead in Europe in developing chemical control laws. Sweden enacted an "Act on Articles Hazardous to Human Health" in 1972, and Germany passed its directive on marketing and use of dangerous substances in 1976. These laws laid the foundation for European-wide legislation that developed with the creation of the European common market. As early as 1967, the European Council adopted a directive on "dangerous substances" that required the classification and labeling of hazardous chemicals (including both pesticides and industrial chemicals). This was followed in 1976 with the Limitations Directive, which was intended to harmonize the process for the European Union's Member States to follow in restricting the use and marketing of the most dangerous chemicals and chemical preparations.

In 1979, the Dangerous Substances Directive was expanded to cover new chemical reviews. Similar to U.S. Toxic Substances Control Act (TSCA), substances produced before 1981 had to be registered with a new European Inventory of Existing Chemical Substances (EINECS), similar to the TSCA Inventory, whereas those produced after 1981 were considered "new chemicals." Any chemical manufacturer or importer wishing to place a new chemical on the European market was required to submit a premarket notification package that included basic chemical characterization, a minimum battery of toxic effects tests, and recommended procedures for safe handling and disposal. By 1988, the lack of progress on testing and regulating existing chemicals—those originally listed on EINECS in 1981—led the European Union to enact the Existing Substances Regulation, which set out specific chemical information reporting requirements for all chemical manufacturers or importers of existing chemicals.[1]

Until this past decade, these laws provided the backbone for regulating industrial chemicals within the European Union. While manufacturers and importers had responsibilities for providing information on their chemicals and preparations, the burden was on the European Commission and the Member States to conduct risk assessments; regulate manufacturing, use, and disposal practices; and restrict the marketing and use of the most dangerous.

In terms of chemicals, European and American authorities have different perspectives on when to act on dangerous chemicals. European regulatory agencies assess the hazards of a chemical, and when there is sufficient evidence to suggest harm, they are likely to take preventive actions, whereas American agencies often wait for substantial evidence that exposure leads to unreasonable levels of harm. Here lie differences over what constitutes sufficient evidence, over how much exposure should be considered, and over what criteria should require governments to act. European authorities tend to base their chemical management policies on the inherent hazards of a chemical, whereas American authorities focus more on the potential for adverse human exposure. This leads European authorities to adopt what is referred to as a hazard-based approach versus the risk-based approach more common among U.S. authorities.[2]

The European hazard-based approach arises from a commitment to a "polluter pays principle" and a "precautionary principle" in guiding environmental health policy. The polluter pays principle, which has long been enforced in European jurisprudence, requires that the party responsible for harm should pay for its damages. This principle, which has many parallels in U.S. law, underlies policies for waste site clean-up, water pollution, and basic negligence law. The precautionary principle requires action on potential hazards, even where the science to substantiate those threats is not conclusive. In 1992, this principle was explicitly incorporated into the environmental section of the Maastricht Treaty of the European Union, and since then, it has been consistently referenced and adopted in many national laws and international treaties.

The polluter pays and precautionary principles provide for a proactive approach to risk management. Uncertainties in studies linking hazards and outcomes are accepted and high standards of proof are waived because the risks of acting are seen to outweigh the risks of not acting. In theory, this means that those responsible for introducing hazardous chemicals should be responsible for potential harms and should intervene early in the face of uncertainty and avoid well-recognized chemicals of high concern. Joel Tickner, a professor at the University of Massachusetts Lowell and one of the convener's of the Wingspread Conference on Implementing the Precautionary Principle (see box 4.1), describes this approach as a "better safe, than sorry approach." Joel goes on, "This is basic common sense policy. It is the same as 'first do no harm' or 'an ounce of prevention is worth a pound of cure.'"[6]

By the close of the 1990s, there was growing pressure in Europe for a major reconsideration of European chemicals policy. A series of dramatic lapses in public health and food protection over the 1990s ("Mad Cow Disease" in the United Kingdom and dioxin in Belgian chickens) and the unaccountable die off of seals in the Baltic Sea led to a heightened public concern over hazardous chemicals. Government and environmental

Box 4.1
The Precautionary Principle

The Precautionary Principle is codified in both European and international environmental policy. First developed in German water protection law, the concept was adopted into the Rio Declaration of the 1992 United Nations Conference on Environment and Development, which stated:

Where there are threats of serious or irreversible change, lack of full scientific certainty shall not be used as a reason for postponing cost-effective measures to prevent environmental degradation.[3]

Such an approach promotes action in the face of uncertainty and encourages thoughtful assessments prior to introducing new technologies. The precautionary principle has never been formally adopted in U.S. law, although the idea is inherent in the nation's air quality standards and in the environmental impact assessment procedures required for federal projects under the 1969 National Environmental Policy.

In 1998, a group of scientists and policymakers met at the Wisconsin Wingspread Conference Center to formally consider the precautionary approach and issued a statement called the "Wingspread Statement on the Precautionary Principle."[4] The statement defined precaution with four components:

• the proponent of an action should bear the burden of proof of safety;
• preventive action should be taken in advance of scientific proof of causation;
• a reasonable range of alternatives, including no action, should be considered; and
• decision making should be open, transparent, and engaging of all appropriate stakeholders.

Critics of the precautionary approach have focused on the caution part of precaution and see it as stifling innovation by requiring a definitive proof of safety. This is too narrow a reading of the concept. Unlike the "permissive approach" of the environmental regulatory laws that is satisfied with compliance with standards or a "reactive approach" that only rushes in regulatory responses after damage is done, the precautionary approach is a driver for anticipatory innovation and proactive prevention. The German origin of the "precautionary principle" lies in the concept of *Vorsorgeprinzip*, which roughly translates as "foresight," a posture that encourages anticipatory action. By "thinking ahead"—"looking before you leap"—precaution avoids the negative and delaying interpretation of caution and supports technological innovations and chemical substitutions that prevent harm and promote safety.[5]

leaders were increasingly frustrated with the ineffectiveness of the myriad of European chemical policies. Their so-called "flow of worry" focused on the lack of progress on chemical testing, the limited amount of public data on chemicals, the asymmetry between new and existing chemical regulation, and the slow response to persistent and bioaccumulative chemicals. Something new and different needed to be done.

4.2 REACH

In June 1999, the European Council of Ministers asked the European Commission to prepare a proposal for a substantial overhaul of the European Union's chemical policies. The European Commission in Brussels plays a significant role in shaping and implementing European law. The Commission typically crafts the initial text for new laws, and these texts are modified through negotiations within the European Parliament and the Council of European Ministers, both of which must agree on and pass the final text. In terms of environmental policies, these laws can be either "regulations" that come into immediate effect as European Union-wide laws or "directives" that provide more flexibility in how they are adopted by each of the Member States.

The European Commission spent more than a year in deliberations and released its *European White Paper on a Strategy for Future Chemicals Policy* in 2001, calling for the replacement of many of the existing directives and regulations on chemicals with a new, more integrated, and precautious chemical policy called "REACH" (Registration, Evaluation, and Authorization of Chemicals). The proposed regulation would require all chemicals to be registered, eliminate the distinction between new and existing chemicals, shift the burden of proof concerning the safety of chemicals from government to industry, require industry to conduct new chemical safety test, and substitute safer substitutes for the most hazardous chemicals. A new European Chemicals Agency (ECHA) would manage the volumes of new information made available by the registering firms and provide for technical backup while the European Commission would retain its policymaking role. The Member States would be responsible for overseeing the generation of new chemical information through various testing requirements and for the overall compliance and enforcement procedures.[7]

It took five years of slow and heavy negotiations before the European Parliament and the European Council would adopt the REACH regulation, and the political lobbying against the proposal was intense. The U.S. State Department and U.S. trade associations actively lobbied against the regulation in Brussels and many of the Member States. The political pressures were effective, and there were many changes and compromises on the journey between the *White Paper* and the final text. However, the basic framework of the *White Paper* held, and the result was a major overhaul of European Union chemical policy.[8]

Although there are many details and complexities, the REACH process basically involves five related steps: preregistration, registration, evaluation, authorization, and restriction.[9]

Preregistration

During the first step, all manufacturers or importers of chemical substances in volumes of one metric ton or more per year were required to notify the new European Chemicals Agency (ECHA). At the deadline in 2008, ECHA received 143,000 chemical substance preregistrations from 65,000 companies.

Registration

The registration process is staged over a ten-year period, with the amount of data and the number of tests required increasing as the volume of chemicals manufactured or imported rises. The first deadline for registration was January 2011, and ECHA reported receiving 28,942 registrations on 5,147 substances.[10] This included substances with annual production of more than 1,000 metric tons, and all carcinogens, mutagens, or substances toxic for reproduction (CMRs); all persistent, bioaccumulative, and toxic substances (PBTs); and all very persistent and very bioaccumulative substances (vPvBs). For these substances, the manufacturer or importer was required to provide a registration dossier that includes both a technical dossier and a chemical safety report. The technical dossier contains information about the identity, chemical-physical, toxicological, and eco-toxicological properties; information about the possible uses; and a newly developed safety data sheet. The chemical safety report presents "exposure scenarios" and includes a special chemical safety assessment. Exposure scenarios provide descriptions of how a substance should be used, what human exposures might be expected during those uses, and what risk management procedures are recommended. The chemical safety assessment has similarities to a conventional chemical risk assessment. For substances with annual production of 100 to 1,000 metric tons, the manufacturer or importer was given six years to register, and the registration dossier includes a similar technical dossier and chemical safety report. Manufacturers or importers of substances of 1 to 100 metric tons per year do not need to register until 2016 and have reduced requirements for their registration dossier. In preparing the chemical safety assessments, REACH provides generous provisions for confidential business information (CBI) protections and encourages collaborations among firms—so-called Substance Information Exchange Forums (SEIFs)—in preparing common dossiers among firms.

Evaluation

There are two types of evaluations: dossier and substance. The dossier evaluation completed by the ECHA provides for a completeness check on the registration dossier and a review of testing procedures to prevent unnecessary animal testing. Each year, the Member States are required to conduct substance evaluations on a select number of registered chemicals. In 2012, ninety substances were identified for the Member State substance evaluations. The substance evaluation permits the government authorities of each Member State to review the registrations and require the manufacturer or importer to obtain and submit more information if the authority believes that the substance poses a risk to human health or the environment.

Authorization

The Commission is responsible for identifying substances of very high concern (SVHCs) and placing them in a special annex (so-called "Annex XIII"). This annex includes the CMRs, PBTs, vPvBs, and other substances of equivalent concern. Once a substance is placed in this annex, the Commission establishes a sunset date after which the chemical cannot be manufactured, imported, or used without special authorization. In 2010, the Commission posted forty-six candidate substances of very high concern to Annex XIII, and by 2014, 155 candidate substances had been posted and thirty-one substances had been approved for authorization.

The Commission is responsible for granting authorizations. In order to obtain an authorization, an applicant must present a chemical safety assessment to demonstrate that the projected risks during production, use, and disposal ("exposure scenarios") are "adequately controlled." Evidence of adequate control is not possible if there is no safe threshold for exposure, and the regulation is clear that this is the case for all CMRs, PBTs, and vPvBs. If adequate control cannot be established, the authorization may yet be granted if the applicant can show that the social and economic benefits of the production and use of the chemical outweigh the projected risks and that there are no other suitable substitute chemicals or technologies available. If the applicant cannot present compelling evidence by either of these two routes, then the authorization application will be denied and the chemical cannot be produced or used.

Restriction

The production, importation, or use of a substance can still be prohibited or subjected to control conditions throughout the European Community if the Commission determines that the risks to human health or the environment are "unreasonable." Where the Commission reaches such determination, it can ask the ECHA to prepare a dossier that includes options for restrictions. Member States can also prepare such dossiers on substances they believe require community-wide restriction; however, only the Commission can make a final decision on a community-wide restriction. Currently, there are fifty-nine categories of restricted substances including more than one thousand substances.

REACH provides a fundamental transformation of European Union chemical policy. As a European Union regulation, it supersedes the individual laws of the Member States, harmonizes chemical policies across Europe, supports the European common market, and promotes innovations in chemical research and development. With some exceptions (e.g., polymers, research chemicals), it covers all chemicals on the European market over one metric ton per year. Foregoing the older directives (and in contrast to the American approach under TSCA), it eliminates the distinction between new and existing chemicals and requires the registration of all chemicals and, thereby, eliminates the advantages that less regulated existing chemicals previously had over new chemicals.

REACH was promoted during its negotiations as a new model of the precautionary approach. The regulation adopts the chemical registration employed in existing American and European pesticides laws and requires that chemical manufacturers or importers generate an extensive dossier of information on the inherent properties, hazards, and potential exposures of each chemical. The regulation puts teeth into this requirement by threatening market prohibition to chemicals without the required information. In the regulation, this is called "no data, no market."[11] In so doing, REACH shifts the burden for presenting scientific evidence on chemicals from the government to the chemical manufacturer and changes incentives in such a way as to make the manufacturer responsible for demonstrating the safety of the chemical. The regulation clearly states that it is "based on the principle that it is for manufacturers, importers and downstream users to ensure that they manufacture, place on the market or use such substances that do not adversely affect human health or the environment."[12]

By stipulating that some substances of very high concern (e.g., CMRs, PBTs, vPvBs) cannot be "adequately controlled," REACH creates a precautionary barrier to easy authorization. In establishing a rebuttable presumption that certain highly dangerous chemicals should not be used and requiring current or prospective users to defend their need for the chemical against the risks, this again shifts the conventional relationship between the private sector and the government. However, this stark reversal of responsibilities is clouded by two somewhat generous routes that REACH left open to the manufacturer or downstream user for presenting their case for authorization. By permitting an option for demonstrating that a substance is "adequately controlled" and providing for a social and economic cost-benefit calculation as a defense for continued use, the law provides a potentially wide "off ramp" for some substances of very high concern to remain on the market without pressure for substitution.[13]

Moreover, once a manufacturer or importer completes a chemical safety assessment to register the chemical, there is no requirement to implement appropriate risk management measures. Nor is there much new in the restriction authority granted to the Commission as a kind of "default" for hazardous substances that otherwise do not get addressed. Here, the burden for proving that a chemical should be restricted falls to the government in the conventional manner consistent with the now replaced directives.[14]

These issues aside, REACH offers a major policy step forward. In its foundation, it is comprehensive and inclusive. It is clearly driving the generation of a significant amount of new chemical information, and it has dramatically shifted the burden to industry to provide that information. Although the regulation does offer a significant confidential business information (CBI) curtain, much of the process is relatively open and transparent. Requiring chemical manufacturers and users to provide new or previously hidden information in a manner that is open to public perusal and use is intended to drive the market toward safer chemicals. However, it is still too early to determine whether the authorization or restriction process will result in phasing out the production and use of many substances of very high concern.

4.3   European Policies on Chemicals in Products

Although REACH provides an overarching framework for chemical policy in Europe, it has not stopped a wave of new legislation regulating chemicals in products. As examples, we consider four of these.

Electronic Products
In 2002, the European Union enacted a Directive on Waste Electrical and Electronic Equipment (WEEE) to divert computers, television, stereos, and other electronic products from disposal by encouraging their collection and recycling. However, during the contentious negotiations, it became apparent that electronic products were difficult to recycle because they contained various hazardous heavy metals and chemical additives. Recognizing that these substances needed to be eliminated, the European Commission drafted a separate Directive on the Restriction of the Use of Certain Hazardous Substances in Electrical and Electronic Equipment (RoHS) that was subsequently enacted by the Parliament and Council and came into force in 2003.[15]

The RoHS Directive sets restrictions on the use of lead, cadmium, mercury, and hexavalent chromium used in the components of electronic products and polybrominated biphenyls and polybrominated diphenyl ether used as flame retardants in the plastic housing of commercial products. Under the directive, all new electronic products in seven product categories put on the market after 2006 must be free of these substances. These restrictions have been impressively successful at driving these chemicals out of European electronic products and forcing the adoption of safer alternatives and, in so doing, causing similar substitutions in electronic products around the world.

Cosmetics
In April 2009, the European Parliament approved a new regulation that overhauls the 1976 Cosmetic Products Directive. The new Cosmetics Products Regulation sets European-wide standards for notification, testing, and labeling of cosmetic products and replaces a patchwork of twenty-seven sets of national rules and fifty-five amendments to the original directive. One of those amendments passed in 2005 required that all chemicals used in cosmetics be tested and that all carcinogens, mutagens, or substances toxic for reproduction (CMRs) be eliminated from cosmetic products. By the end of 2006, the list of substances restricted from cosmetics (the "negative list") had grown to 1,100. In addition to this negative list, the Cosmetic Directive establishes a list of conditional use substances and specific lists of permitted colorings, permitted preservatives, and permitted ultraviolet filters. The new regulation also regulates nanomaterials in cosmetics requiring notification, safety assessments, and effective labeling and prepares the way for regulating endocrine disrupting chemicals once an authoritative list is established. Today, scientific committees

established by the Health and Consumer Protection Directorate are over-seeing the testing of hundreds of ingredients found in cosmetics.[16]

Biocides

European community legislation on pesticides dates back to a 1976 directive that specified maximum levels of pesticide residues on certain fruits and vegetables. In 1991, the Biocidal Products Directive was adopted by the European Council to create a cross-national framework for authorization, use, and control of plant protection products and has led to harmonized Maximum Residual Levels for some 1,100 pesticide products. In addition, the directive classified active ingredients as either "existing" (on the market by 1993) or "new" and required that all (920) existing active ingredients be tested to determine whether they should be placed on a special annex of "acceptable" pesticides or, if not, phased out. Over the past fifteen years, the evaluations have been conducted by government agencies in each of the Member States and have been staged in four groupings of chemicals, beginning with the most widely used and progressing over time to more marginal compounds. As of 2009, 889 of the 920 active ingredients have been evaluated, and only 194 have been posted to the acceptable substance annex. New active ingredients have been emerging during this period at a rate of about eight per year. Each new compound must go through an evaluation by a Member State authority similar to the evaluation required of the existing ingredients. To date there have been eighty-two new active ingredients posted to the acceptable substance annex.[17]

Toys

In 2009, the European Union passed a Directive on the Safety of Toys that replaces and updates existing European Union and Member State laws. This new directive provides for a comprehensive coverage of all the safety features of all children's toys, ranging from physical and mechanical hazards to chemical hazards, and prohibits a range of fragrances and CMRs except for those that are inaccessible to a toy user, such as the nickel used in steel parts. A new set of harmonized safety standards have been created so that toy manufacturers can self-certify their products if they are in conformance with the requirements of the standard. The part of this safety standard addressing hazardous substances has yet to be completed.[18]

Each of these new product directives is focused on products in specific economic sectors—electronics, cosmetics, agriculture, toys—and on accepting or phasing out chemicals based primarily on the inherent hazards of the chemical and only secondarily on how it is used or how exposures might occur. The RoHS Directive is particularly impressive because, in its efforts to regulate the environmental aspects of electronic products, it eliminates the use of several hazardous chemicals that provided risks to workers in fabrication plants and problems in end-of-life recycling processes. By focusing directly on several chemicals of high concern, the RoHS Directive provides a telling example of how a hazard-based approach shifts the focus from managing human exposures to catalyzing product redesigns that lead toward inherently safer products.[19]

4.4   Chemicals Policy in Asia

Asian countries have also been active in crafting chemicals policies.[20] Early in 1973, Japan enacted a framework Chemical Substances Control Law that requires reporting and classification of new and existing chemicals and provides for the regulation of chemicals found to be unreasonably harmful. Japan established a pollutant release and transfer registry (PRTR) in 1999 similar to U.S. Toxic Release Inventory to track the annual releases of a small set of hazardous chemicals. During this past decade, Japan enacted laws on poisonous and deleterious substances and the control of household products containing hazardous substances and ministerial ordinances on chemical testing and risk assessment.

Korea passed a Toxic Chemicals Control Act in 1990. It provides a framework for distinguishing between new and existing chemicals, offering a separate reporting requirement for both. This law further provides authority for the creation of lists for toxic chemicals, "observational chemicals" (chemicals to be avoided) and severely restricted or banned chemicals, rules for hazard and risk assessment, and the establishment of a PRTR for tracking chemical releases.

In China, chemical management authorities were originally laid out under the National Environmental Protection law of 1979 and implemented by several agencies according to chemical uses (industrial chemicals, agricultural chemicals, medicines, food additives, disinfectants, cosmetics, etc.). National regulations enacted during the 1980s established criteria for the classification of toxic and hazardous operations and the safe management of hazardous chemicals. An Inventory of Existing Chemicals Substances was established in 2001 that today lists 45,612 chemicals. In 2008, China elevated its State Environmental Protection Agency into a Ministry of Environmental Protection, which currently maintains the Inventory of Existing Chemicals and separate inventories for highly toxic chemicals, highly poisonous chemicals, and dangerous goods. However, the regulation of existing chemicals occurs under a complex web of regulations administered through an ill-coordinated arrangement among some eleven agencies ranging from the Ministry of Health and the Ministry of Commerce to the National Development and Reform Commission. Those chemicals not appearing on the national chemical inventory are considered new chemicals and are now regulated under the Measures on the Management of New Chemical Substances and the Hazard Evaluation of New Substances both enacted in 2008.[21]

Asian countries have been particularly sensitive to European initiatives. The RoHS Directive enacted by the European Union was followed by the Chinese Management Measures for the Prevention and Control of Pollution from Electronic Information Products in 2006 and similar versions modeled on RoHS in Japan, Korea, Thailand, and Taiwan. The China measure has two phases. The first requires that the packaging for all electronic products must be labeled as to whether the product contains any of the substances restricted under the RoHS Directive, and the second, sets specific restrictions for those chemicals. The other "Asian RoHS" measures have similar labeling requirements but are not mandatory although "strongly recommended."

The passage of REACH further spurred Asian chemical policy. In 2009, Japan significantly amended its chemical substance control law to expand the PRTR and establish a tiered procedure for hazardous chemical regulation. Chemicals found to be severely dangerous to human health or flora are posted to a special Class I or Class II list. Chemicals on the Class I list (currently 16) require permission for further manufacture and use, whereas chemicals on the Class II list (currently twenty-three) require annual production reports, special handling, and the labeling of products in which they appear. Another amendment in 2011 requires notifications on all chemicals produced or imported in volumes over 1 metric ton and any chemical on a list of Priority Assessment Chemicals (PACs) (expected to grow to 1,000 chemicals) for which a special risk assessment must be conducted.[22]

Hazardous chemicals in Korea are regulated under the 1991 Toxic Chemicals Control Act. In 2009, the Korean Ministry of the Environment issued a chemicals action plan called GreenSHIFT (Safety, Health, Information, and Friendship Together) based on amendments to Korea's Toxic Chemicals Control Act. Green SHIFT requires new testing requirements for new chemicals, mandatory submission of hazard and exposure information on existing chemicals, data collection on high-production-volume chemicals, and the introduction of risk assessment and socioeconomic analysis tools in chemical policymaking.

In 2010, China amended its Measures on the Management of New Chemicals to lay the framework for a PRTR and create a classification of new chemicals into three categories: general new chemicals, hazardous new chemicals, and priority hazardous new chemicals. Regulations under the amendment now require annual reporting on the use of hazardous and priority new chemicals and the phase-out of 120 priority substances through new chemical substitution. A regulation on prohibiting the import and export of highly dangerous substances (currently 158) was passed in 2011.[23]

Although Japan has an early history of chemical policy dating back to the 1970s, the experience of other Asian countries has more recent beginnings. Enforcement and compliance are strong in Japan, although there remains a gap between policy adoption and local-level compliance in the other countries. However, there have been significant chemical policy initiatives in Asia, many driven by a strong desire to harmonize with the policies of the European Union.

4.5   International Chemical Policies

As Europe and Asia have moved forward with new chemical policies, a growing body of international law has emerged regarding chemical management.[24] Under sponsorship of the United Nations Environment Program, new international treaties on the management of industrial chemicals have been drafted and ratified. The 1985 Vienna Convention for the Protection of the Ozone Layer set conditions for the phase-out of ozone-depleting chemicals, the 1989 Basel Convention on the Control of Trans-boundary Movement of Hazardous Wastes and Their Disposal established regulations on the global shipment of hazardous wastes, and the International Labor Organization's 1992 Convention Concerning Safety in the Use of Chemicals at Work recommended guidelines for the workplace use of chemicals. Of these the convention on ozone protection has been the most effective with a well-implemented international agreement under the Montreal Protocol to phase out the production and use of several ozone-depleting substances, including the ubiquitous chlorofluorocarbons.

The United Nations Conference on Environment and Development (UNCED) convened in Rio de Janeiro in 1992 provided a significant catalyst for international chemicals policy. *Agenda 21*, the strategy adopted at the close of the conference, set the stage for a broad international effort to transition to a sounder and more sustainable economy. In chapter 19, *Agenda 21* laid out a set of six program areas to encourage the "environmentally sound management of chemicals." To carry this out, the World Health Organization and the International Labor Organization joined the United Nations Environment Program in establishing the Intergovernmental Forum on Chemical Safety (IFCS). One of the first tasks of the new IFCS was to convene government and nongovernment organizations from around the world to recommend a new international treaty to restrict persistent organic pollutants (POPs). Negotiations to create a global treaty on POPs began in 1998, and after several rounds of contentious negotiations, a formal treaty was signed in Stockholm in 2001. The Stockholm Convention identified twelve organic chemicals of common international concern and set in place guidance by which national governments agreed to work to eliminate them. Among the chemicals were eight pesticides, polychlorinated biphenyls and dioxins, and furans. Since the treaty came into force in 2004 with more than 150 signatory countries, another nine new substances have been added to annexes for elimination or restriction.[25]

A second treaty, the Rotterdam Convention on Prior Informed Consent, focused on the exporting of hazardous chemicals to developing countries proceeded in parallel with the Stockholm Convention and came into force in 2004. This treaty requires that exporting enterprises in a signatory country must provide advanced notice (prior informed consent) to importing countries when shipping any of the hazardous chemicals listed in a special annex. Today, that annex lists forty-three substances (thirty-two pesticides and eleven industrial chemicals).[26] Pesticides are more fully addressed under the United Nations Food and Agriculture Organization International Code of Conduct on the Distribution and Use of Pesticides. This voluntary standard covers pesticide testing; distribution and trade; use; labeling, storage, and disposal; information exchange; and regulation with special attention to developing countries.[27]

In 2009, negotiations began on an international treaty on mercury, a neurotoxin considered highly toxic to fetuses. Some three years later, the negotiations were concluded, and in 2013, the treaty was signed in Minimata, Japan, where industrial mercury discharges polluted a bay and caused serious physical and reproductive damage to local residents. The treaty addresses several of the leading sources of mercury exposure, such as mercury emissions from coal-fired power plants, worker exposure in small-scale artisanal gold mining, mercury use in medical devices (thermometers), and mercury in dental amalgams. However, mercury use in vaccines was exempted from the treaty in deference to the need for cheap vaccines.[28]

In 2003, the United Nations Economic and Social Council adopted the Globally Harmonized System of Classification and Labeling of Chemicals (GHS). The GHS establishes a single, international standard for the classification of hazardous chemicals and the labeling, nomenclature, and communication about their hazards. The GHS standardizes the hazard statements, warning words, pictograms, precautionary statements, and supplier identification for communicating about chemicals for every country and every language. It draws on the hazard terms and symbols and the means of communication that a few industrialized countries had developed independently and creates a uniform system for all countries. In addition, the GHS set an international standard for the Safety Data Sheets (called "Material Safety Data Sheets" in the United States) that chemical manufacturers prepare to provide chemical identification, proper management, and safety instructions to chemical users. The GHS is not a binding regulation but an international standard. The motivation was to advance global safety and global trade by harmonizing national nomenclature and standards on chemicals in commerce. As the GHS is adopted, it is intended to increase the level of corporate and governmental chemical information transfer throughout the world.[29]

The European Union has moved rapidly to adopt the GHS, and in 2006, the Regulation on the Classification, Labeling and Packaging of Chemical Substances and Mixtures (CLP) was adopted by the European Commission. The CLP replaces past European labeling requirements with a system harmonized with the GHS, which provides new classification criteria, hazard symbols, and labeling phrases. The regulation requires companies to classify, label, and package all chemical products appropriately and to notify the European Chemicals Agency (ECHA) of the GHS classification of each chemical on the market. By 2011, some 3.1 million chemical classification notices had been received.[30]

This rapid proliferation of international and regional chemical treaties and bodies with chemical responsibilities has led to inconsistencies and gaps. For instance, there are no international conventions for addressing hazardous chemicals that are not POPs, no authorities for managing the trafficking in illegal and banned chemicals, and no programs for promoting safer chemicals in products. This has raised concerns about a lack of coordination among the international agreements, and the 2002 World Summit on Sustainable Development (WSSD) called for the development of a global strategy on the management of chemicals. In so doing, the WSSD renewed the *Agenda 21* commitment to the sound management of chemicals and pledged:

to achieve, by 2020, that chemicals are used and produced in ways that lead to the minimization of significant adverse effects on human health and the environment, using transparent science-based risk assessment procedures and science -based risk management procedures, taking into account the precautionary approach, as set out in principle 15 of the Rio Declaration on Environment and Development.[31]

The idea was to create an international strategy that could serve to unify and coordinate the emerging international treaties and standards and provide for their effective implementation particularly in the industrializing countries. In 2006, after three years of development, an International Conference on Chemicals Management (ICCM) held in Dubai adopted the Strategic Approach to International Chemicals Management (SAICM). Rather than a separate treaty, SAICM provides a global strategy made up of three core texts: the *Dubai Declaration on International Chemicals Management*, an *Overarching Policy Strategy*, and a *Global Plan of Action* designed to facilitate a comprehensive, worldwide approach to promoting sound chemical management.

The Dubai Declaration laid out a clear goal modeled on the 2002 WSSD goal:

The overall objective of the Strategic Approach is to achieve the sound management of chemicals throughout the life cycle so that, by 2020, chemicals are used and produced in ways that lead to the minimization of significant adverse effects on human health and the environment.[32]

The *Global Action Plan* includes thirty-six work areas with some 273 specific implementation activities that range from cleaning up contaminated chemical waste sites to encouraging industries to seek safer chemical processes, increasing education about the health effects of chemical exposure, and building regulatory capacity in developing countries. With funds from the United Nations Global Environmental Facility, SAICM set up a secretariat in Geneva and now provides competitively awarded funds to government initiatives under its "Quick Start Program." SAICM provides a nonbinding strategy rather than a regulatory treaty. However, the importance of the agreement lies in the legitimacy that it provides all participants, including small, developing countries and nongovernmental organizations, in equal access to global chemical policy agenda setting and program development. Five global initiatives have been established, including heavy metals (mercury, lead, cadmium, and hexavalent chromium), nanotechnology, information on chemicals in products, electronic product waste, and endocrine disrupting chemicals, and workgroups have been established in each area to collectively set goals and design international implementation programs.

The United States has been hesitant and sometimes resistant to participating in negotiations on these conventions, arguing that these treaties are a veiled attempt at trade barriers and domestic market protection. The U.S. delegation worked hard to prohibit the Stockholm Convention from covering more than the original twelve substances. During the SAICM negotiations, several countries, including the United States and Australia, worked to narrow the strategy. Moreover, the U.S. Congress has failed to ratify most of these treaties, and U.S. participation by the Department of State in the ensuing meetings of the treaty members has been limited and subdued.

4.6   The International Effects of Chemical Policies

The European chemical policies of the past decade are shifting the design and manufacture of internationally traded goods and reframing the regulatory requirements on the global chemicals market.

The RoHS Directive banning six hazardous substances from most electronic products sold in Europe has driven electronic manufacturers in Korea, Taiwan, China, and Japan to eliminate these same chemicals from products sold throughout the world. As noted above, Japan, Korea, and China adopted measures to promote the RoHS Directive among their national electronics industries. Ted Smith, the founder of the Silicon Valley Toxics Coalition, the leading environmental advocacy organization in the United States focused on the electronics industry, calls the RoHS Directive the "most important regulatory driver for changing the chemicals used in electronic products since the Montreal Protocol." With his help, in 2005, California passed a law phasing out the same six substances and requiring that any additional substances added for phase out to the RoHS Directive would automatically be phased out in California as well.[33]

Several recent policy reviews have pointed to the range of impacts that REACH requirements are having on the international chemicals market, in the United States and globally.[34]

First, the EU requirements for chemical information disclosure have become the international standard for chemical and product ingredient information. All of the major U.S.-based chemical corporations are generating chemical safety reports conforming to REACH registration requirement for most of their products regardless of whether those products are sold in Europe. Henrik Selin and Stacy VanDeveer note the effects of these policies on multinational firms:

If non-EU firms want to continue to sell their products in the EU, they will have to comply with EU product rules and standards. Most firms operating in multiple markets prefer to produce their products to as few different standards as possible and they often follow the highest regulatory standard. This is particularly likely for products where major producers compete across markets.[35]

Years ago, David Vogel, a professor of Business Ethics at the University of California, Berkeley, termed this the "California effect," whereby the more demanding policies of jurisdictions with the "greenest policies" become the dominating policies throughout the market. In this way, "market incentives can serve to promote the ratcheting upward of regulatory standards."[36]

Second, the new chemical information generated under REACH requirements is having broad impacts on product manufacturers, retailers, and consumers. Product manufacturers around the world look to these initiatives in making chemical selection decisions as the chemical information presented in the registration safety dossiers is driving new demands from retailers and consumers in selecting safer products.

Third, REACH-generated information has become useful in non-European countries when they are implementing their own environmental programs and regulations. The U.S. EPA is now using REACH registration data to support new testing requirements and pursue "substantial risk" information revealed by firms manufacturing existing chemicals.

Fourth, the bold ambitions of these new international initiatives, particularly the enactment of REACH and RoHS, have already become a model for emulation by activists and policymakers in Asia and the United States. State legislators in Maine, Minnesota, Washington, and California have referred to the successful campaign for REACH as a positive example in promoting their own new state chemical policies, and REACH has become a catalyst for a recent activist-driven campaign to reform TSCA here in the United States.

The international chemical treaties and active United Nations bodies are also driving local and regional chemical policies. The National

Implementation Plans required under the Stockholm Convention and the periodic surveys conducted under SAICM have spurred national governments to assess domestic chemical management practices and plan for improved chemical policies. The GHS has become the global standard for chemical hazard assessment and funding from the United Nations agencies and various national development funds underwrites new government chemical initiatives within developing countries. With parallel funding, the International POPs Elimination Network, (a nongovernmental organization) offers small grants to hundreds of small community-based organizations around the world that are working on local chemical management programs.[37]

There are important lessons in these new chemical policy initiatives:

• The policy shift ushered in by REACH from a chemical-by-chemical universe to a holistic approach that covers the entire chemicals market provides a comprehensive platform for collecting and presenting chemical information, setting priorities, and strategically targeting regulatory responses.

• The precautionary approach that shifts the burden for proving the safety of chemicals creates a normative "duty of care" onto the suppliers of chemicals.

• The European RoHS, Cosmetics, Biocide, and Toy Directives illustrate the potential opportunities for policy approaches focused on chemicals that are produced for and used in specific economic sectors.

• The goals of SAICM to promote the sound management of chemicals globally suggest the importance of broad and ambitious goals and the value of a defined strategy for managing chemicals in a truly global market.

Whether by default or conscious intention, these national, regional and international government initiatives are setting the global standards for chemical policies. The pressures created by these programs drive higher levels of environmental performance throughout the world and feed the desire for more and better government and corporate chemical policies.

4.8   Chemical Policies for a New Chemical Market

Although little progress has been made in changing federal chemical policies in the United States, the past twenty-five years have witnessed continued development outside U.S. borders. European environmental policy that developed in tandem with the U.S. policies up through the 1970s has continued to develop independently in the decades that have followed. Significant new concepts such as the polluter pays and precautionary principles have developed at the national level while REACH has introduced government registration for all chemicals and shifted responsibility for providing chemical information to industry. The European RoHS Directive and Cosmetics Directive have advanced prohibitions stricter than anywhere in the world on the use of certain high hazard chemicals in commercial products.

These new European policies owe much to the structure of the European Union. The EU is new, with a relatively open and flexible set of agencies that are still evolving and lack the rigidity of decades of formalized structure building. The expertise, knowledge, and commitment of Sweden, Denmark, the Netherlands, and Germany, eager to protect their own high standards, have dominated Europe in shaping chemical and product policies. In particular, the Green Parties of these northern countries have provided political support for chemical policy innovation, and without their determination, these new directives and regulations would never have been so bold or precautious.

Europe's progressive steps are reshaping the global chemical market. With more than 40 percent of the worldwide chemical sales now required to meet new European standards, these regulations have become *de facto* international standards, reshaping the U.S. chemical market as well as the global market. In some cases, chemical health and safety protections can simply float upward, buoyed by the rising global standards. In other cases, and this is more likely with regard to chemical constituents of products, the United States may be consuming more hazardous products than those sold in Europe, and the more dangerous products made in China and other industrializing countries that are now banned from European markets may now be entering the U.S. market.[38]

The new chemical policy directions set by the international chemicals treaties and conventions and the SAICM commitment to the sound management of all chemicals further shape the requirements of the global chemical market. The refusal of the U.S. government to ratify most of the international chemical treaties or to formally acknowledge SAICM means that the federal government can play no more than a nonvoting observer in international chemical policy decision making. This makes U.S. resistance to new international chemical policies increasingly irrelevant. The global chemical market is now responding to the more progressive chemical policies beyond U.S. borders. However, this presents significant domestic costs. The inability to participate in international chemical policymaking has meant that the U.S. government is increasingly unable to set the health and environmental standards of its own chemical market, and the United States is decreasingly a significant influence in the global governance of chemicals.

The outdated chemical control policies of the United States and their disappointing performance can be contrasted with the innovative and ambitious chemical policies arising internationally. In the face of an increasingly globalized chemical market, with environmental and consumer safety standards now dictated by foreign governments, the United States cannot long remain a hold out if it is to remain competitive internationally. Were the United States to update its chemical policies, it could begin with minor adjustments to its existing statutes to harmonize them with international requirements. However, there is a bolder opportunity here to build on these international standards and create an innovative, new approach to chemicals management that not only reclaims U.S. leadership but erects scaffolding for a truly sustainable system of chemicals production and consumption. However, this would require some fundamental rethinking about the chemicals problem.

5

Reframing the Chemicals Problem

An old adage holds that solutions too often fail, not because they are not good solutions, but because they are addressing the wrong problem. To this point, we have largely reviewed the various solutions that have been put forward during the twentieth century for addressing the risks of hazardous chemicals. We have not considered with equal scrutiny the nature of the problem that these solutions were trying to solve. Before we move on to look for new solutions, we will consider more carefully "the chemicals problem" as it has been framed and how it could be reframed more effectively.[1]

5.1 Rethinking the Chemicals Problem

Those who came together to draft the chemical control policy framework of the 1970s were leaders from the environmental, consumer protection, and labor movements. They had a strong sense of their mission. The problems that they faced were fairly clear. Toxic and hazardous chemicals were polluting the environment and threatening the health of workers and consumers. They believed in the efficacy of government and saw government regulations as a means to reduce and control the risks. These advocates were well grounded in the social movements they represented. They were also pragmatists willing to negotiate in the halls of Congress for remedies that fit within the structure of Congress and the ideology of the time. Their grandest goal was protection of public health and the environment. What they were not were chemists who knew how to make chemicals or how to use chemicals to make products.

Had the leaders of the environmental movement of the 1970s been more focused on chemical production, they might have recognized the complex ways in which chemicals are synthesized and processed and understood that, in order to shift to safer chemicals, fundamental changes might be required in the structures of the chemical market and the practice of chemistry. They might have decided to draft laws that promoted new innovations in chemistry and new ways to shape and redirect the production of chemicals. But the proponents of the 1970s were not trying to make safer chemicals or reorganize the chemical industry. For them, the challenge was to provide government with the authority and mandate to identify the most dangerous chemicals and require a reluctant industry to restrict their use. The trajectory was defensive and protective, distrustful of industry and reactive to the dangers that appeared self-evident. They were writing chemical control policy not industrial policy.

However, much has changed during the past thirty years.

• The number and volume of synthetic chemicals on the market has grown significantly. In the 1970s, the TSCA Inventory included some 67,000 substances; today, it is more than 85,000 substances. Of these chemicals, an estimated 30,000 are in wide commercial use.[2] The pre-registrations submitted under REACH in 2008 identified 143,000 substances, and this figure does not include polymers or chemicals used under 1 ton per year. Not only, the number of chemicals, the volume of chemicals is staggering. The EPA's 2012 Chemical Data Reporting program identified 9.5 trillion pounds of chemicals manufactured or imported into the United States. Of these chemicals, some 2,800 are manufactured or imported in volumes over 1 million pounds per year and another 14,000 over 10,000 pounds.[3]

• The chemical industry and its markets are now truly global. The manufacture of products is a global enterprise, with chemicals manufactured in most industrialized countries and, increasingly, in many developing countries. The chemical sectors of Brazil, China, Russia, and India are steadily expanding. China has recently become the world's largest national chemical supplier. The purchase and use of chemicals in the manufacture of products ranging from foods, cosmetics, and household products to durable goods and vehicles occur throughout the world. The majority of workers who are exposed to dangerous chemicals during production processes are now widely dispersed about the industrializing world, where the protective regulations of the industrialized countries often have no reach.

• Scientific understanding of the environmental and health effects of many chemicals has advanced and confirmed what could only be previously assumed from medical records and daily experience: many commonly used chemicals display one or multiple hazard traits of reasonable concern. There are more than a hundred confirmed and probable human carcinogens, more than a hundred substances classified as reproductive toxins, and 1,045 chemicals classified as "very toxic to aquatic life," and the number of chemicals listed on the federal Registry of Toxic Effects of Chemical Substances rose to 153,000 in 2001, when it was transferred to a private subscription service.[4]

• Some chemicals, including some previously thought to be benign, are capable, at extremely low exposure levels, of interacting with biological systems and altering how genes of living organisms behave. Such changes are implicated in learning and behavioral disorders, infertility, diabetes, heart disease, and cancer. These low-dose exposures can cause effects that are difficult to predict and different from the effects of larger dose exposures, which have conventionally formed the scientific basis of current regulatory standards.

• Although substantial information on the biological effects exists for some of the most studied chemicals, for many other chemicals, that knowledge is absent or insufficient. The EPA's High Production Volume Chemical Program and the REACH registration dossiers have generated significant information on the health and environmental effects of several thousand chemicals, but there remain many data gaps where no studies have been completed or the research is inconclusive.. For most small- and moderate-level production chemicals, there is even less information. However, even where quality research is available, the singular-substance, chemical-by-chemical advances of this science only partially reveals the effects of real-time chemical exposures that occur from chemical mixtures in complex environments involving multiple chemicals with synergistic effects.

• The potential hazards of the novel chemistries emerging from advances in nanotechnology and synthetic biology are largely unknown. In 2007, the global market for products containing nanotechnologies was about $147 billion, and it is expected to grow to $2.5 trillion by 2015. Although much smaller today, the synthetic biology research market is expected to grow to $3.5 billion by 2020. Research on the health and environmental effects of these substances lags behind the rate at which they are being introduced into commerce. Of the $2 billion federal budget for nanotechnology research, less than 5 percent is allocated for research on environmental, health, and safety.[5]

• For some chemicals of concern, the most serious exposures now come from consumer products. The thousands of personal care products, domestic cleaning agents, textiles and garments, and home furnishings that are available on domestic markets throughout both industrialized and industrializing countries are composed of many chemicals that lack sufficient testing. Direct testing of contaminates found in human tissue and fluids reveals that people in advanced industrial countries carry within them many synthetic chemicals found in commercial products. Internationally produced products often carry with them a legacy of processing chemicals used to make them and the promise of a future of pollution once they are disposed. Some 20 to 50 million tons of electronic product waste worldwide contribute to the hazardous properties of many country's domestic wastes and waste exports.[6]

• Public awareness and expectations about the safety of synthetic chemicals are expanding globally. The consuming public in industrialized countries—those who buy and waste most of the products of the global market—have grown wealthier and more demanding about the health impacts and safety of the products they consume. Similar public attention is arising in fast-growing industrializing countries such as Brazil, India, and China. Just in the past decade, a growing consumer movement in China has been raising concerns about industrial pollution and the harmful chemicals in electronics and personal care products.

• National government authorities alone are not capable of managing the vast number of chemicals in a manner that ensures environmental protection or human safety. With no upper bound on the number of chemicals of concern, the *de facto* limit on the number of regulated chemicals has become the capacities of government agencies. Even with more resources, government agencies cannot overcome the information asymmetry, whereby chemical manufacturers have superior access to information on chemical composition and potential effects; and, even if they did, they could not effectively regulate thousands of chemicals.

So this is the state of the current chemicals problem. Scale, globalization, scientific uncertainty, public expectations, and the limits of government resources work against a comfortable resolution. It should be clear from this review that modest and incremental changes alone will not solve the chemicals problem. Put simply, the chemicals problem cannot be solved one chemical at a time. Indeed, government regulations alone cannot solve the problem. The needed changes are broad and structural. The chemicals problem derives from the systemic way in which chemicals are synthesized, processed, used, and disposed within an economy that requires a never-ending stream of chemically based products and services.

The problem has grown from a few recognized toxic chemicals to an economy rich in hazardous chemicals and fundamentally built on them. The earlier chemical control laws were an important step, but they are not expansive or penetrating enough to address the broader chemicals problem.

The trajectory of industrial development has resulted in the increasing chemical intensification of national economies, whereby today most all economies, from the richest to the poorest, are heavily dependent on synthetic chemicals. If this massive infusion of synthetic chemicals into every corner of the world economy from product manufacturing to food production, transportation, construction, and health care were based on relatively safe and benign chemicals, then this intensification would not be of much concern. But public concern has grown because the chemical intensification of national economies has ushered in a great bath of toxic and hazardous chemicals, some well recognized and many unstudied, and there is growing apprehension that such a soak is not a safe one.

In this view, it is the economy generating and consuming millions of pounds of chemicals that is now burdened by the hazards that these chemicals present. The costs begin with the heavy costs on industry for special chemical handling, lost worker productivity due to injury and illness, waste management, pollution control, liability insurance, and regulatory compliance. There are costs borne by workers in terms of compromised health, medical treatment, and lost work days; costs borne by governments in terms of agency staffing, emergency response, and municipal waste management; and costs borne by society in terms of disability insurance and public health care costs. There are the difficult to measure costs borne by the environment in terms of damaged ecosystems and loss of biodiversity. Although markets perform many functions well, they have been poor at incorporating these costs into the full costs of hazardous chemicals. Instead, such costs are externalized onto workers, consumers, governments, and the environment. Reducing these costs and lowering the risks of hazardous chemicals have become one single goal.

Achieving this goal requires transitioning these chemically intensive economies to equally productive but safer economies. Elsewhere, I have argued that the process of moving from a high to a low chemical hazard economy could be referred to as detoxification. By detoxification I mean:

…reducing the toxicity of the materials used in industrial production while maintaining or improving product effectiveness. It means seeking substitutes for the most toxic materials, or developing production processes or products that do not require them, or shifting to new sources of materials that are less likely to leave chemically active and biologically threatening residues.[7]

The detoxification of the global economy requires a transition from products and processes dependent on chemicals of concern to safer ones. This could be accomplished by phasing out dangerous chemicals one chemical at a time—the international phase out of chlorofluorocarbons provides the world's best example, but this will be long and costly. Instead the detoxification of global economies could be accomplished by redesigning production processes and products. Production processes could shift to safer technologies using less hazardous chemical intermediates and generating less hazardous chemical wastes. Products could be redesigned to substitute chemicals of high concern with safer and better performing chemicals. Paints could be reformulated without volatile compounds; perfumes, without benzyl acetate; sunscreens, without benzophenones; plastics, without phthalates; cleaning supplies, without triclosan; inks, without heavy metals; nonflammable textiles, without fire retardants; and electronics, without lead and cadmium.

A chemically intensive economy need not be a dangerous one. However, it will take substantial effort to transition current economies where hazardous chemicals are common to more sustainable economies where safer chemicals are used more respectfully. Hazardous chemicals are not the fundamental problem; instead, they are a symptom of the broader problem of a largely misdirected chemical economy. The chemicals problem has its roots much deeper in the practice of chemistry, the drivers of the chemical market, the objectives of the chemical manufacturing industry, and the structure of the economy. Resolving the chemicals problem requires addressing these underlying determinants.

5.2 Chemicals as Systems

Chemicals do not stand alone. They are constituents of materials, parts of products, and embedded in systems of production and consumption (see figure 5.1).

Modern synthetic chemicals are manufactured in complex and highly integrated systems of production. They are processed and distributed into sophisticated chemicals markets from which they are formulated into mixtures and assembled into the myriad of functional products fundamental to modern systems of consumption. Throughout these processes, they are continuously interacting and reacting with other chemicals,

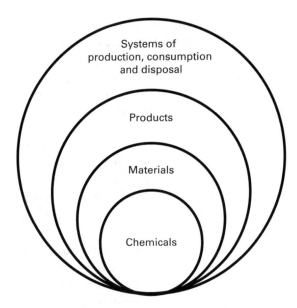

Figure 5.1

Chemicals in Systems

materials, and products. Addressing the problem of hazardous chemicals at this more fundamental level suggests focusing on the systems of chemical production and consumption.

A system is a bounded complex made up of interrelated parts whose behavior is determined by numerous information "feedback" loops. The chemical market and the chemical industry form the core of a gigantic economic production and consumption system. Nearly two hundred years of industrial development have produced a tightly integrated formation of production technologies, commercial corporations, and government authorities that make up a complex system of chemical production, consumption, and disposal.

The manufacture of polyethylene bags provides a good example. The production of polyethylene bags takes place within a subsystem of the petrochemical industry that converts petroleum into naphtha, which is converted into ethylene and then polymerized into polyethylene. However, when naphtha is processed to make ethylene, other co-products are made, such as propylene, butadiene, toluene, and benzene. When the ethylene is processed and polymerized to make polyethylene, co-products are also made that are polymerized to make polystyrene and polyester. These integrated production processes producing products and co-products have been refined over many years to maximize efficiencies and economics. If a decision is made to phase out polyethylene bags, then the polyethylene that was to go into making plastic bags needs to go into other polyethylene products or the production of polyethylene needs to decrease and then the ethylene will need to go into the production of other products. Much of this would be accomplished through the normal functioning of the chemical market because the market is a sophisticated system of interdependencies linking buyers and sellers and constantly adjusting the flow of commodities through price-setting processes. However, the market is only one of the systems that shape patterns of production and consumption. Government policies, technological innovations, social values, ecological dynamics, political campaigns, and sometimes simple personal interventions can dramatically transform or stubbornly protect production and consumption systems. Chemicals, the chemical industry, and the broader economy are all intertwined in a complex set of systems and subsystems, such that changes, even quite minor ones, such as phasing out a chemical product, can be complex and broadly resisted.

This is what Rachel Carson came to understand. She originally set out to determine whether pesticides were causing the demise of songbird populations. However, as she followed the trail of pesticides up to the farmer, she learned how industrial farming, based on single crop production, drove the use of insecticides and, as insects became more resistant, how farmers increasingly needed more and more insecticides. To meet this demand, she found a loosely linked formation of pesticide manufacturers, farm industry organizations, and government extension services all promoting pesticide sales, and then to her great dismay, she found another such formation made up of public health scientists, physicians, and government regulators effectively functioning to calm public concern about the risks of pesticides. In the center of it all, she came to focus on DDT, not because it was the only dangerous pesticide, but because it was such a central product in the midst of a far-flung and multifaceted system. Rachel Carson did not describe these findings in terms of systems, but she did see with prophetic eyes that the cause of the songbirds' demise lie not in suburban backyards but quite far away in highly integrated complexes of ecological, economic, and political dynamics.[8]

Considering chemical production, consumption and disposal requires seeing chemicals in terms of integrated systems that function along the life cycles of chemicals, from chemical synthesis through production and consumption to disposal and dissipation. The linear flow of chemicals

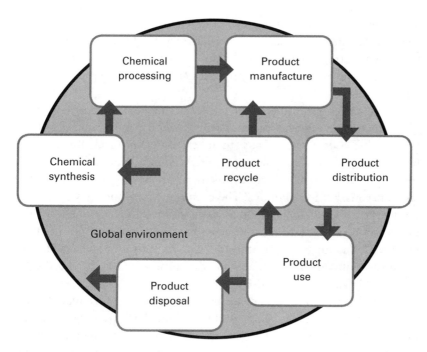

Figure 5.2

Life Cycle of Chemicals

through these multiple stages makes up a life cycle that can be diagramed to present a time-lapse picture which can be analyzed using existing life cycle assessment tools to identify the various impacts that a chemical may have over time. Seeing chemicals dynamically across their life cycles introduces a perspective on the temporal relations in a chemical system and the many points in time where a given chemical may create threats to human health or the environment (see figure 5.2). Careful attention to chemical life cycles also reveals the significant amount of energy, raw materials, and ecological services needed to produce and use chemicals and the large amounts of wastes and emissions generated throughout the production, use, and disposal of those chemicals.

Take nitric acid. Nitrogen is separated from air by liquid air distillation, and hydrogen is conventionally derived from the production of coke. The two elements are reacted over a catalyst at high temperatures to generate ammonia, which is oxidized to produce nitric dioxide. Nitric dioxide is then absorbed in water to produce nitric acid, which is reacted with ammonia gas to produce ammonium nitrate. Ammonium nitrate is used primarily as agricultural fertilizer, where it is dissipated into soils decomposing into nitrogen, oxygen, and water. The life cycle of nitric acid begins with nitrogen from air and returns to nitrogen in air, but along the way, it involves risky oxidative processes and several hazardous processing chemicals.

If chemicals—hazardous as well as safer chemicals—are always parts of production and consumption systems, then it may be fruitful to look within those systems to find ways to increase safer chemicals and decrease hazardous ones. If the problem lies in the chemically intensive economy, then it makes sense to look more directly at the chemical industry and the consumer market for solutions. The chemical control policies developed during the 1970s were based on an assumption that government regulation alone could generate a safer economy. Government policies remain important, but nongovernmental strategies that drive the chemical industry toward safer production and drive the consumer market toward safer chemical consumption will also be critical.

This suggests that the chemicals problem is not best formulated as an environmental or public health problem alone. Such framing limits the options for solutions. Instead, the chemicals problem emerges directly from the way in which systems of chemical production and consumption are organized. These systems make up an economy—a chemically intensive economy—an economy that includes the chemical industry and the chemical market but is also larger, encompassing how chemicals are used to make products and how they are treated once products are disposed. This economy is in need of a redesign. The systems in which the chemicals problem has been conventionally defined lie within broader systems wherein better solutions may appear. Therefore, the chemicals problem can be seen as a problem for which a combination of industrial policy, science policy, and environmental and public health policy may offer the most promising approach to solutions. This is more than a chemical control strategy; it is a strategy to convert the chemical economy to safer chemistries—it is better defined as a chemical conversion strategy.

5.3   A Chemical Conversion Strategy

By reframing the chemicals problems in terms of flaws in the chemical economy, it is possible to move beyond the limitations of the chemical control policy framework that emerged in the 1970s. That framework was focused on the control of hazardous chemicals rather than the conversion of the systems of chemical production and consumption. A strategy that focuses on conversion is one that digs more deeply into the structures of the economy to make system changes that will result in safer chemicals.

While a conversion strategy for changing the chemical economy needs to identify opportunistic intervention points, it must also provide a broader context into which government interventions, industry initiatives, and consumer activism can be coordinated in mutually reinforcing campaigns that add up to pressures and enticements large enough to change systems. Today, in the United States, there are several strategic forces or "strategic fronts" that could collectively be assembled to form campaigns of a broadly integrated conversion strategy for the chemical economy. These strategic fronts are composed of a wide diversity of initiatives, sometimes duplicative and seldom broadly coordinated, but often thoughtfully conceived, enthusiastically led, impressively engaged, and all directed at reducing chemical hazards. These include forces directed at converting the chemical market, transforming the chemical industry, and designing greener chemicals (see figure 5.3).

Converting the Chemical Market

First, there is a broad assortment of consumer education programs, advocacy campaigns and retailer and institutional procurement initiatives that are focused on driving the chemical market toward safer chemicals. This includes government and private product labeling programs, consumer-oriented product rating databases, retailer chemical screening programs,

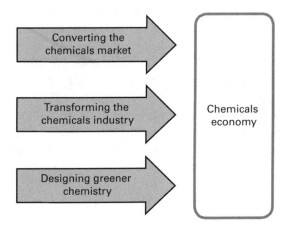

Figure 5.3

Three Strategic Fronts for Converting the Chemical Economy

and institutional preferred purchasing programs. Additionally, the nongovernmental advocacy community has become adept at using market pressures to drive out hazardous chemicals used in products such as lead and phthalates in children's toys, mercury in thermometers, and brominated flame retardants in household furniture.

Transforming the Chemical Industry

Over the past decade, a second strategic front has emerged where multiple initiatives within specific firms or collaborations of several firms in certain sectors of the product manufacturing industries have arisen to shift these industries toward less hazardous chemical products and production. These initiatives are often led by pioneering corporate managers eager to meet customer expectations for safe products. Many of these initiatives have begun to create lists of restricted chemicals, assessment tools, and rating systems for classifying chemicals and exerting pressures on chemical supply chains that are leading toward safer chemicals.

Designing Greener Chemistry

The rapid and global emergence of green chemistry within the academic and applied chemical fields makes up a third strategic front. First appearing some twenty years ago, this campaign by a few dedicated chemists has now expanded among academic chemists and become an accepted section of chemistry conferences, with green chemistry research programs and centers in universities, green chemistry awards programs, and green chemistry textbooks and curriculum appearing around the world.

Activity on these three strategic fronts now makes up a broad, albeit emergent movement toward safer systems of chemical production and consumption. Those involved in the many diverse initiatives are seldom aware of all that is ongoing among others with similar objectives, and indeed they are unlikely to see that all the work even in their own areas makes up something that is here loosely called a "strategic front." However, there is much that is common in the objectives of many of these initiatives, and although it remains too early to call this a clear social movement similar to the civil rights or human rights movements, the makings of a broadly defined safer chemicals movement is a possibility.

5.4   Safer Chemical Policy Framework

The chemical conversion strategy needs a new policy framework—a safer chemical policy framework that can address the entire system of chemical production and consumption. To change that system, this policy framework must be comprehensive, hazard-based, and transformative. Such a framework may involve government or corporate policies. The REACH regulation is comprehensive, as well as some corporate chemical policies. Leading corporations such as Nike, Herman Miller, Volvo, and Seventh Generation have adopted safer chemical policies, some covering chemicals from product ingredients to processing chemicals to cleaning and office supply chemicals.

What is a safer chemical policy framework? A simple definition would be:

An integrated collection of universally inclusive, hazard-based, and transformative policies designed to achieve the development and use of safer and more sustainable chemicals in production and consumption systems.

Comprehensive

If it is to address the entire chemical economy, a safer chemical policy framework must provide a broad and inclusive approach to all chemicals. The early assumption that the chemicals problem could be solved by regulating a modest number of truly dangerous chemicals led to a focus on only part of the chemical production and consumption system and ignored the broader chemicals market and the interrelations among the many firms that operate in the chemical economy. The chemicals initially identified for regulation were interlaced with many other chemicals of concern and could be substituted with other chemicals that might be of equal concern. A systems perspective invites consideration of all of the chemicals commonly used in industry and commerce.

Rather than focusing only on the most dangerous chemicals, a safer chemical approach gives equal consideration to all chemicals ranging from the most hazardous to the least. FIFRA and the FFDCA do this by covering all pesticides and all pharmaceuticals. The European Union's REACH does the same for all chemicals manufactured or imported into the European Member States. SAICM takes this approach internationally, creating a global strategy for the sound management of all chemicals in all countries. Thus, a universal approach creates a comprehensive, system-wide landscape of chemicals ranging from the safest to the most concerning and from the best studied to the least studied.

A safer chemical policy framework would not provide a new, single, unified policy but rather a framework for many related policies. These policies need to provide roles and responsibilities for government agencies, chemical manufacturers, product manufacturers, chemical and product users, and product recyclers and end-of-life managers. However, a safer chemical approach dives deeper beneath the chemicals on the market to address the systems of chemical production and consumption. A systems orientation encourages a broad perspective on chemicals across all phases of a chemical's life. Chemicals are manufactured from basic molecular building blocks, mixed and reacted into complex compounds, processed into useful formulations, assembled into commercial products, used in widely diverse applications, discharged into complex waste streams and recycled, reused or released into the ambient environment. A comprehensive approach considers all of these life cycle stages.

Hazard-Based

Although the chemical control policies drafted during the 1970s were effective in curbing some of the worst chemical abuses, they did less to bring about fundamental change in the chemical economy. In hindsight, it is easy to see why. The chemical control framework that developed during the 1970s and 1980s was a narrow, confrontational, risk-focused approach destined to be tough to implement and limited in its broader impacts. The focus on risk turned out to be particularly problematic. As a decision-assisting tool, risk assessment created a well-documented and orderly procedure for evaluating the hazards of chemicals and the likelihood of exposure. However, judicial determinations, business pressures, and administrative practices locked much of the chemical control policies into a rigid, risk-based approach that encumbered decision making with costly animal tests and complex exposure modeling and favored exposure control over hazard reduction.

The risk-based approach that emerged assumes, as given, that hazardous chemicals must be used and then focuses risk management attention on managing exposures through barrier technologies, exposure management practices, and safety standards. Such efforts reinforce existing production and consumption systems and always add costs to production. Exposure standards and controls never achieve perfect protection, often fail in practice to meet expectations, and always leave opportunity for accidents and mismanagement. Today, the risk-based approach is fundamental to the way in which government regulates, business operates, and science is employed. It has become the conventional way of understanding the problems with chemicals and responding to them.

A hazard-based approach offers a different way to understand the problem of chemicals and a more effective, direct, and innovative means for building safer chemical policies. Under a hazard-based approach, the primary focus is on the inherent hazards of a chemical and, only secondarily, on exposures. All chemicals have some hazards and risks associated with them. As noted earlier, a hazard is the chemical's innate ability to cause adverse health effects (cancer, birth defects, nervous system effects, etc.). How likely it is that one of these effects will occur (how much risk) in any given situation is determined by how exposure occurs—how much, how often, and how long a person is exposed—and other factors such as age, gender, health, medication, and nutrition. These exposure factors are complex, varied, uncertain, and difficult to model and measure, particularly where exposures are repeated and cumulative and occur in complex environments where there are many chemicals of concern interacting synergistically. Instead of using science to estimate all of these factors to determine what level of chemical exposure is acceptably safe, a hazard-based approach encourages the development and adoption of inherently less hazardous chemicals. The goal of a hazard-based approach is not chemicals that are "safe enough" but "safer chemicals." This is a journey, an exploration, a quest. The objective is the safest chemicals not the least exposures.

Because no chemical can be assumed to be completely safe, the concept of a "safe chemical" is misleading. Instead, the degree of concern or degree of preference across a range of hazard traits serves as the criteria for calling a chemical "safer" or "preferred." For a chemical to be considered safer than a chemical of concern, it must be relatively safer than the chemical of concern in terms of a baseline of factors that includes:

- its inherent toxicological profile;
- its physical characteristics in terms of acute hazards, flammability, corrosiveness, and explosive potential;
- its potential for persistence and accumulation in organisms and food chains; and
- its metabolic, ecologically transformative, and degradation characteristics.

In addition, it is useful to consider other (life cycle) factors that may directly affect the hazards of a chemical:

- its fate and transport in the environment,
- the characteristics of its synthetic production processes and the intermediary chemicals used in those processes, and
- its energy and resource consumption during production, use, and disposal.

Under a hazard-based approach, the focus of safer chemical policies is on the inherent hazards of a chemical. Inherent safety is a term often used in the chemical industry to describe "primary prevention" or elimination of a hazard. This differs from "secondary prevention," which includes processes that seek to reduce the risk of a chemical, often by the use of technological barriers or managerial controls. With regard to chemicals, primary prevention seeks to reduce or eliminate the inherent hazards of a chemical at the chemical design or selection stage prior to use and exposure. The focus is on substituting safer chemicals for chemicals of concern and converting to cleaner and safer technologies, thereby reducing the need for exposure controls. Toxics use reduction is a hazard-based approach as are inherently safer technology, green chemistry, and green engineering. These strategies avoid protracted efforts to control exposure and instead seek to prevent harms by developing and substituting safer alternatives to hazardous chemicals. Whereas a risk-based approach tends to reinforce the existing systems of chemical production and consumption, a hazard-based approach more directly seeks to change the chemical market.

Transformative

The narrow focus of the chemical control laws provided little incentive for considering how to change chemical production and consumption systems. The focus on regulating chemicals was ambitious. The initial standards set for air, water, and workplace exposures were pioneering, but after the 1980s, the struggle to set tighter standards in keeping with the increasing number of toxicological and epidemiological studies slowed. Every effort by OSHA to create new substance-specific standards was resisted through court challenges. The primary air quality standards did not grow beyond six whereas the hazardous air pollutant standards and drinking water Maximum Contaminant Levels only emerged where lawsuits forced the EPA to meet the requirements of the law. As noted in chapter 3, chemical regulations under TSCA and CPSA simply ground to a halt.

The chemical control policies would certainly have been more effective without the active resistance of the regulated community. The confrontational nature of the government regulatory approach engendered its own well-constructed and well-resourced opposition. The control metaphor proved too easily stopped. Instead of a control approach, a safer chemical policy framework needs a more multifaceted, government and market-based, transformative approach that promotes conversion rather than commands compliance. It needs to provide both positive and negative incentives for change, encouraging and rewarding leaders as well as identifying and penalizing laggards. A transformative approach needs to marshal a collection of instruments, methods, and tools that can be employed by governments, corporations, and nongovernmental organizations to gradually and incrementally change the chemical market and reform the chemicals industry. A big, integrated system like the chemical economy does not change quickly, but the accumulation of many small changes can, over time, bring about significant changes, and unlike the government-focused chemical control laws, these changes cannot be easily stalled.

Corporations must play a central role in a chemical conversion strategy. Chemical manufacturers make chemicals; chemical distributors market chemicals; product manufacturers use chemicals to make products; and retailers sell products. These corporations are optimally placed to shift chemical production and use to safer chemicals. This requires a new vision of a corporation; one that sees it as a valued agent of environmental and health protection. Transferring the responsibility for chemical safety from government agencies to the private sector places the burden for selecting safer chemicals on those who have or should have the most chemical information and technical expertise and integrates the cost of chemical safety into the market price of chemicals. Government regulations are still important in a transformative approach, but they must be integrated with economic instruments, technical assistance services, consumer education, civil society campaigns, and new analytical tools, such as life cycle assessment, alternatives assessment, technical assessment, total cost assessment, and social and economic impact analysis.

5.5 Building Blocks of a Safer Chemical Policy Framework

A safer chemical policy framework needs to include the existing chemical control policies, but it is new and different from those conventional approaches to chemicals in both scope and action. It must be broader and more comprehensive, addressing the entire system of chemical production and consumption and dynamic and creative in setting priorities and compelling action. It must promote the generation of chemical information and increase its access and flow, and it must specify a set of tools and methods and legal instruments for encouraging or compelling change. Although many features are important to a safer chemical policy framework, it is useful here to refer back to the three common building blocks of the existing chemical control policy framework described in chapter 2:

characterize and prioritize chemicals, generate and make accessible chemical information, and regulate the manufacture and use of chemicals of concern. Here, building blocks modeled on that earlier framework could be constructed to design a safer chemical policy framework as depicted in figure 5.4. The first two blocks are essentially the same while the third block is divided into two blocks: selecting safer alternatives and developing safer chemicals.

Characterize and Prioritize Chemicals

For a chemical to stay on the market, every chemical manufacturer or importer would need to create a chemical profile that fully characterizes the chemical by its physical and chemical properties, hazard traits, and energy and resource requirements. This chemical profile should provide enough data for classifying the chemical by its inherent hazards. The classification could be constructed in a hierarchy of tiers ranging from the highest to the lowest level of concern. By establishing a set of benchmark criteria for differentiating chemicals by their inherent properties, well-characterized chemicals could be slotted into these tiers and grouped by level of concern ranging from the least preferred to the most preferred.

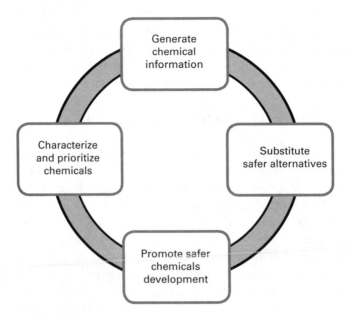

Figure 5.4

Building Blocks for a Safer Chemical Policy Framework

Such a classification framework would provide a valuable landscape from which governments, corporations, or other interests could identify priority chemicals for regulation, substitution, or further research. Prioritization could be guided by a formal protocol that includes both the hazards of chemicals and the likely exposures to people and ecosystems. Where the prioritization process considers exposure and risk, the classification of chemicals into the classification framework is based on the inherent hazards of the chemical alone.

Generate and Make Accessible Chemical Information

Information on chemicals is a fundamental requirement for an effective chemical policy. This includes information of the health and environmental effects of each chemical, its uses, and its potential for exposures during those uses. Unlike a half-century ago, significant advances in toxicology and pharmacology have expanded the range of knowledge about chemical effects, and there is now a host of databases and inventories for identifying many well-studied chemicals.

It is not just information on the health and environmental effects of chemicals that is needed; information on the production, use, and disposal of chemicals as well as information on the environmental releases and potentials for exposure is needed. This information may be developed by government laboratories, universities, professional associations, and corporate research facilities, but the responsibility for ensuring that this information is generated and made available needs to rest with the chemical supplier—the manufacturer, importer, distributor, or vendor. By making the supply of sufficient information the responsibility of the chemical supplier, the cost of generating that information can be factored into the price of the chemical rather than the cost of government.

Transparency and broad public access to chemical information is critical to ensuring that such information is valid and useful, not only where consumers make decisions but throughout supply chains such that product manufacturers can be well informed in determining what chemicals go into products and retailers can confidently assure customers about product safety and safe product use.

Substitute Safer Alternatives for Chemicals of Concern

Safer chemical policies need a well-integrated set of instruments and tools for selecting the most dangerous chemicals and driving their substitution with safer alternatives. Those who drafted the chemical control laws of the 1970s relied heavily of the regulatory powers of government to restrict the threats of hazardous chemicals. The use of regulations made sense if the chemicals problem was defined as eliminating a modest number of chemicals of high concern. However, if the problem is framed as a broader systems problem, then regulations will not be enough. New instruments and approaches are needed that tap the power of consumers, scientists, educators, businesses, and nongovernmental organizations in promoting the selection and adoption of safer alternatives.

Substitution of chemicals of concern can be driven by governments using legal authorities to ban the manufacture and use of chemicals, by companies voluntarily seeking to avoid chemicals of high concern, or by consumers, retailers or government agencies selecting products that contain safer chemicals. Governments and companies might set up programs for identifying, evaluating, and selecting safer alternatives to chemicals of concern and creating chemical action plans for guiding the adoption of the selected alternatives. This will require new decision-supporting methods and tools such as life cycle assessment, comparative chemical hazard assessment, and alternatives assessment. Life cycle assessments provide the broad perspective on how chemicals impact the broader temporal context in which they function. Alternatives assessment and various forms of comparative chemical hazard assessment are tools for ensuring that alternatives are safer and sounder than the chemicals they replace.

Develop Safer Chemicals

Developing tools and methods for converting to safer chemical alternatives increases the demand for safer alternatives. Green chemistry and engineering and new chemistries and green chemical companies will be central to generating safer chemicals and chemical processes. This could be encouraged by government or corporate policies and the campaigns of nongovernment organizations.

Conventional chemistry is a dynamic and innovative enterprise, and new chemicals and chemical processes are continuously emerging. The most recent advances in nanotechnology, synthetic biology, and biochemistry are generating waves of new chemicals that could provide safer alternatives. Human health and environmental protection considerations need to be integrated into the design of these compounds to ensure that they are truly safer alternatives.

What is needed here is a formal protocol for defining the design space for chemicals in such a way as to progressively ensure safer and then safer chemicals. Such a design protocol needs to begin with fundamental rules and progress though a set of criteria and principles that promote novel chemicals and chemistries that are safer and likely to be adopted into the mainstream chemical economy as transformative innovations.

5.6   Developing a Chemical Conversion Strategy

The reframing of the chemicals problem lies at the heart of a chemical conversion strategy and opens opportunities for new chemical, material, and product solutions. A strategy like this is big—big and complex—and it needs to organize a host of disparate parties. The challenge is to orchestrate an assortment of diverse tactics among a group of important parties that are all generally committed to transitioning toward a safer and more sustainable chemical economy. The first step, and certainly one of the hardest, is to get the many important forces to accept this new definition of the chemicals problem and agree to work together to find transformative solutions.

The basic structure of such a strategy already exists today. SAICM, the global strategy for the sound management of chemicals, is a strategy. With SAICM, the United Nations did not just create an international policy, it launched a global strategy on chemicals. SAICM sets out a broad declaration, a strategic plan, and a long list of specific tasks. This is a strategy—a comprehensive and ambitious strategy for shifting the international economy toward the sound management of chemicals.

A similar chemical conversion strategy could be developed in the United States. Such a strategy would include a safer chemical policy framework with the building blocks identified here—the characterization and classification of chemicals, increased chemical information, a processes for identifying safer chemicals, and a means to promote their development—and if there were then some goals, some benchmarks, and a list of tasks, the course could be set for converting the chemical economy to safer and more sustainable chemistry. This is not difficult to envision. Over the past decade, government chemical policy has begun to shift as corporate and private sector institutions are assuming more responsibility for chemical management. Increasingly, nongovernment organizations are playing more aggressive roles in promoting safer chemicals, sometimes working directly with corporations to assist them in transitioning to safer chemicals. Starting with REACH and SAICM, new chemical policy frameworks are appearing across Asia, and several U.S. states have enacted new safer chemical policies for phasing out hazardous chemicals and adopting safer alternatives. This is leading to a resurgence of interest in reforming federal chemical policies.

Admittedly, the journey suggested here appears too bold and too broadly stretching of the imagination. However, initiatives already in place suggest how effective each of the strategic fronts of a conversion strategy might be and how a range of chemicals policies could be stitched together to generate a comprehensive framework. So the task here, then, is to identify those initiatives, evaluate them, draw lessons from them, and place them into the broader perspective of a national safer chemical policy framework.

In the sections that follow, each of the three strategic fronts of the chemical conversion strategy will be examined in more depth by reviewing many of the current initiatives, and the building blocks of a safer chemical policy framework will be developed in more detail. However, before we begin, it is valuable to take a closer look at the system that we are trying to change: the chemical economy.

6

Understanding the Chemical Economy

If you visit the Shell Chemicals petrochemical plant in Norco, Louisiana, it is difficult not to be overwhelmed (figure 6.1). The plant extends over a mile along the Mississippi River. It appears as a gigantic plumber's toy with pipes running over and under each other and a forest of exhaust stacks pointing a hundred feet into the air. There is an eight-foot-high chain link fence that extends the length of the facility. This is a highly productive facility with both an ethylene furnace and a petroleum refinery on site allowing for significant economies of scale. The plant was expanded once in the 1990s and again in 2000.

The refinery manufactures 220,000 barrels of petroleum gases and heavy liquids per day, and the chemical plant has an annual capacity of 1.5 million tons of ethylene. This makes the Norco facility Shell Chemicals' largest ethylene production facility in the United States. The plant also manufactures a broad spectrum of petrochemicals that range from propylene and butadiene to butyl alcohol and aromatic hydrocarbons. Plants like this one at Norco make up the basic subsystems of the world's system of chemical production.

However, the facility has been the subject of a long history of local concern as the residential community of Diamond, which lies just outside the plant gates, fought the company's expansion for years. The Concerned Citizens of Norco contended that pollution from the plant led to a litany of health problems and a host of respiratory ailments. Following the death of two community residents during a chemical accident in 1973, the citizens waged a long struggle to convince Shell to buy out their homes and provide relocation expenses. Finally, in 2002, Shell negotiated an agreement and bought and demolished houses over a four-block area.[1]

A broad, comprehensive chemical strategy needs to address not just chemistry but the chemical industry itself. Transitioning the chemical industry from its current unsustainable structures to a more sustainable form opens up significant opportunity for a safer and greener chemical market. Here, it is worth taking a closer look at the chemical economy and consider how the chemical industry produces chemicals today and how those chemicals are used to make commercial products.

6.2   The Chemical Manufacturing Industry

The chemical manufacturing industry is one of the largest industries in the world. It produces some 100,000 to 140,000 chemical substances. A relatively small number of products (some 2,500) account for the largest proportion (roughly 95 percent) of the production volume. With annual sales at some $4.12 trillion, the global industry makes up nearly 5 percent of worldwide gross domestic product (GDP). While the industry is spread across the world with production facilities in more than 100 countries, the largest share of production comes from Europe, Japan, China, and the United States. The European chemical industry is the world's largest, with sales over $690 billion a year and 1.9 million employees scattered throughout 31,000 firms. However, the United States maintains the largest single nation industry, with annual sales of nearly $540 billion generating nearly 4 percent of the U.S. national gross domestic product (GDP) or about 17 percent of the total contribution to GDP from the manufacturing sectors.[2]

The chemical industry is a system—a highly complex and diverse system. It is composed of chemical manufacturers that make primary tier production chemicals; chemical processors that refine those chemicals to make secondary tier intermediary chemicals; and chemical formulators that produce commercial grade chemical products that are used to make the brand-name products that appear on the commercial market. The synthetic organic side of the industry takes a few basic carbon-rich resources such as petroleum, coal, and natural gas and makes thousands of chemical derivatives, ranging from plastics, resins, and fibers to dyes, coatings, and drugs. The inorganic side starts with the nonfuel minerals of the planet and produces a wide range of acids, metals, salts, and gases that are then processed to make products such as pigments, ceramics, fertilizers, glass, and a myriad of construction materials. Conventionally, the industry is divided into four subsectors: bulk chemicals, specialty chemicals, agricultural (or life science) chemicals, and consumer products (pharmaceuticals are sometimes included as a fifth category or included within the life science chemicals).[3]

Figure 6.1

Norco Shell Facility (photo credit: Candie Wilderman)

Bulk Chemicals

Bulk or commodity chemicals include olefins, aromatics, industrial gases, inorganic compounds, synthetic rubber, and resins, most of which are sold within the chemical industry as feedstocks or intermediaries in the production of other chemicals. These large volume chemicals are made in a highly decentralized industrial system, with leading companies making up no more than 5 percent of any commodity market. Production takes place in big, integrated facilities using large amounts of energy and continuous—as opposed to batch—operations. Bulk chemical production is considered a mature industry, with the fundamental products, processes, and technologies little changed over the past 50 years.

Specialty Chemicals

Specialty or fine chemicals are made by hundreds of chemical producers, and the chemical products are made by thousands of large- and small-scale formulators. These include a diverse array of high-value, low-volume chemicals with applications in chemical additives, dyes and inks, surfactants, coatings and sealants, electronics, and advanced polymers. Production may take place in continuous or batch operations. Specialty chemicals make up a relatively dynamic, growth-oriented sector of the industry with higher than average rates of return. Research and innovation on products and processes are often keys to success.

Life Science Chemicals

The life science chemicals sector, composed largely of fertilizers, pesticides, soil amendments, and pharmaceuticals, are produced by both large and small manufacturers in batch and continuous operations. The largest volume chemicals involve nitrogen- and phosphorus-based fertilizers while herbicides comprise the largest share of pesticides. Although the sector as a whole is considered low margin and mature, pharmaceuticals make up one of the fastest growing and most profitable branches of the entire chemical industry.

Consumer Product Chemicals

The consumer product sector of the chemical industry manufacturers the vast array of brand-name commercial products that fill the shelves of retail stores and range from personal care products to soaps, detergents, cleaning products, and fragrances. These chemicals are formulated "end" products typically made in batch processes by thousands of large- and small-scale producers making modestly different versions of similar products that compete against one another in the commercial market.

6.3   The Structure of the Chemical Industry

The chemical manufacturing industry is large and profitable. In the United States, the chemical industry is a $720 billion enterprise generating some 70,000 products and producing about 12 percent of the nation's manufacturing revenue.[4] Today there are approximately 13,500 chemical manufacturing facilities in the United States owned by more than 9,000 companies and employing more than 790,000 employees. Since the 1990s, U.S. chemical industry production has been growing at about 4 percent per year (the national average for all manufacturing is about 3 percent); however, during this period industry employment has fallen by some 300,000 jobs. Although bulk chemical production is largely located in Texas, Louisiana, New Jersey, Pennsylvania, and California, the specialty chemical industry is widely dispersed and composed of small companies, with nearly 90 percent employing fewer than 500 employees.[5]

Up through the 1950s, the majority of the chemical manufacturing industry was concentrated in Europe, Japan, and the United States; however, since that time, chemical production has been gradually relocating. The basic chemical industry and much of the agricultural chemical industry have been moving from highly industrialized countries to developing regions of the world, attracted by lower labor costs, reduced tariffs and trade barriers, availability of feedstocks, and growing consumer markets in developing countries. Only about 20 percent of global pesticide production remains in the United States. Today, almost no production of methanol or ammonia occurs in the United States as Trinidad, Chile, and Saudi Arabia have become key suppliers. Nearly a third of bulk chemical production remains in Europe today, whereas another third of the industry is now based in Asia. This trend is expected to continue. Half the growth in ethylene production in the next five years is expected to be in the Near East, with an additional one third of growth expected in Southeast Asia.[6]

However, production of specialty chemicals and pharmaceuticals has remained strong in Europe, Japan, and the United States. Europe provides the largest volume of specialty chemicals, and the United States maintains substantial production; however, the specialty chemicals industries in China, India, and the Near East are growing rapidly. In 2011, China became the largest producer of chemicals in the world, and today China maintains the fastest growing chemical market. Japanese and Western multinationals have been rapidly expanding facilities in China, and new Chinese firms are emerging as well. Although much of this growth is in bulk and agricultural chemicals, China is also investing heavily in specialty chemicals.[7]

The chemical industry is composed of some of the largest corporations in the world. Ranked by international sales, BASF is the largest chemical corporation, followed by Sinopec of China and then Exxon/Mobil, Dow, SABIC, and Shell (see table 6.1).[8]

During the 1940s, major corporations in the petroleum industry began investing in chemicals, buying up chemical divisions or merging with chemical firms. Firms such as Shell, Chevron, and Mobil invested heavily in chemicals, thereby vertically integrating energy and chemical production while DuPont bought Conoco for much of the same reason. Over the past thirty years, the chemical industry has been restructuring through mergers, divestments, and acquisitions. Several firms have transitioned into the life sciences. Monsanto shed its chemical business to concentrate on biotechnology and downstream seed markets. Hoechst and Rhone

Table 6.1
Largest Global Chemicals Corporations by Global Sales

Company	Home Country	Global Sales, 2012 (in millions)
BASF	Germany	$95,100
Sinopec	China	$64,894
Exxon/Mobil	USA	$60,885
Dow Chemical	USA	$56,786
SABIC	Saudi Arabia	$50,390
Shell	Netherlands	$45,757
Lyondell/Basell	USA	$45,352
DuPont	USA	$34,833
Mitsubishi Chemical	Japan	$32,782
INEOS	UK	$29,908
Bayer	Germany	$27,370
Total, S.A.	France	$26,050

Poulenc combined to form Aventis and concentrate solely on the life sciences. Dow Chemical has announced plans to sell nearly $5 billon of its chlorine business. Nowhere is this corporate restructuring more pronounced than in pharmaceuticals, where Sandoz and Ciba-Geigy merged divisions to form Novartis and ICI spun off its pharmaceuticals to form Zeneca, which then merged with Astra AB to form AstraZeneca.[9]

The chemical industry has also been shedding research and development (R&D). The industry has long been known for basic research and technological innovation with breakthrough products like nylon and polyethylene and process technologies that dramatically improved production yields. However, with the exception of pharmaceuticals, most firms have significantly reduced investments in R&D. Although R&D expenditures in the pharmaceutical sector in 2010 was 13.1 percent of sales, it was 1.5 percent of sales for the remainder of the industry or less than 45 percent of the average for U.S. manufacturing as a whole.[10]

The economic strength of the chemical manufacturing industry is tied to the cost of raw materials, price of energy, structure of the market, and government policy. The chemical industry is the United States' largest industrial consumer of natural gas, with about 80 percent of olefin production based on natural gas liquids. Because the industry relies on fossil fuels as both a feedstock and an energy source, the price of energy has direct effects on the industry. For instance, when natural gas prices nearly quadrupled in 2000, 50 percent of the methanol capacity, 40 percent of the ammonia capacity, and 15 percent of the ethylene capacity were idled or shut down, and the costs of these chemicals rose.[11]

The internal dynamics of the chemical market has significant effects on the industry. The manufacture of bulk and agricultural chemicals is a low-margin business. Downward price pressures due to many equally matched competitors and upward cost factors due to material and energy volatility leave little room for long-term capital surpluses. To counter the effects of a relatively flat market and increasing global competition, many chemical manufacturers began during the 1980s to cut costs, consolidate operations, reduce research investments, and increase automated and information-based production processes. The result has been increased efficiencies but also short periods of growth followed by dramatic cutbacks, delayed capital investments, and active periods of corporate restructuring. After many such years, the infrastructure of many U.S. basic chemical production facilities is aging and no longer state of the art.

Finally, government policy helps to shape the industry and its market. Government regulations, particularly on hazardous waste treatment and pollution control measures, generate significant costs to the industry. Regulations leading to high costs for the siting and construction of new facilities in industrialized countries are one of the factors favoring developing country investments. However, the chemical industry benefits significantly from the large subsidies provided to oil and gas companies that significantly reduce the full costs of energy and chemical feedstocks. These include tax credits for producers (Expensing of Exploration and Development Costs Credits, the Percentage Depletion Allowances, and Alternative Fuel and Production Credits), reduced royalties for oil leases on federal lands, grants for oil and gas research and development, safety programs for pipelines, and maintenance of the nation's Strategic Petroleum Reserve. In 2006, these subsidies totaled some $3.5 billion. In addition, the chemical industry, like other industries, enjoys research and development tax credits and accelerated depreciation credits for new equipment installation and investments in pollution control equipment.[12]

6.4   The Manufacture of Chemicals

The production of bulk chemicals takes place in complex, large-scale manufacturing plants. Such plants typically begin with a few simple inorganic or hydrocarbon feedstocks and, using a mix of chemical intermediates (reagents, solvents, catalysts, acids, and bases), combine, heat, react, and refine them into a diverse pallet of basic chemical products. These highly integrated production facilities produce many co-products with many different applications and market values. The chemical products are transferred or sold to downstream chemical processors and formulators, where they are combined with other organic or inorganic compounds, separated and purified, and supplemented with a wide range of additives to meet various product grades and performance requirements.

Petrochemical production provides an instructive illustration. ure 6.2 displays the common production sequences in the petrochemical industry that generate many of the industry's primary products. The petrochemical production system begins with the refining of oil, natural gas, or coal to make the starting feedstocks. In the United States, natural gas and natural gas liquids make up the largest feedstock (more than 65 percent) for petrochemicals, with oil converted to naphtha and other heavy liquids making up most of the rest. The largest production stream "cracks" natural gas to produce methanol and the olefins—ethane, propane, and butane— from which primary tier chemicals—ethylene, propylene, and butadiene— are produced. These are the three largest building blocks for making secondary tier intermediaries and derivatives. Three quarters of ethylene production goes into making polymers such as polyethylene, polyvinyl chloride (PVC), and polystyrene, with the remainder consumed in producing products such as antifreeze, synthetic rubber, solvents, and detergents. Forty percent of propylene is used to make polypropylene, whereas the remainder goes into isopropanol, acrylonitrile, propylene oxide, and cumene (isopropylbenzene)—all valuable compounds for making consumer products. The largest share of butadiene goes into synthetic rubber, with a small share going into latex, resins, and nylon. About 40 percent of methanol is used to make formaldehyde, which is used in the production of adhesives, mastics, paints, explosives, and plastics, and the remainder goes toward acetic acid.[13]

Inorganic chemical synthesis can be much simpler. Consider ammonia, which is used in large quantities in the manufacture of agricultural fertilizers, whereas smaller amounts go into explosives. The production of ammonia through the Haber-Bosch process involves the reaction of gaseous hydrogen and nitrogen. Large volumes of hydrogen are derived from the methane generated from petroleum, which must be separated from sulfur and then reacted with carbon monoxide and carbon dioxide to form purified hydrogen. The nitrogen is stripped from air by either absorption or membrane separation, before it is reacted with the

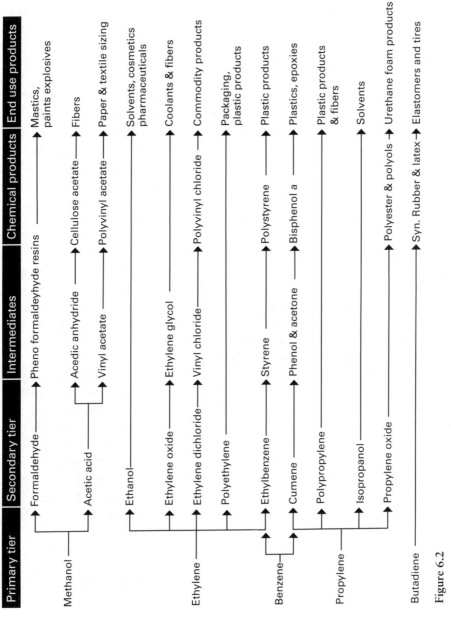

Figure 6.2

Petrochemical Production, Intermediaries, and Products

hydrogen and a catalyst to produce anhydrous liquid ammonia. This is a simple, straightforward, and mature technological process.

The significant integration and interdependencies of the production systems in the manufacture of bulk chemicals are critical to the efficiencies and economics of the industry. In many of the chemical production branches, a small number of the large production products are lead products, and the profitability of the entire manufacturing operation may be carried by these few commodity chemicals, whereas other products are just breaking even. However, these interdependencies also create rigidities, such that singular changes in the systems (such as elimination of a commercial grade product) can cause multiple consequences rippling up and down the production line.

For example, during the 1990s, large volumes of methanol and isobutylene were used to make methyl tertiary butyl ether (MTBE). MTBE was introduced in the United States during the 1970s to maintain gasoline octane ratings when lead additives were being phased out. It was an attractive choice. MTBE was inexpensive, blended well with gasoline, and could be made in existing refineries with little structural change. The use of the additive grew rapidly following the 1990 Amendments to the Clean Air Act requiring fuel oxygenates that could reduce the production of carbon monoxide and ozone. By 1998, 10.5 million gallons of MTBE were produced per day, making it the fourth highest production volume chemical in the country. However, MTBE began showing up as a contaminant in groundwater, and municipal water wells in California and Maine were forced to close because of high MTBE levels. At the same time, evidence began to appear that MTBE might be carcinogenic and associated with the prevalence of asthma. In 1998, the EPA convened a "Blue Ribbon Panel on Oxygenates in Gasoline," which determined that some 5 to 10 percent of the water supplies in high MTBE use areas were contaminated with the compound and called for a "substantial and rapid reduction in MTBE use." In 2000, the EPA began to take steps to phase out the use of MTBE and shift fuel oxygenates to ethanol. The production of MTBE fell dramatically. With the reduction in MTBE production, huge volumes of methanol and isobutylene needed to be diverted to other uses.[14]

6.5   The Hierarchy of Chemicals

The production of the highest volume bulk chemicals has remained fairly constant for the past forty years, with well over forty of the top fifty high-

Table 6.2
High-Volume Bulk Chemicals

High-Volume Inorganic Chemicals	High-Volume Organic Chemicals
Sulfuric Acid	Ethylene
Nitrogen	Polyethylene
Oxygen	Propylene
Lime	Ethylene Dichloride
Ammonia	MTBE
Chlorine	Vinyl Chloride
Phosphoric Acid	Polypropylene
Diammonium Phosphate	Benzene
Sodium Carbonate	Polyvinyl Chloride
Sodium Hydroxide	Ethanol
Sulfur	Ethyl benzene
Carbon Dioxide	Styrene
Nitric Acid	Formaldehyde

est production volume chemicals in the 1970s still in the top fifty today.[15] Table 6.2 lists the highest production volume chemicals worldwide.

Many of the largest production volume production chemicals are crucial "platform chemicals," which serve as central feedstocks in the production of many other chemicals, whereas other large-volume chemicals are "intermediates" used in the production of other chemicals, and still others are finished chemical "end use" products. Benzene, ethylene oxide, and phthalates provide examples of each of these types of chemicals.

Benzene is a platform chemical. It is an aromatic hydrocarbon derived from crude oil and manufactured as a co-product with toluene and xylene. Until World War II, benzene was primarily used in the rubber industry. The war created a heavy demand for toluene for use in the manufacture of explosives. This generated a surplus of benzene, which found use as an additive in high-octane gasoline and in the production of pesticides, adhesives, paint removers, detergents, and plastics. In 1977, benzene was found to be a carcinogen, so its use in commercial products was gradually reduced; however, it remains used extensively as a production feedstock. In 2008, benzene consumption worldwide was just under 40 million metric tons. Half of all benzene produced is used to make styrene while the remainder is used to make adipic acid and caprolactam, both intermediates important in the manufacture of nylon and phenol, which is used to make bisphenol A, a critical production chemical used in the manufacture of polycarbonates and epoxies[16] (see figure 6.3).

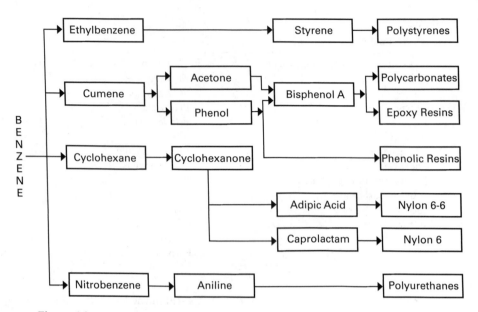

Figure 6.3

Benzene and Its Derivatives

Although benzene is a highly hazardous chemical, it is a critical building block in the production of important industrial, agricultural, and pharmaceutical products. Other such hazardous platform chemicals are chlorine, formaldehyde, ethylene dichloride, butadiene, caustic soda, and ammonia.

Ethylene oxide, a recognized carcinogen, is an intermediate in the production of many other chemicals and chemical products. It is conventionally manufactured by the direct oxidation of ethylene in the presence of a silver catalyst. Although small amounts of ethylene oxide are used as sterilants in hospitals, the largest volumes of ethylene oxide are used as production intermediates in the production of detergents, adhesives, solvents, antifreeze, pharmaceuticals, and polyurethane foam. World production of ethylene oxide was 20 million metric tons in 2009, with just over 4 million metric tons produced in the United States largely by Dow Chemical. Nearly 75 percent of global production goes into the synthesis of ethylene glycols. These are used in the production of antifreeze, solvents, coolants, and polyester and polyethylene terephthalate (PET, used in the majority of plastic beverage bottles). The polyethylene glycols are used to make cosmetics, lubricants, paint thinners, and plasticizers, whereas the ethylene glycol ethers are used to manufacture brake fluids, detergents, solvents, lacquers, and paints.

Unlike benzene or ethylene oxide, phthalates are intended to end up in end-use, plastic products because, as additives in polymeric compounds, they provide important functional characteristics. Indeed, in some plastic products, phthalates may make up some 40 to 60 percent of the weight. Phthalates are esters of phthalic acid, which are generated by a catalytic oxidation of naphthalene to produce phthalic anhydride, which then can be reacted with various compounds to produce useful derivatives. High-molecular-weight phthalates such as di(2-ethylhexyl) phthalate (DEHP) and butylbenzophthalates are used as plasticizers in a large number of vinyl products ranging from floor and wall coverings to toys and medical devices. Low-molecular-weight phthalates, including diethyl phthalate and dibutyl phthalate, are used in many personal care products (e.g., shampoos, cosmetics) and as dispersants, lubricants, gelling agents, binders, and emulsifiers. However, phthalates also exhibit hazardous properties. Low-dose exposure to some phthalates has been implicated in lowering sperm counts in adult males and retarding male sexual development in laboratory animals and some human studies. However, phthalates remain widely used in commercial applications because of their desirable performance characteristics and low cost.

6.6   Energy, Resources, and Wastes

The chemical manufacturing industry consumes large amounts of chemical feedstocks, water, and energy. The industry uses about 5.15 quadtrillion Btu of energy per year in the United States or about 24 percent of the energy used by the manufacturing sector as a whole and 7 percent of total U.S. energy consumption. The production of bulk chemicals (e.g., propylene, polypropylene, cumene, and butadiene) and certain specialty chemicals (particularly gasses) are the most energy intensive, whereas pharmaceuticals are among the least energy intensive. The petrochemicals sector in the United States made up 43 percent of the industry's energy consumption in 2010, followed by the plastics, synthetic rubber, and fibers sectors, which combined made up another 33 percent.[17]

Not only is a large share of the nation's fossil fuel sources used as energy for powering the chemicals industry, the industry uses some 9 percent of the nation's fossil fuels as feedstocks for making chemicals. The natural gas and petroleum feedstocks used by the U.S. chemicals industry are equivalent to 3.4 quadrillion Btu of energy per year, accounting for approximately half the chemical industry's use of petroleum-based fuels and 3 percent of the total U.S. natural gas consumption. Indeed, the chemical industry is the largest consumer of natural gas within the manufacturing sector. Ethylene, ammonium, and chlorine production are some of the largest energy-consuming processes. For some energy-intensive products, the price of fuel for power and feedstocks makes up 85 percent of the cost of the finished product. Until recently, the high and unstable costs of the fossil fuels, particularly natural gas, have contributed to the general decline in U.S. bulk chemical production.[18]

Because of the effects of energy prices on the economics of the chemical industry, the industry has invested in energy efficiency. Since 1974, the industry has cut its energy consumption per unit of output roughly in half. Whereas the thermodynamic constraints of various high-energy reactions set limits on some of the largest production processes, many firms have cut energy consumption in other functions such as transportation, storage, and intermediary production.[19]

Because these energy-intensive processes are fueled by carbon-based fuels, they generate large amounts of greenhouse gas emissions. The U.S. chemical industry generates more than 250 million metric tons of greenhouse gas equivalents annually. This represents roughly 4 percent of total U.S. greenhouse gas emissions. Some 230 million metric tons of this is carbon dioxide from fuel consumption and production processes.[20]

The chemical industry generates large amounts of waste. In 2012, the U.S. chemical industry reported to EPA's Toxics Release Inventory (TRI) the release of 100 million pounds of waste to the air, water, and land. Another 9.8 billion pounds of production-related wastes were reported to be treated, recycled, or land disposed. That same year, the chemical manufacturing industry reported 168 million pounds of air emissions to the TRI. This represents about 15 percent of the hazardous waste and emissions generated nationally. Nitrate compounds, manganese, ammonia, and acetonitrile make up the largest volume of disposed wastes. The largest volume air pollutants released included ammonia, methanol, ethylene, and carbonyl sulfide (box 6.1). Over the past two decades, the total releases of toxic chemicals to the air and water have been reduced by 80 percent. Because much of the hazardous waste is treated on site at production facilities, the amount of wastes sent for treatment or disposal has also declined but more slowly.[21]

6.7   The Chemical-Using Sectors

Although a substantial share of the chemicals manufactured by the chemical industry is used internally, the chemical market is largely driven by demand from those who manufacture products. A formulated or

Box 6.1
Top TRI Air Emissions from the Chemical Industry, 2005 (in millions of pounds)

Ammonia	50.4
Methanol	19.2
Ethylene	16.6
Carbonyl sulfide	12.8
Carbon disulfide	10.1
Toluene	6.9
Sulfuric acid	3.6
Xylene	2.8

assembled product is simply an ordered assortment of chemicals. The product manufacturing industry is a huge and highly diverse system of facilities and linkages that marshal labor, energy, and technologies to convert chemicals from the chemical industry into "finished" products.

In the United States, the manufacturing industries make up $5.3 trillion in annual sales, include some 288,000 companies, and employ about 13.4 million people. The 2008–2009 financial crisis and subsequent recession led to a tightening credit market and a large pool of unemployed consumers, which resulted in significant reductions in consumer spending on products and services. Aggregate consumption fell by 0.6 percent in 2008 and 1.9 percent in 2009. This spending reduction was not evenly distributed. The North American Industry Classification System divides manufacturing into twenty-one subsectors that range from the transportation and telecommunications industries to electronics,, fabricated metal, textile, apparel, furniture, and food and beverage industries. Consumption of durable consumer products declined faster than services and non-durable goods. Sales of automobiles, fuels, and transportation services declined sharply during this period but gradually increased during 2011 and 2012.[22]

About half of the volume of chemicals manufactured or imported into the United States goes into the manufacturing sector. About 7 percent of total chemical sales go into health care, 6 percent go into agriculture, and the construction industry is responsible for about 2.2 percent.[23] The automotive sector is another major market in the United States. An estimated $3,000 of the cost of every manufactured automobile goes toward chemical processing and products, with automotive paint making up the highest value. The American Chemistry Council estimates that nearly 80 percent of the material value of laminate and vinyl floor coverings and 70 percent of the value of carpets come from chemistry.[24]

Agriculture consumes a large amount of chemicals as well. Nearly 21 billion tons of nitrogen, phosphorus, and potash were spread on U.S. agricultural fields in 2010. In addition, the agriculture sector uses some 440 million pounds of herbicides per year, as well as 65 million pounds of insecticides and 44 million pounds of fungicides. The shift to large industrial agriculture based on mono-cropping of grains, fruits, vegetables, fibers, and animal feed creates an ever-increasing demand for synthetic fertilizers and pesticides. But farms also require a host of synthetic chemicals, such as livestock hormones, pharmaceuticals, and nutrient supplements; and, once the produce has left the farm, huge amounts of synthetic chemicals are used in the storage and processing of foods and fibers for the commercial market.[25]

According to the World Trade Organization, the United States is the largest importer of goods in the world, with an estimated 2012 value of $2.33 trillion accounting for one eighth of all global imports. China is the largest source of these imports, followed by the European Union, Canada, and Mexico in that order. Imports are sensitive to a broad set of factors, including trade policies, transportation costs, currency exchange ratios, and household consumption. Although the United States is a large importer of chemicals, it is a major exporter of products and services. With an export trade estimated at $1.54 trillion in 2012, the United States is the second largest global goods exporter, second only to China, making up 8.4 percent of the world's total export economy.[26]

The manufacture of articles in such a global economy requires chemicals that flow along a material supply chain and become increasingly integrated to form mixtures, materials (e.g., polymers, ceramics), components (e.g., switches, semiconductors), and, finally, finished and assembled products. The supply chain is composed of the various firms and organizations through which chemicals flow from their synthesis through product processing and formulating to their ultimate consumption (see figure 6.4).

The globalization of the chemical economy has created long supply chains with substantial strains. There are often many tiers of suppliers in international supply chains, and it is often difficult for product manufacturers to know who makes up their suppliers, much less what chemicals are in the components they purchase from suppliers.

The product manufacturing industries are largely dependent on external sources of research and product development. Historically, the

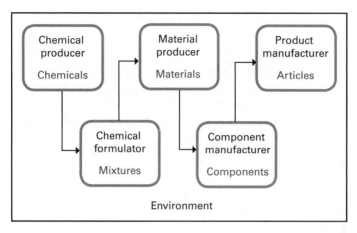

Figure 6.4

A Product Supply Chain

government contributed heavily to research and development (R&D), supporting more than half of all R&D during the 1950s. This proportion has been decreasing since then, dropping to 30 percent during the 1980s, and it is now less than 10 percent. In 2009, the national manufacturing sector spent some $195 billion in domestic R&D. In contrast, the chemical industry investments in R&D have been relatively small and on a steady decline. The U.S. chemical sector (without pharmaceuticals) spent $53 million in 2009 whereas the pharmaceutical industry spent another $44 million.[27]

Formulated and assembled products add to the chemical intensification of the economy. Nearly 80 percent of U.S. households own a computer, and this is near a saturation level, such that total sales have substantially slowed over the past several years. In contrast, sales of cosmetics, personal care products, and domestic consumables have been on a general incline for years. A consumer economy, such as the U.S. economy, is driven by a constant flow of products that creates large demands for the production of products and, therefore, large amounts of waste. In order to keep markets from saturating (consumers having enough products), products are made to be easily consumed and disposed (e.g., tissue paper, product packaging, personal care products), products are designed to be frequently replaced (e.g., cell phones, laptops, toys), fashion changes require new products (e.g., footwear, cosmetics, clothing and apparel), and mass advertising artificially generates demand (e.g., television and billboard ads). As long as consumers have the income to spend, product manufacturers will supply that demand with products and thereby create increasing demand for chemicals.

As the growing share of global product production shifts out of the United States and the chemical industry that supplies that production shifts as well, the United States is increasingly becoming predominately a large consumption engine converting products to waste. The consumption of chemicals depends on this economic pump, and to the degree that hazardous chemicals provide an inexpensive and effective feedstock for consumer products, this pump delivers health and environmental risks. Many U.S. citizens will continue to be exposed to chemical hazards where they work or in the communities that are near chemical or product production facilities. This is particularly true for lower income residents and recent immigrants who work on farms or in industrial facilities or who often live near them. However, for an increasing share of middle-income citizens who live in suburbs or well-scrubbed urban centers and who work in the professional service sectors, the primary exposure to hazardous chemicals comes from the products they purchase and consume. The near obsessive drive among some well-off consumers to buy organic foods and shop for healthy and safe products attests to their concern about their most vulnerable routes of exposure.

6.8   The Waste Management Sector

The production of chemicals and chemically intensive products generates large volumes of chemical wastes. Some 34 million tons of hazardous wastes were reported to the EPA in 2011. That year the chemical industry reported 4.8 million tons of on- and off-site releases of hazardous chemicals, of which 4.6 million tons were production-related wastes, making the industry the third largest hazardous waste-generating sector in the country. A large proportion of this waste is managed by the industry at its production sites. An additional share of the nation's hazardous waste is generated by the product manufacturing industries, and a large percentage of this waste is sent off site for treatment and disposal. Because these hazardous wastes are regulated and the costs of treatment and disposal are high, over the past 25 years, the chemical and product manufacturing industries have steadily decreased the rate of hazardous waste generation per unit of sales. The industries proudly tout this as evidence of increasing efficiency. However, the shift in the production of chemicals and product manufacturing to other countries has meant that increases in hazardous waste generation that might have occurred domestically had these industries expanded in the United States do not now add to this nation's waste-generation profile. Instead, these wastes are generated in China, India, Brazil, and other industrializing countries, even though the manufactured products are sold and consumed domestically in the United States.[28]

Because thousands of products are consumed annually in the United States, large amounts end up in the nation's waste and disposal treatment facilities, and because the products (plastics, packaging, formulated chemical products, etc.) are largely composed of synthetic chemicals, the nation's commercial and domestic wastes (so-called municipal wastes) are filled with these chemicals. During the 2000s, the cumulative volume of such municipal wastes hovered around 250 million tons. Municipal waste authorities are responsible for managing these wastes, although an increasing share of this waste is shipped to foreign facilities.[29]

However, a much larger volume of chemical wastes remains that is less visible. This is the so-called nonhazardous industrial wastes of the manufacturing industries. Although this waste is not legally considered hazardous, it is not benign. It includes everything from detergents and cleaners to paints, plastics, photographic film, waste paper, grindings, shavings, lubricants, rubber goods, construction wastes, textiles, wasted containers, furniture, and electronics and electrical equipment, most of which is rich in petrochemicals and other synthetic chemicals. The total volume of non-hazardous industrial wastes was about 6.5 billion tons in 1992, the last year that the federal government attempted to estimate it.[30]

6.9 Finding Leverage in the System

So this is a quick profile of the chemical economy. Considering the chemical economy as an integrated system extending from chemical synthesis to final product disposal reveals several observations that have important implications for a chemical conversion strategy.

• The chemical economy is global. Globalization means longer supply chains, less transparency along the supply chain, more difficulty ensuring that products meet health and safety standards, unregulated and unsafe production conditions, and more difficulty in gaining recourse against responsible parties. Policy changes in one country (even a country as large as the United States) may not have significant effects on chemical production in other parts of the world. Achieving a safer chemical economy will require international coordination among nations and the active participation of international institutions.

• Chemical production is a highly integrated system. Phasing out the use of some large-volume hazardous chemicals (particularly the platform chemicals) can have significant impacts on the entire chemical production system, and because of this would most likely be heavily resisted. Government regulations or market shifts that change the sales of even small-volume chemicals must be careful to consider the substitutes in order to avoid unintended consequences.

• Chemical production is highly dependent on the oil and gas industries for raw materials and energy. The two industries are tightly integrated, with many investment linkages between them. Some of the largest chemical companies (e.g. ExxonMobil Chemical, Shell Chemical, and Chevron Phillips Chemical) are owned by petroleum companies. Changing the chemical economy will require addressing these large and powerful petrochemical corporations.

• The investment return on many chemicals is marginal. The economic trends of the chemical industry generally track the larger economy, and many of the bulk chemicals only make a profit when the price of consumer end-use and specialty chemicals is high. Inexpensive ocean transport makes it possible to produce most chemicals overseas, and the United States is increasingly a large chemical importer. Placing taxes or regulations on domestic chemical producers may affect their competitive position and lead to further domestic industry disinvestment.

• Product and chemical value chains are long and sheltered. Supply chains are of particular importance because the interdependence along supply chains often leads to resistance to change. Chemical and production information often flows poorly from chemical suppliers to downstream product manufacturers and product retailers. Improving the generation and flow of reliable chemical information in supply chains could have positive impacts on chemical selection.

• The chemical economy is partitioned such that important dialogues are limited. Consumers with chemical concerns are many steps removed from chemical manufacturers. Chemical manufacturers are not getting broadly based market signals from suppliers and retailers concerning undesirable chemicals. Municipal waste managers handling hazardous products have no effective communication with chemical suppliers. Information does not easily flow along supply chains, and there are many missing linkages, such as those between chemists and health scientists or between government regulators and investors. Communication systems need to be developed such that valuable information generates more awareness and accountability.

Considering the chemical production and consumption system from afar makes it appear monolithic and impenetrable. However, a closer examination reveals many drivers for safer chemicals and points where interventions could be made to promote safer and more sustainable chemical production and use.

First, there are economic drivers. The costs and liabilities of hazardous chemicals in terms of purchase price, processing costs, handling and storage costs, waste treatment costs, compliance costs, spill and accident costs, future cleanup costs, and injured workers all have the potential to encourage firms to seek safer substitutes. However, many product manufacturing firms are unaware of the magnitude of these costs and therefore miss the compelling signals. Revealing the true costs of hazardous chemicals could create the needed economic leverage.

Second, shifts in social values and expectations are taking place. A growing segment of trend-setting consumers is creating demand for safer, cleaner, greener, healthier products. As product manufacturers compete to attract these customers, they create a demand on chemical suppliers for safer chemicals. However, pressure from individual consumers is too sporadic and unorganized to encourage significant changes. Mobilizing consumers through campaigns and intermediary organizations could create more effective market leverage.

Third, leading firms in the product manufacturing sectors are actively seeking to replace chemicals of concern. Within these firms, production managers, marketing directors, environmental officers, and others are pressing their companies toward more sustainable practices, including the elimination of hazardous chemicals. Rewarding the leading companies and penalizing the laggards could create competitive leverage.

Fourth, organizations ranging from the European Union and the United Nations to national governments in Asia are developing new chemical policies, enacting new laws, and negotiating new agreements that are putting international pressure on global chemical markets to commit to the sound management of chemicals. Encouraging state and nongovernmental safer chemical initiatives in the United States could create political pressures.

Each of these drivers within the chemical economy points out opportunities for promoting changes in chemical production and use. To be effective, these forces need to be more broadly scaled, better integrated, and more visible to end users. The energy for change is there, and a careful examination of the chemical industry reveals promising points of leverage.

6.10   The Chemical Economy

Transforming the chemical economy is a huge and ambitious project. The chemical industry is large and complex, and much of its technical foundation is mature and rigidly structured. Globalization has created extensive spatial differentiation, with much of the bulk chemical production now in Asia. The United States and Europe continue to lead only in the innovative specialty and pharmaceuticals sectors, and even this lead is vulnerable.

Chemical consumption and disposal is also being restructured by globalization. The trend since the 1970s is geographically separating large sectors of product production from the large volumes of product consumption of Western Europe and North America. The low cost of energy means that chemicals and products are cheaply transported around the planet. The result is a chemical life cycle pump drawing chemical flows from China and the Near East to product production centers in Asia, Eastern Europe, and Latin America and onward into Western Europe and North America for consumption before pushing product disposal on into Asia, Africa, and Latin America. Add to this an increasingly significant flow of chemically intensive products (paints, adhesives, textiles, lubricants, and chemically complex articles such as cell phones and laptops) from European, American, and Asian economies into the developing world, and an intensive global chemical economy is now worldwide.

It is no longer effective to address chemicals singularly and out of the context of their production and consumption systems. What is needed now is a chemical conversion strategy that focuses on the chemical market, the chemical industry, and the chemical-using sectors that make up the chemical economy. From a systems-changing perspective, this requires a strategy that addresses market conversion, industry transformation, and an expansion of the science and practice of chemistry. Many such initiatives are ongoing today and the sections that follow describe and assess them.

III

A Chemical Conversion Strategy

The overall objective of the Strategic Approach to International Chemicals Management is to achieve the sound management of chemicals throughout their life-cycle so that, by 2020, chemicals are used and produced in ways that lead to the minimization of significant adverse effects on human health and the environment.

—*Overarching Policy Strategy*, Strategic Approach to International Chemicals Management (2006)

7

Driving the Chemical Market

If you shop for hair shampoo you know there are scores of products on the shelves. There are products made for oily hair and products made for aging hair; there are products for fighting dandruff and products that are gluten-free; there are products offering herbal fragrances and products with no fragrance at all. However, a careful reading of the labels will tell you little about the ingredients that make up the shampoo. Some products will be labeled as composed of "natural ingredients," some will claim to be "organic," and others will claim to be free of toxic chemicals and good for the environment. In truth, a shopper has little information to judge the chemical safety of hair shampoos or most any other product.

Every day thousands of people make decisions that affect the chemical market. Most decisions are about the costs, availability, and performance of products; few involve consideration of human health or the environment. If the consumer market offers an important opportunity for promoting safer chemical production and consumption systems, the amount of chemical information in the market must expand, and the number of products that take health and environment into account must increase.

A chemical conversion strategy needs to address the chemical market. Driving the market is the first of three strategic fronts that we examine here. Many current initiatives are seeking to shift the chemical market toward safer chemicals. Some drive the market by providing more consumer and buyer information, some educate consumers, some change purchasing programs, some alter marketing and retail practices, and some organize consumers to demand change. A focus on commercial products provides important leverage in shifting to safer chemicals because products are so central to a consumer economy and so accessible to decision making by an informed public. Consumers, at the point of purchase, can select products with safer chemicals; retailers and institutional buyers, when negotiating supplier contracts, can specify products with safer chemicals; product manufacturers, when designing products, can specify safer chemical ingredients; and consumer advocacy campaigns, when targeting specific chemicals, can recommend products to buy or avoid.

7.1   Market Drivers

The chemical market is driven by prices that consumers will pay for products and by the expectations they have for product performance and quality. Market research and direct experience demonstrate that some consumers, when provided a choice, will select safer and more environmentally friendly products if the costs are not significantly higher. However, in the absence of appropriate information, consumers cannot make such choices. Price is readily available, and people learn about performance from advertisements, past experience, and friends. It is much more difficult for consumers to access information on the health, safety, and environmental impacts of products. Indeed, it is not easy for retailers or, even, product manufacturers to acquire such information. Providing more and better chemical information, therefore, is an important step toward using markets to promote safer chemicals.

Ensuring accurate and appropriate information for consumers is conventionally considered a basic requirement for an effectively functioning market. Providing consumers with health and environmental effects information could encorage consumers to select safer and more sustainable products.[1] Some studies suggest that among consumers in the United States, more than half will consider health or environmental values when making purchase decisions. A recent study of 6,000 shoppers interviewed as they exited retail stores found that 47 percent reported that they looked for and saw products labeled as good for the environment; however, only 22 percent reported actually buying those products. This study found that, although some 18 percent of shoppers consciously shop for "green products," a mere 2 percent reported making their shopping choices primarily on perceived environmental attributes. Impressive numbers, but they are not enough to swing markets.[2]

This focus on consumer behavior can be misleading. Markets may be driven by customer desires; however, in large complex societies, individual consumers have limited influence over what they can purchase. Their knowledge about what is on the market is largely shaped by media advertising, and their purchases are determined by what they find when they shop. The range of products that domestic customers can buy is predetermined by intermediary institutions such as retailers, distributors, and large institutional buyers. "Big box," chain retailers and a limited number of distributors select small numbers of products to put before customers, and large institutional buyers exert significant purchasing power on markets and tend to narrow product diversity. The real forces of consumer choice are located not with "mom and pop" harassed by media advertising but in the decisions of large retailers, group purchasing organizations, and government agencies such as Walmart, Target, Home Depot, Staples, CVS, Kroger, Carrefour, Consorta, Novation, Univar, Ashland Distributors, and the Pentagon.

Consumer advocacy organizations have become more sophisticated in recognizing how effectively market information can shape product markets, and they have developed programs, designed tools, and mobilized consumer forces to drive markets toward safer and more sustainable products. Generally, these initiatives can be grouped into four categories: consumer education initiatives, retailer initiatives, institutional procurement initiatives, and consumer campaigns.

7.2   Consumer Education

Product Certifications and Labels

The growth of environmental awareness among consumers has driven a parallel growth in the number of product labels declaring environmental benefits. Today, there are well over 430 such "eco-labels" awarded by private and government organizations in 147 countries covering products ranging from coffee to building products. Only some of these programs address the chemical constituents of products or the chemicals used in manufacturing products; most do not.[3]

Under the requirements of the early federal chemical control laws, all containers of pesticides, prescription drugs, and over-the-counter drugs are required to carry a product label revealing active chemical ingredients, safety warnings, and instructions on proper use. The Nutritional Labeling Education Act of 1990 set standards for nutrition labels on food packaging. The "nutritional box" label is carefully designed to present the product's nutrient ingredients and food additives along with the percentage of daily allowances of several important food characteristics (fats, carbohydrates, vitamins, etc.). Today, the names of the manufacturer, place of origin, date of expected durability, and special instructions for use are also required on food product packages. Under the Federal Hazardous Substances Act, formulated household products such as paints, glues, alcohols, and cleaning fluids are required to provide on the container package, information on chemical ingredients, name and location of manufacturer, safety warnings, and directions on safe use and disposal.

Beginning in the 1980s, the number of products making environmental claims started to mushroom, with hundreds of product packages claiming to be green, safe, biodegradable, ozone-safe, or environmentally friendly. This led the Federal Trade Commission in 1992 to issue guidelines for responsible green marketing that covered general statements of environmental benefits and terms that suggest a product is recyclable, compostable, or degradable. As these guidelines became accepted by major product suppliers, certified labeling programs ("eco-labels") began to replace misleading environmental claims.[4]

These environmental labeling programs have been provided some degree of standardization through the efforts of the International Standards Organization (ISO). The ISO 14025 standard on environmental labels and declarations, published in 2006, identifies three types of labels: type I (multi-criteria symbols developed by third-party organizations), type II (self-declaration symbols developed by a product producer), and type III (labels presenting several attributes based on a life cycle inventory and verified by a third party).[5] Table 7.1 identifies some of the type I programs based in the United States and Canada.

Only some of these type I product labels address the use of hazardous chemicals. For instance, the U.S. Department of Agriculture (USDA)'s organic certification program makes use of a National List of Allowed and Prohibited Substances (mostly pesticides, fertilizers, and processing agents) in determining products permitted to carry the federal organic label. To achieve certification, an agriculture product producer or handler must submit a plan to a USDA-approved private or state certification program ensuring that all production or processing steps are free from chemicals on the National List. Each certifying program is responsible for periodic on-site inspections of certified entities to guarantee certification conformance.[6]

Green Seal, a private nonprofit organization, offers a type I seal-of-approval label for nearly 200 product and service categories that "cause less harm to the environment than other similar products" (figure 7.1). In awarding its certification mark, Green Seal evaluates each product against other products in the product's product category using life cycle assessments, existing technical studies, and independent testing laboratories. Once a standard has been established, Green Seal staff conduct an evaluation of the data supplied by applicants and visit their production

Table 7.1
Examples of Type I Eco-Labels

Label	Products Covered	Chemicals Specifically Covered	Life Cycle Considered
U.S. EPA Energy Star	Electronic products	No	No
Marine Stewardship Council	Fisheries products	No	No
Sustainable Forests Initiative	Forest products	No	No
Forest Stewardship Council	Forest products	No	No
USDA Organic	Organic ingredients	Pesticides	No
EPEAT	Electronic products		Yes
EcoLogo	Cleaning, office, electronic, and building products	Carcinogens, reproductive toxins, PBTs*	Yes
U.S. EPA Design for Environment	Cleaning products	Carcinogens, reproductive toxins, PBTs*	Yes
Green Seal	Cleaning, office, and building products	Carcinogens, reproductive toxins, PBTs*	Yes

*Persistent, bioaccumulative, toxics

facilities before awarding the use the Green Seal mark. A broad range of criteria are used for determining environmental performance, including chemical hazards such as the presence of carcinogens, reproductive toxins, persistent, bioaccumulative, and toxic substances, ozone-depleting substances, and lethal toxins.[7]

In Europe, national governments set up the first eco-labeling programs such as the German Blue Angel and the Nordic Swan label. In 1992, the European Union established a European-wide *Ecolabel* (figure 7.2) that, like Green Seal, uses life cycle assessments to evaluate products against standards set for similar products in a product category and provides a flower logo mark to those meeting or exceeding the standard. Most of the product categories include criteria setting limits on the use of some

Figure 7.1

Green Seal mark

Eco LABEL

Figure 7.2

European-wide *Ecolabel*

hazardous chemicals. With some 750 companies awarded *Ecolabels* for 17,000 products in 22 product categories, the *Ecolabel* is becoming the European standard. Indeed, a survey by Eurobarometer found that nearly 80 percent of European respondents say that they pay attention to environmental attributes when shopping, 47 percent report that they are inclined to pay attention to product labeling, and 40 percent recognize the *Ecolabel*.[8]

While type I and type III labels require some third-party certification or audit, type II labels based on self-declaration provide little assurance that the information underlying the labels is accurate. There is widespread concern that many of these environmental claims on product labels are misleading or simply false—a condition described as "greenwashing." Terra Choice, a Canadian-based environmental marketing firm, conducted a survey in 2010 of producers of more than 5,000 products making nearly 12,000 green marketing claims and found that for 95 percent of the products the company had made some meaningless claims or could not provide appropriate verification. The largest numbers of questionable claims were found on children's toys, cosmetics, and cleaning products. Responding to concerns like these, in 2011, the Federal Trade Commission issued a revised set of standards ("Green Guides") that update and strengthen the guidance for environmental marketing claims.[9]

Some environmental labels such as EPA's Energy Star logo used to rate energy-efficient products are well recognized by consumers and appear to have a direct effect on purchase decisions. Studies indicate that roughly 50 percent of consumers recognize the logo, and somewhere between 30 and 50 percent of respondents say that the logo influenced their purchase. EPA's Design for Environment (DfE) Program created the Safer Product Labeling Program for formulated household and commercial products. If manufacturers can demonstrate that their cleaners and detergents meet the DfE Program's requirements for chemical ingredient safety and sign an agreement to continuously improve their product, they can use the DfE logo on their product packaging. Today, more than 2,800 products carry the DfE label.[10]

The effectiveness of the environmental logos that include chemical safety in shaping consumer behavior has been little studied. Evaluations of the German Blue Angel credit the program with 30 percent reduction of sulfur dioxide, carbon monoxide, and nitrogen oxides from home heating furnaces and significant reductions in solvents emitted from household paints and varnishes. Studies of the Nordic Swan label in the Nordic countries credit the program with considerable reductions in chlorinated pollutants in the Swedish pulp and paper industry and reductions in the volume and toxicity of laundry detergents released to sewers.[11]

Product Declarations and Ingredient Disclosure

A more direct means for addressing the chemical ingredients of products involves product declarations. Environmental product declarations are type III eco-labels that disclose the environmental performance of products over their life cycle in a standardized, accessible format. Early efforts to develop product declarations by Volvo, Siemens, and others were standardized under the ISO 14025 standard that sets out methods for presenting information on environmental life cycle impacts and recycled content.

The ISO standard requires the establishment of a product category rule (PCR), which defines a discrete product group and sets the criteria and metrics by which products within that category are to be evaluated. Each product manufacturer then generates its own product label and contracts with a third-party for validation. Although there has been little interest in the United States, environmental product declarations are increasingly used in Japan and the European Union. Japan lists nearly 200 completed product declarations. France has passed national legislation making environmental product declarations mandatory for all high-volume consumer products. In United States SCS Global Services, Green Standard and the Sustainability Consortium are each developing industry-wide product category rules and seeking clients for product declarations.[12]

The ISO 14025 standard addresses chemicals primarily through the methods set out for life cycle assessments under ISO 14040 and does not require that specific chemical ingredients be listed on the product declaration. However, the Healthy Building Network, an environmental advocacy organization, which promotes safer chemicals and materials in building construction materials, has developed a *Health Product Declaration* (HPD) that is designed to complement an environmental product declaration by generating the chemical hazard information typically missing from life cycle assessments. The HPD, which is increasingly accepted among many building material and product suppliers, provides a standardized template for suppliers to present information on the chemical ingredients, emissions, and potential health impacts of the products they market.[13]

Although product labels and declarations have been effective in reaching some consumers, they have often been more effective in influencing product manufacturers that respond to such labels by reformulating or redesigning products to conform to the label or certification requirements. This was demonstrated by the range of products that were reformulated to remove carcinogens and reproductive toxins following the passage of California's Proposition 65, which required product warning labels. For instance, Gillette reformulated its typing correction fluid in order to avoid the "Prop 65" requirement to label the product as containing a carcinogen.

The large number of eco-labels and certification programs is quite diverse in terms of scope, criteria, and design, and this diversity undermines the very objectives of these programs as too many environmental labels tend to confuse rather than direct the market. A national government "eco-label" program was proposed in a Senate bill in 2008; however, opposition from national trade associations and those currently operating their own product labeling programs suggests that this and similar legislation face a steep uphill struggle.

Consumer Guides

Consumer guides offer a third means of informing consumers about the chemical composition of products. Consumers Union has long provided a publically trusted product evaluation service that offers product ratings on a range of indices, including quality, performance, price, and safety. In 2005, the organization launched *Greener Choices*, an Internet-based service that provides information on the environmental attributes of products in five product categories: appliances, cars, electronics, food, and home and garden products. Another trusted institution, the National Geographic Society, provides its own consumer guide to environmentally friendly products, *Green Guide*, which offers recommendations and shopping tips to consumers seeking green products. However, neither of these guides provides information on the chemical constituents of products or their chemical hazards.

The Internet provides a range of searchable databases that provide more information on product chemical hazards (see table 7.2). The National Library of Medicine maintains an easily searchable database, the Household Products Database, covering some 9,000 commercially available household products. This database and other government databases encourage chemical ingredient disclosure but make no effort to rate products or provide recommendations to consumers. However, because the source of these data is often manufacturer reports and manufacturer-generated Material Safety Data Sheets, the data provided on chemicals is

Table 7.2
Sources of Consumer Information on Chemical Hazards

Name	Products Covered	Comments
Household Products Database (National Library of Medicine)	9,000 consumer products	Health effects information taken from suppliers Material Safety Data Sheets
Consumer Product Information Database (Delima Associates)	8,000 consumer products	Same database as Household Products Database
National Pesticide Information Center (Oregon State University and U.S. EPA)	Pesticides	Fact sheets based on government health information databases
Consumer Products Inventory (Woodrow Wilson Center, Project on Emerging Nanotechnologies)	1,015 consumer products	Database of products from 485 company reports containing nano-scale materials
Consumer Action Guide to Cars, Consumer Action Guide to Toys, (Ecology Center)	200 vehicles and 1,500 toys and children's products	Database derived by physical and chemical test results
Skin Deep (Environmental Working Group)	53,000 cosmetic and personal care products	Pooled data from more than 50 government, industry, and academic databases
GoodGuide	210,000 products in nine product categories	Database built from more than 1000 professional, scientific and government sources

likely underreported, particularly for low-volume chemicals and unintended chemical ingredients (contaminants).

In 2004, the Environmental Working Group launched a website called *Skin Deep* to provide an Internet-based, searchable database for identifying the health and safety attributes of cosmetics and personal care products. Today, this rigorously assembled resource includes nearly 53,000 commercially available products and provides toxicity and hazard data on more than 8,800 ingredients from some fifty government databases. Each product is presented by an Internet product page that identifies the chemical ingredients and indicates whether the product contains chemicals of high concern (carcinogens, reproductive hazards, etc.). On the basis of the ingredient hazards, each product is scored with a preference number (from 1 to 10) and a color-coded button. With about one million page views per month, *Skin Deep* has come to fill a vital consumer need for personal care product information.[14]

As useful as the *Skin Deep* website is, it only covers cosmetics and personal care products. Such an interactive consumer information service could cover a broad range of product categories. The premier example of such a shopping tool is the *GoodGuide*, created by Dara O'Rouke, a professor of environment and labor policy at the University of California at Berkeley. "Product labels are a good start," O'Rouke says, "but they don't provide enough information or they provide it in a way that is hard for consumers to use. Consumers want to know is the product safe and will it harm the environment, but they don't want to struggle to find that information. They want simple rating codes, tailored to their needs and available where and when they make shopping decisions."[15]

The Internet-accessed side of *GoodGuide* is similar to the *Skin Deep* service. By naming a product category, the service lists the products in that category in order of the scoring. By clicking on the product name, scores for the health, environmental, and social attributes of the product are displayed by a set of color-coded (red, yellow, and green) ratings and a ten-point scale. Once consumers have reviewed the information and selected a product for purchase, they can add it to a printable shopping list or order it directly through a link with Amazon. Consumers can also "drill down" from the product pages to identify those attributes that determine the ratings or, further down, to review the scientific information on which the ratings are based. Today, *GoodGuide* covers some 210,000 products in nine product categories: foods, personal care products, baby and kid's products, appliances, pet foods, automobiles, apparel, electronics, and household products.

Because the *GoodGuide* service is accessible through a cell phone, it is interactive at the point of sale in a retail store. Once customers download the *GoodGuide* "app" to their cell phones they can photograph a product's Universal Product Code (UPC) bar code with their cell phone, and the *GoodGuide* screen can immediately display the product information and rating. For instance, a cell phone image of Seventh Generation's Dish Washing Liquid is quickly augmented with a graphics bar that displays an overall green rating code of 8.3 (out of 10) plus three subsidiary ratings for health, environment, and society. Clicking on the health box (the lowest of the ratings—8.0) reveals a "medium concern" warning arising from the inclusion of 1,2 benzisothiazolin-3-one, and clicking on the chemical name provides a drop down box with toxicological information. Then by clicking on a side button, a screen appears identifying dish washing liquids that do not contain 1,2 benzisothiazolin-3-one and their aggregate ratings.

Dare O'Rouke notes that *GoodGuide* provides a platform for displaying product ratings, but as a platform, it could be developed to provide for much more. For instance, with a few cell phone key clicks, the consumer could send a message to the manufacturer about their response to the product's rating. Or such information could be sent out by email or Twitter to friends. Such "real-time" and direct information can be valuable to both customers and retailers.

Such an impressive amount of information should have the power to change markets. This is what sociologist Daniel Goleman calls "radical transparency," a flood of product and market information that has the potential to "radically" transform the consumer product market. "Radical transparency means tracking every substantial impact of an item from manufacture to disposal…and summarizing those impacts for shoppers as they are deciding what to purchase," Goleman writes, "transparency's power comes from providing key information that changes customers' choices, which in turn creates new incentives for businesses to align their practices with the public's priorities."[16]

By marrying life cycle information on the health and environmental impacts of a product to a consumer's need for information at the point of a purchase decision, a simple mobile phone becomes a powerful tool for changing the chemical market. "We're in the Dark Ages now," Goleman quotes Dara O'Rourke as saying. "We know brand and price, and we think we know quality. But no one knows what's behind the label: what the product actually does to us or the planet. We want to pull back the veil of brand and go far beyond what the company tells you. What ingredients are health concerns? How far did it travel? How were workers treated?"[17]

The potential here is to open up vastly more information to consumers on the health and environmental impacts of products and, indirectly, on the chemical ingredients in those products. It is not necessary for a large majority of consumers to become highly knowledgeable, expert shoppers: the existence of such public information in the market and the possibility that it could affect consumer's purchasing behavior provides a potent driver for product manufacturers and retailers to seek safer chemicals. It is not so much that consumers become the drivers for change as it is that the availability of public information creates positive information feed-back loops that become the drivers of change. Here lies a potentially significant lever for changing the chemical production and consumption system.

7.3   Retailer Initiatives

As consumers increasingly seek more sustainable products, retailers—especially large retailers—have become a new voice for the chemical safety of the products they sell. In the absence of strong federal regulations, retailers have begun to set their own chemical policies that include requirements for their suppliers and lists of chemicals that should be avoided.

For some time, specialized retailers such as Patagonia, REI, H&M, and the Body Shop have had product supplier programs that screen hazardous chemicals out of their products, but these are niche enterprises serving health-conscious customers, and their total inventory has typically been too small to directly shift a market. However, because these firms are often market leaders, they can have broad indirect effects on their own economic sector by introducing environmentally superior products into the sector, encouraging customers to experiment with new, health-sensitive products, and setting high standards for other firms within the sector.

Patagonia provides an illustrative example of a small apparel retailer with a history as a pace setter in environmentally conscious marketing. In 1996, Patagonia's founder, Yvon Chouinard, switched the majority of the apparel line to organic cotton as a means of raising customer awareness. Although Patagonia's organic cotton t-shirts, "Beneficial Ts," became a national market leader, Chouinard recognized the limits of a small retailer in shaping the market and turned his attention to the broader apparel market. Working together, Patagonia and several similar retailers engaged the Switzerland-based bluesign Technologies to develop a product standard for apparel. With the motto "if you don't know, you don't care," bluesign built a standard around five values: consumer safety, resource consumption, water quality, air quality, and occupational health and safety. To be certified by bluesign, a supplier of dyes, coatings, finishing agents, and other textile constituents must submit to an on-site facility audit and prepare a chemical screening using special "bluetool" software that converts a list of chemical ingredients into a color-coded score indicating whether a substance can be used without restriction or only under specific conditions.[18]

H&M, a larger clothing retailer with operations in 22 countries, has created its own chemical management screening process. The firm first restricted azo dyes during the 1990s in response to proposed German regulations. This early initiative soon turned into a full-fledged list of restricted substance that by 2007 contained more than 170 chemicals and chemical categories. To supply H&M, clothing manufacturers must present tests from H&M-approved labs demonstrating that these chemicals are not present or are below defined thresholds.[19]

Whole Foods Market with some 400 grocery stores in the United States has built a reputation selling high-end, "natural" foods. Its self-created quality standards screens for minimally processed foods that are free of hydrogenated fats, artificial flavors, colors and sweeteners, and ingredients listed on its "Unacceptable Food Ingredients" list. The firm offers a "Premium Body Care" seal for products that suppliers demonstrate do not include synthetic fragrances or any of a list of 250 synthetic chemicals.

On the other end of the retail market, the huge purchasing contracts of "big box" retailers such as Walmart, Target, K-Mart/Sears, Home Depot, Staples, Kroger, and Safeway can more directly affect a manufacturer's market share. Within the retail sector, there has been a steady rise in corporate concentration, with a few large retailers in some sectors now accounting for up to 50 to 70 percent of the market. Because they have much greater leverage on suppliers, these firms requirements can put significant pressure on upstream chemical suppliers.[20]

In the home improvement and hardware retail sector, Home Depot and Lowes together control more than a third of the U.S. market. Both of these firms require suppliers to meet various environmental requirements, including agreeing to seek lumber certified by the Forest Stewardship Council (FSC). Since working with the FSC, Home Depot has sold more than 630 million pieces of FSC-certified wood, shifted out of lauan-finished doors, moved to second-growth cedar forests for wood shingles, and phased out sales of lumber from forty endangered trees.[21]

Two of the three largest office products retailers, Staples and Office Depot, representing nearly 50 percent of the sector's sales, now impose environmental requirements on their suppliers. Staples is the world's largest supplier of office products, with annual sales of $27 billion. During the early 2000s, Staples came under pressure from a nationwide campaign organized by consumer and environmental advocacy organizations that resulted in Staples ending sales of paper made from old-growth forests and increasing sales of recycled paper. In 2009, Staples purchased

Corporate Express, a leading supplier of business office supplies. Corporate Express had a line of Sustainable Earth commercial cleaning products developed by Roger McFadden, then the chief scientist at Corporate Express. To ensure high standards for these products, Roger had developed a three-part Sustainable Product Design Standard (SPDS), which he brought along to Staples during the merger. The first screen of the standard is composed of criteria for nine chemical hazard traits (e.g., CMRs, endocrine disruptors, volatile organic compounds) that a supplier's products must meet in order to sell through Staples. The second screen includes criteria for twenty-two environmental, and health, and safety attributes that a product is scored on, while the third (optional) screen encourages continuous improvement. While products with low scores may still be sold through Staples, they cannot be marketed as environmentally preferable. Roger sees such corporate policies as driving the market: "Being a market leader is a big responsibility. We are not just improving the environmental profile of our suppliers we are making customers aware of how their purchases have real consequences on the health and safety of the offices that they work in and the places they go home to."[22]

In terms of scale, Walmart is the world's heavyweight retailer. With 7,000 retail stores in 15 countries and annual sales of $410 billion, this one retailer has an unparalleled capacity to shape commercial markets. Walmart launched its sustainability program in 2005, with goals to use 100 percent renewable energy, achieve zero waste, and "sell products that sustain our resources and environment." In 2004, Walmart began tracking hazardous chemicals in its stores to ensure regulatory compliance. New policies were established for Walmart buyers focused on preferential purchasing of garments made with organically grown cotton, toys that did not contain lead or phthalates, and electronic products compliant with the European RoHS Directive. To gain advice from outsiders, the retailer set up a series of fourteen advisory groups called Sustainable Value Networks that focused on various environmental impacts ranging from energy efficiency, waste reduction, and packaging reduction to "chemical-intensive products." This last group identified twenty chemicals of high concern and recommended restricting their presence in products, leading the retailer to ban pesticide products containing propoxur and permethrin and cleaning products containing nonylphenol ethoxylates.

In 2008, Walmart changed course by moving away from a discrete list of restricted substances to a third-party-based screening process. The process uses a screening tool, called the *GreenWERCS Chemical Screening*

Tool, designed by The Wercs, a private firm with a twenty-five-year history of assisting corporations with chemicals compliance. The *GreenWERCS* tool evaluates data on some 2,400 chemicals from thirty authoritative lists of chemical hazards. To sell a product through Walmart, a supplier must present a confidential list of all intentionally added chemical ingredients to Wercs, and Wercs then uses its screening tool to develop a color-coded "green score" for the product. Before recommending the product to Walmart, Wercs reviews the product scores with the supplier and allows the supplier to adjust the chemical ingredients in order to achieve a higher score. Since its launch, Walmart has used Wercs to score more than 150,000 formulated products before placing them on Walmart display shelves.[23]

In 2013, Target, another "big box" retailer, also took steps to rate products on environmental attributes. Target took a different course from Walmart by working with *GoodGuide* to set a standard that product suppliers must use to rate their products, with scores that range from 0 to 100 based on points that can be earned in five categories: chemical hazard, transparency, animal testing, packaging, and water quality protection. Currently, the standard is being piloted on 7,500 products in household cleaners and personal care and baby care products.[24]

None of these retailers has a perfect record on all features of sustainability. Many have been criticized for problems in their foreign supply chains, and some have been criticized for their domestic labor relations. Such criticisms may be warranted; however, they do not negate the progress that these firms are driving in terms of chemical hazards. Put bluntly, it is possible for retail firms to see a competitive advantage in offering safer products, particularly to business customers that they do not see in their labor relations. Thus, if the efforts of retailers on eliminating hazardous chemicals are to be seen as universally valued, they need to be balanced with incentives for promoting other values such as labor and community relations.

The role that these retailers are assuming in product supply chains presents several opportunities for shifting the chemical market. First, these retailers are broadening the definition of a quality product by making consumers more aware of the health and environmental values of products, particularly chemical hazards. Second, the retailers are further defining their mission as a product screening and testing agent, suggesting a new role as a guardian of public and environmental health. Third, these retailer initiatives are sending assertive messages up the supply chain, pressing accountability for product safety and the burden of proof for chemical safety on to product manufacturers and, potentially, on up the supply chain to chemical manufacturers.

Conventionally, retailers have been seen as agents of product producers; however, some retailers are defining a new role as agents of consumers and, even more, as educators and partners of consumers. Such a reorientation creates a new voice for safer chemicals and provides important leverage for changing the chemical economy. Chemical manufacturers have not looked with much favor on this new activism by product retailers, arguing that restricting chemical ingredients in products should be the task of governments. However, Richard Dennison from the Environmental Defense Fund praises this new corporate role as "regulation by retailer" and quips, "Chemical manufacturers need to be reminded that good business practice involves delivering what the customer wants—and product retailers know that voice best."[25]

7.4   Government and Institutional Procurement Initiatives

In terms of scale, government purchasing operations compete with large retailers in shaping commercial markets. The federal government occupies nearly 500,000 buildings, operates more than 600,000 vehicles, and purchases some $350 to $500 billion per year in products and services. The Department of Defense (DOD) has an annual budget of $420 billion and purchases more than $30 billion of commercial products each year. The impact of the government's massive procurement budget has a significant effect on the economy and could play a major role in promoting safer chemicals.

The federal government has several agency-wide programs for environmentally preferred purchasing (EPP), including programs for selecting energy-efficient products, products made from recovered and biobased materials, and products that avoid ozone-depleting substances. Many of these programs promote interagency collaboration. For instance, an early Cleaning Products Pilot Project brought the EPA together with the Government Services Administration (GSA) to develop a matrix of performance and environmental attributes for ranking building cleaning products. Executive Order 13514 issued by President Obama's administration in 2009 follows upon previous presidential orders by setting goals for the federal government in the purchase of environmentally superior products and requiring that 95 percent of federal purchasing agreements must meet those goals.[26]

Responsibility for Executive Order 13514 falls heavily on the GSA and the DOD. The GSA, which provides procurement services to non-defense agencies, publishes an Environmental Products brochure listing 150 green products and maintains an Environmental Specialty Category in its global catalog and online purchasing services. This includes special attention to safer paints, coatings, and cleaners. The DOD has its own Green Procurement Program that sets a framework for educating procurement officers in each of the military services, increasing the purchase of green products, promoting energy and resource conservation, reducing solid wastes, and promoting markets for green products and services. Although there are explicit goals for GSA and DOD on energy, water, and waste, there are only general goals for reducing the quantity of toxic and hazardous chemicals acquired. The GSA is now working to create criteria for identifying chemicals to avoid in federal purchasing.[27]

The DOD, the Department of Energy, the National Institute of Standards and Technologies, and several other federal agencies also shape markets through issuance of specification standards. For instance, there are some 25,000 to 30,000 military specifications ("mil-specs") that define in detail the design, performance, and makeup of thousands of parts and processes. There are many examples where these specifications specify by name the use of specific substances or materials although seldom on the basis of health or environmental attributes. These specifications have served to harmonize equipment and process compatibilities for years, but they also have been criticized for limiting flexibility and inhibiting innovation. Because of this, the DOD has made various recent efforts to shift its standards toward private standard-making bodies.

For instance, the federal government purchases billions of dollars' worth of information technology equipment. In 2007, the EPA and the electronics industry established the Federal Electronics Challenge (FEC) to drive greener equipment purchases and reduce the environmental impacts of these products during use and disposal. To facilitate the procurement of products with preferred environmental attributes, the FEC has developed the Electronic Product Environmental Assessment Tool (EPEAT) to assist institutional purchasers in comparing computers, printers, and monitors based on a clear set of environmental criteria. Products are rated as gold, silver, or bronze based on a scoring protocol made up of environmental attributes. EPEAT provides a Type I eco-label and registers products that meet minimum performance standards for energy efficiency, end-of-life product management, and corporate responsibility. The certification is based on data supplied by vendors that are periodically verified for accuracy. EPEAT addresses the chemicals restricted under the European RoHS Directive; however, there has been resistance to including the reduction of other hazardous chemical as an aspect of the standard.[28]

Many state and local governments also maintain environmentally preferred purchasing programs. Most of these programs focus on attributes such as recycled content and energy efficiency, but a few directly address hazardous chemical ingredients in products. The Minnesota purchasing program identifies potential products that are "less toxic," the City of Buffalo has an ordinance that encourages the elimination of PBTs in municipal agency purchasing, and San Francisco maintains an Approved Green Products List containing more than a thousand products selected because they present low environmental impacts. The City of Berkeley and Alameda County in California have established EPP programs that "require the purchase of products and services that minimize environmental and health impacts, toxics, pollution and hazards to workers and community safety." A 2012 Oregon Executive Order requires the revision of the state procurement policy to reduce the use of toxic chemicals in state purchasing. In New York, an Interagency Committee on Sustainability and Green Procurement has proposed the creation of a "Chemical Avoidance List" that includes ninety-five "priority toxic substances" that would be phased out of state purchases when safer alternatives are clearly available. Massachusetts has a well-established EPP program that has been recently directed by an Executive Order requiring purchases that avoid products containing toxic chemicals if safer, comparable products can be purchased instead. The Responsible Purchasing Network, an international network of more than 400 government and public institutional purchasing organizations, tracks and encourages these green purchasing programs.[29]

Notwithstanding these last examples, it is notable how few government procurement programs prioritize chemical hazards in directing purchasing decisions. Energy efficiency and waste reduction are attributes not likely to be controversial; toxicity in products is a more contested attribute. It would appear that the public accountability of government agencies would make them ideal institutions for promoting safer products; however, too often the reverse is true. The public nature of governments makes agencies reluctant to rate products or draft lists of preferred brands due to concerns about political pressures from affected industries and potential liabilities that could be exploited by trial lawyers. Early efforts by the GSA to prefer the purchase of chlorine-free photocopy paper brought intense pressure from paper manufacturers and resulted in the abandonment of the effort.

Some large private procurement organizations are now playing a role similar to these government agencies in demanding safer chemicals in products. More than a decade ago, the environmental advocacy organization Health Care without Harm led a campaign to press large hospital group purchasing organizations (GPOs), such as Consorta, Premier, and Novation, to eliminate mercury thermometers and take mercury sphygmomanometers "off contract." Success in that campaign led the environmental organization to work with these large GPOs to develop lists of restricted chemicals that became part of the GPOs' conventional purchasing criteria.

Like the retailers, these government and private procurement organizations have significant power in the chemicals markets to demand the elimination of targeted chemicals of concern. Whereas the fraction of individual consumers willing to shop for safer products is limited, and therefore often not scaled to change chemical markets, "big box" retailers and government and institutional procurement organizations offer more robust and potentially effective capacities.

7.5   Consumer and Environmental Advocacy Campaigns

Since the early part of the twentieth century, nongovernmental organizations (NGOs) have mobilized public pressures for banning various chemicals of concern. Early efforts focused on dangerous drugs, food additives, and pesticides. Following the publication of Rachel Carson's *Silent Spring* in 1962, a national campaign arose spearheaded by the United Farm Workers (UFW) to phase out the use of the pesticide, DDT. Together with public health professionals and young lawyers recruited by Ralph Nader, the UFW campaigned for and, in 1972, won a ban by the EPA on use of the pesticide. Although this campaign was seminal in the history of environmental protection, there have been other well-publicized advocacy campaigns organized to reduce the use of specific chemicals. A national movement of professionals and citizen activists to restrict the use of lead won passage of the national Lead-Based Paint Poisoning Prevention Act in 1971 and laid the basis for government regulations that phased out leaded gasoline.[30]

These early campaigns demonstrated how mobilizing consumers, workers, or public health advocates could lead to government restrictions on chemicals of broad public concern. The 1980s campaigns against pesticides used in agricultural fields and orchards or brought in on imported fruits and vegetables resulted in government restrictions on aldrin, dieldrin, 2,4-D, and toxaphene. The broader campaign for organic farming was driven by food and small farm advocacy organizations opposed to industrial forms of agriculture that required massive amounts of synthetic fertilizers and pesticides. Originally organized around subscribers to the Rodale magazine, *Organic Gardening and Farming*, the movement grew to include consumer organizations and organic food producers and through the 1980s achieved state-based standards in California, Oregon, and Maine. Not satisfied with a myriad of state regulations, the campaign lobbied for a national organic certification standard and in 1990 achieved passage of the Organic Foods Production Act, which required the USDA to establish such a standard. It took a serious campaign and some 275,000 public comments to ensure a tightly defined standard. Today, there is a well-recognized national organic food standard that covers fruits, vegetables, meats, and the ingredients in many packaged foods, as well as various natural fibers and personal care products.[31]

Over time, these public advocacy campaigns shifted from a single focus on government restrictions to a more direct focus on corporate decision making. For example, the computer and electronics industries have been the target of an international campaign focused on eliminating highly hazardous chemicals. The campaign first emerged during the 1970s in California with the formation of the Santa Clara Center for Occupational Safety and Health (SCCOSH) and the Silicon Valley Toxics Coalition (SVTC), a coalition of environmental and labor activists mobilized to confront the pollution and waste generated by the Silicon Valley computer and semiconductor industry. Following several campaigns focused on forcing IBM and Fairchild to clean up contaminated groundwater, the coalition shifted its focus to press for the substitution of glycol ethers, an occupational health hazard linked to miscarriages, which were used in semiconductor production facilities. When the electronics industry transferred most of its production to foreign countries, the coalition set up the International Campaign for Responsible Technologies to focus on the workplace hazards of foreign production centers and the electronic waste—"e-waste"—generated by the foreign disposal of the industry's products. By focusing on the global occupational and environmental health impacts of the expanding industry, these campaigns made the link between hazards uncovered in Silicon Valley and similar problems in developing countries. The coalition has gone on to develop a consumer "score card" for revealing differences among computer makers in terms of their environmental performance. Periodic versions of the scorecard and the participation of Greenpeace have led to the development of the "Guide to Greener Electronics," with expanded criteria that consider each firm's use of targeted hazardous chemicals (e.g., polyvinyl chloride, brominated flame retardants) and the effectiveness of their chemical management procedures.[32]

Health Care without Harm (HCwH), noted earlier, began as a campaign to reduce the use of mercury and polyvinyl chloride (PVC) in health care services. The organization arose during the late 1990s as community activists struggled to shut down local hospital trash incinerators that were identified as leading sources of dioxin emissions. The early success of this campaign led the HCwH to broaden its focus to address the chemicals used in health care facilities that were the cause of dangerous emissions. PVC products and packaging were identified as major sources of the dioxin generated in hospital waste incineration, and mercury-containing medical devices were identified as a major source of mercury waste water contamination. Mobilizing local community campaigns around the country, HCwH pressured many local hospitals to examine their medical supply purchases. Soon major hospital groups, such as Kaiser Permanente and Catholic Healthcare West, with big purchasing inventories, were requiring suppliers to find substitutes for PVC-based intravenous bags and tubing and mercury-containing thermometers and sphygmomanometers. Beginning in 2004 with a mercury thermometer take back campaign, HCwH put pressure on national pharmacies to eliminate mercury-containing products. Within ten years, all of the nation's pharmacies had stopped selling mercury-containing thermometers, some 1,200 hospitals had signed a pledge to phase out the use of mercury-containing products, and 400 health care facilities had gone "mercury free."[33]

While the initial focus on single chemicals proved successful for HCwH, during the 2000s, the campaign shifted to a broader focus on the many hazardous chemicals used in hospitals, ranging from surface-cleaning disinfectants to facility building materials and pesticides used on foods. In 2001, HCwH convened "CleanMed," an annual conference for hospital administrators on purchasing safer health care products, and in 2004, the campaign published a "Green Guide for Health Care" that provides guidance on hospital facility construction with a clear focus on avoiding chemicals of high concern. Working with the American Hospital Association and the EPA, the campaign has helped to launch Hospitals for a Healthy Environment (recently renamed Practice Greenhealth), which by 2006 had 1,300 partners representing more than 7,000 health care facilities.[34]

The Campaign for Safe Cosmetics was established by several women's, environmental, consumer, and health advocacy organizations in 2004 to encourage safer chemicals in personal care products. As part of the strategy, the campaign developed a voluntary pledge called the Compact for Safe Cosmetics and pressed cosmetic manufactures to commit to its goals. This included compliance with the European Union's Cosmetics Directive, disclosure of all product ingredients to the Environmental Working Group's *Skin Deep* database, substitution of a list of chemicals of concern, and active participation in the campaign. By 2011, 1,500 companies had signed the pledge, and 322 had fully met the goals of the pledge. Working with these suppliers, the campaign has now begun a new Safe Cosmetics Business Network.[35]

Consumer and environmental advocacy campaigns have proved their mettle in targeting hazardous chemicals used in specific consumer-facing economic sectors. NGOs such as Greenpeace, Healthy Building Network, Pesticide Action Network, Clean Water Action, the Center for Environmental Health, the Center for Health, Environment and Justice, Washington Toxics Coalition, and the Ecology Center have all developed successful campaigns focused on hazardous chemicals. These activist organizations have become increasingly sophisticated in using the media, science reports, government investigations, and direct action to raise awareness around specific chemicals and to press product manufacturers and retailers to substitute chemicals or phase out targeted products. The 2009 NGO-led Internet campaign against bisphenol A provides a good example. This campaign was so effective that purveyors of plastic sports water bottles stumbled over themselves to become "BPA Free" long before any government response was taken. These so-called "market campaigns" are often "media-grabbing," confrontational, and effective. While their direct effects can be quite impressive, the in-direct, deterrent, and catalytic effects on firms concerned that they might become targets of such campaigns are harder to assess but potentially of even greater impact.

7.6   Consumer-Driven Market Transformation

Shifting the market toward safer chemicals can be driven by strategies ranging from informing consumers, engaging retailers, directing institutional purchasing, and mobilizing activists. Products are the central focus of these strategies. Shifting the product market can be one of the most powerful levers for converting the chemical economy. Product labeling, product declarations, and consumer guides increase the information transparency of the chemical market. However, the process of informing and mobilizing consumers is slow and gradual, and there may be an upper bound to the fraction of consumers willing to purchase on the basis of chemical safety.

But informed consumers are not the only driver in the market. The powerful forces of high-volume retailers can amplify consumer desire for safer product ingredients and can provide an even faster and potentially more effective lever for change. Although government procurement agencies may be too vulnerable to supplier pressures to maintain substantial lists of unacceptable chemicals, dominant group procurement institutions and large-volume retailers can play an effective role in "de-selecting" products containing chemicals on such lists. This requires that some of the largest retailers and institutional purchasers come to see value in adopting a public and environmental health mission beyond product price and performance.

Successful consumer campaigns highlight the important role that non-profit and public interest NGOs have in creating social change. Voluntary, special interest organizations have long been a part of American society. These organizations are rapidly emerging throughout both industrialized and developing countries, and they are seen as making up a broadly based, decentralized "civil society sector" that has become a powerful force in promoting social change.[36] The civil society organizations, advocating for environmental protection, public health, and consumer and worker rights in the United States, today play a significant role in raising public awareness, providing technical and professional assistance, writing and negotiating legislation, and influencing changes in social norms and behaviors. Where these organizations have shifted from seeking to change government policy to working to change market dynamics, they reveal the potential leverage in directing market pressures. The growth of these organizations and their emerging sophistication suggest the power they may have in supporting a broader social movement for shifting the chemical market toward safer chemicals.

The effectiveness of these NGO, retailer, and institutional initiatives has led some analysts to argue that these strategies are eroding the role of government and government regulations. By taking regulatory responsibilities out of the cautious hands of politically accountable government agencies, these initiatives place health and environmental protection in the allegedly fickle hands of unelected activists and business agents. However, the history of these initiatives reveals the opposite story; the increasing failure of government regulators has impelled the activism of nongovernment, retailer, and institutional procurement activists. The stagnated condition of the U.S. federal government's chemical policies has led to increased nongovernment direct action. Richard Dennison of the Environmental Defense Fund sums it up shortly: "Retailers and consumers can set their own chemical regulations. If consumers want products without dangerous chemicals, the market can aggregate that demand and drive it up the supply chain."[37]

Potentially, such civil society, retail and institutional organizations can be more effective than government action because, unlike government initiatives, their initiatives do not need to be compromised through negotiation with reluctant business interests. Indeed, if these organizations and their activist members could identify and co-align with those inside the product manufacturing and chemical supply industries who are also pushing for safer chemicals, then they might together amass pressures powerful enough to achieve significant change. The next chapter considers this possibility.

8

Transforming the Chemical Industry

In 1997, the *New York Times* exposed dismal working conditions and low wages in the Nike Corporation's Asian-based network of suppliers. As word of the conditions at these so-called "sweatshops" spread, labor and student activist groups organized protests at Nike retail stores around the country. Less well known in the United States was a scandal over organotins found in the dyes of the Nike sports shirts worn by the German national soccer team that ignited a European consumer boycott of Nike products. Footwear and apparel makers that succeed on the basis of fashion are among the most sensitive manufacturers to public opinion—an entire product line can be killed by a single bad press article.[1]

Nike responded to the revelations about its social and environmental transgressions with an aggressive reframing of its business model. Looking at the full life cycle of its products, Nike moved to introduce sustainability objectives into its product design and manufacture procedures. In 2005, Nike launched its "Considered Design Program" to reduce waste, use environmentally preferred materials, and avoid toxic chemicals. Since then it has become a corporate leader in promoting sustainability and safer products.

If the market-directed initiatives of consumer activists, retailers, and institutional purchasers can be seen as a "demand-side" approach to shifting the chemical market, then activism inside the chemical- and product-producing industries could be seen as a parallel "supply-side" approach. The many ongoing initiatives inside these industries make up a second strategic front in a chemical conversion strategy. Whereas consumer activism focuses on products, the initiatives inside the industry focus on both products and production processes. From a systems perspective, these inside industry initiatives are linked together through the chemical supply chain. Today, many product manufacturers and even some chemical suppliers in these supply chains are developing programs to transition to safer chemicals. Some of these firms have adopted corporate chemical policies and maintain lists of chemicals they avoid, and others have formed economic sector collaborations to set standards for chemical management as part of broader sustainability programs.

Unlike the initiatives for shifting the consumer market, the inside industry initiatives present a more tenuous and guarded area for work. The public and often the government have little access to activities that go on inside firms, industries, and supply chains. However, the Internet and the new values placed on openness and transparency are diminishing the veils of protection that corporations have conventionally maintained. This is positive because the market-directed initiatives outside industry can be more effective where they are informed by and coordinated with efforts inside industry.

8.1   Chemical Industry Initiatives

The idea that private corporations have a responsibility to protect social values is not without critics. Milton Friedman and other economists at the University of Chicago have argued that, when corporations take on social commitments that involve costs, they pervert the singular function of the corporation to maximize shareholder value, and when such costs are transferred on down to the consumer, the increased costs are essentially a hidden tax.[2] However, corporations have long found that a certain amount of charitable philanthropy and civic involvement has paid off in terms of community good-will, political favors, and comfortable neighborhood relations.

The steady rise in activist campaigns against corporations over working conditions, environmental pollution, plant closings, and weapons production during the 1960s provided a fertile environment for the emergence of a wave of "corporate social responsibility" (CSR) programs. These early programs largely involved firms in the energy and manufacturing sectors; the chemical industry, which was focused heavily on resisting chemical regulations, was slow to jump onto the CSR train. However, the tragic gas release at Bhopal, India, in 1984 led chemical industry's leaders to change strategy. While the traditional opposition to regulation persisted, a new, softer, more voluntary strategy emerged that started as public relations but soon became more substantive. This first appeared as several corporations adopted detailed chemical safety "codes of conduct." In 1988, the U.S. Chemical Manufacturers Association (renamed the American Chemistry Council [ACC] during the 1990s) launched an industry-wide program, first developed in Canada, called "Responsible Care," which was explicitly intended to improve the health, safety, and environmental performance of the chemical manufacturing industry.[3]

Today, Responsible Care offers a broad set of corporate guidelines on everything from community education and emergency response to workplace safety and pollution prevention. A special section called the Global Product Strategy (GPS) addresses chemicals management and includes a focus on increased transparency, a tiered approach to risk characterizations, improved product stewardship, and expanded engagement in public policy processes. It requires that participating member companies prioritize their chemicals according to recognized risks and designate high-priority chemicals based on uses, exposures, toxicity, and production volumes. Detailed risk characterizations are to be developed for each high-priority chemical, such that chemical safety summaries and responsible risk management plans can be developed. Currently, GPS Safety Summaries for some 1,600 chemicals are available through the International Council of Chemical Association website.[4]

Through its trade associations, the chemical industry has struggled to respond to a public that has grown skeptical and concerned about it and its products. Surveys of public attitudes toward the U.S. chemical industry show that the industry is among the least trusted of the economic sectors. The periodic revelations of dangerous chemicals in domestic products has added to the legacy of badly managed hazardous chemical wastes and created a gray cloud that hangs over the industry's reputation. Responsible Care and well-funded public relations campaigns to brighten the industry image have not been enough to change public attitudes. The industry has been slow to understand that the public and, increasingly, downstream firms in the product supply chain are asking not just for assurances of chemical safety but actually for safer chemicals. While the chemicals industry may be slow, others in the supply chain are getting this message and responding accordingly.

8.2   Product Manufacturer Initiatives

Just as retailers, institutional purchasers, and others are putting pressure on product manufacturers to eliminate the use of chemicals of public concern, product manufacturers are putting pressure on their component, material, and chemical suppliers to reduce the use of hazardous chemicals that they perceive may present business risks. Firms such as Herman Miller, Timberland, New Balance, True Fabrics, Seagate, Construction

Specialties, Shaw Carpet, Johnson and Johnson, Seventh Generation, Method, Levi Strauss, and Nike are among the leaders. The programs at Nike provide a good example.

Nike is a $24 billion company with headquarters near Beaverton, Oregon.[5] However, with a broad range of products in footwear, sporting goods, and apparel, the company does not own or operate the facilities that manufacture its products. Instead, it contracts out to more than 600 factories in 52 countries. So when Nike sought to develop a corporate sustainability program, it needed to develop a sophisticated business model based on supplier responsibility. Two of the primary elements of the Considered Design Program that Nike launched in 2005 are the "Considered Index" and "Considered Chemistry" programs. The Considered Index is a product design tool for predicting the environmental footprint of a product prior to commercialization. Products are assigned a score based on a set of metrics covering chemical use, waste generation, material life cycle impacts, post-assembly treatment, and innovation.

Through Considered Chemistry, Nike is committed to reducing or eliminating chemicals harmful to human health or the environment. Toward this goal, Nike utilizes a chemical review process to screen chemicals for hazardous properties, a list of restricted substances for suppliers, a strategy to reduce the use of hazardous chemicals in production operations and a research investment program to develop safer chemicals. The restricted chemical substance list currently has sections that apply to finished product materials, odor management chemicals, intermediary manufacturing chemicals, and packaging materials. To ensure conformity with this list among its suppliers, Nike requires that they be willing to support product tests whenever asked. Nike also has a process to select and test samples and asks that suppliers test their products at Nike-approved laboratories. In addition, Nike conducts random tests on products.

Nike's chemical evaluation program is based on hazard categories. Each chemical identified for consideration is given a rating based on several scientific or government studies. If the chemical presents a high hazard score, then it is screened through a specially developed exposure assessment model to determine likely exposure to consumers, workers, or the environment. A chemical that scores high on both hazard and exposure is prioritized for elimination. John Frazier, Nike's Senior Director of Chemical Innovation, calls this "a process for systematically evaluating our materials and helping to stay on the right track to using better chemicals."[6]

Like Nike, many large manufacturing corporations have drawn up lists of substances to avoid, often called "black lists" or "restricted substance lists" (RSLs). Daimler Chrysler, Volvo, Cannon, Sony, Steelcase, and Ben and Jerry's (ice cream) maintain lists of chemicals to avoid in their manufacturing processes. These RSLs are an increasingly common element in corporate chemical policies. The Lowell Center for Sustainable Production conducted a study of nineteen restricted substances lists from fifteen firms and four industrial sectors. With assurances that the source lists would remain confidential, the Lowell Center pooled the lists to create a master list composed of those chemicals appearing on at least one list. By aggregating the lists, some 650 individual substances were identified. In many cases, these chemicals were listed because they were already on some government restriction list; however, many other chemicals were listed because of specific properties (carcinogens, reproductive hazards, persistence, etc.) that firms were seeking to avoid.[7]

S.C. Johnson, one of the nation's largest manufacturers of consumable household products, with some $8 billion in annual sales, has its own unique approach to ensuring that alternatives to hazardous chemicals are not equally hazardous. The firm launched its chemical classification framework in 2001. The framework called "Greenlist" is designed to provide environmental and human health impact ratings of raw materials such that when the company's chemists design products, they can assess and compare the health and environmental ratings of potential chemicals. The Greenlist covers nineteen material categories that range from surfactants and waxes to dyes, thickeners, colorants, and resins and uses qualitative ratings ("Acceptable," "Better," "Best," etc.) covering seven environmental and health criteria, including biodegradability, human toxicity, and aquatic toxicity to rate each material against an average score for its material category.

The chemical hazard data are provided by hundreds of material suppliers in the form of "toxicological summaries," which Greenlist protects through confidentiality agreements. If a supplier is too reluctant to release the chemical information, then it may use the Greenlist criteria to rate the chemicals and submit the scores directly to the Greenlist. David Long, the company's now retired sustainability officer, notes that, although initially some major suppliers were resistant to provide information, today 98 percent of the raw materials used in the company's products have Greenlist scores. Indeed, some suppliers have used the scores as improvement targets and changed their formulations in order to receive higher scores.[8]

Hewlett Packard (HP), also a leader in setting corporate chemical policy, uses an alternative assessment process to screen and rank alternatives before approving any substitution of a chemical of concern (see box 8.1). Helen Holder, the Corporate Materials Selection Manager, explains, "From a business perspective, it is undesirable to face future restrictions for the same application due to a poor choice of replacement materials. Chemical substitutions can be costly, and the required changes can be disruptive to product releases. In light of the trend towards more chemical regulation and substance restriction, there is a growing risk of multiple substitutions unless potential replacement technologies are properly assessed against environmental and human health criteria in advance of their widespread adoption."[9]

Box 8.1
Materials Restricted from HP Products

- Asbestos
- Brominated flame retardants*
- Cadmium*
- Certain azo colorants
- Chlorinated hydrocarbons
- Chlorinated Parafins
- Formaldehyde
- Halogenated diphenyl methanes
- Hexavalent chromium
- Lead*
- Mercury
- Nickel
- Ozone-depleting substances
- Perfluorooctane sulfonates*
- PCBs and PCTs
- Polychlorinated naphthalenes
- Polycyclic aromatic hydrocarbons
- Polyvinyl chloride (in packaging)
- Radioactive substances
- Tributyl and triphenyl tin

*with some exceptions

HP is one of the world's largest information technology companies, with annual revenues of more than $112 billion in 2013. HP also has one of the industry's most extensive supply chains. Comprising more than

1,000 production suppliers and tens of thousands of nonproduction suppliers, HP's supply chain spans six continents and more than forty-five countries and territories.

In the mid-1990s, HP developed a preliminary RSL and set a timeline for phasing out a series of chemicals of concern. The firm began with a pioneering company-wide Design for Environment program that considers environmental impact in the design of every product and solution, from the smallest ink cartridge to an entire data center. Over the past 20 years, the program has led to innovations in material selection for products and their packaging, as well as product transportation, and return and recycling capabilities. More than fifty environmental product stewards work alongside HP's design teams in a concerted effort to improve product performance, measure progress, and communicate results. The company has also developed a General Specification for the Environment (GSE) standard to guide its many suppliers, and, today HP maintains a robust training program for all of its first-tier suppliers. In addition to restricting chemicals through the GSE, HP collects product and component chemical content information from suppliers for more than 240 substances considered emerging chemicals of concern.[10]

Since 2007, HP has used a set of integrated assessment tools to assist in its chemical and material selection. The approach begins with a comparative chemical hazard assessment to help rule out alternatives that are of equal or greater concern than the substances they would replace. Since the program began, HP has completed assessments on the materials that make up 80 percent of the weight of their products, including materials used in low-halogen power cords, brominated flame-retardant alternatives, general plastic resins, as well as cleaners used in the manufacturing process.[11]

The effort to pursue strong chemical policies at Nike, S.C. Johnson, and HP has been hampered by the same problems experienced by other firms, similarly engaged. Obtaining chemical information from suppliers can be difficult and complicated. Some report fully, but some only report on those chemicals listed on specific government lists. A recent study of corporate chemical classification frameworks found that "upstream" suppliers often do not know the chemical ingredients of products. Suppliers may not have the technical capacity to track chemical ingredients or they fear that disclosing the information will result in losing the contract to a competitor or reveal their own liabilities.[12]

The chemical safety programs at Nike, S.C. Johnson, and HP are not unique. Many other large product manufacturing firms, including Steelcase, Dell, Apple, Toyota, Honda, Panasonic, Mitsubishi, Proctor and Gamble, Sony, Boeing, Pfizer, GlaxsoSmithKlien, Ford, General Motors, and Nokia, have such programs as do hundreds of smaller firms manufacturing environmentally friendly products. Major architecture firms such as Perkins+Will and construction companies such as Skanska maintain restricted substance lists and take proactive steps to specify and use products with safer and more sustainable profiles.

These are company-specific programs. Trade associations and industry collaborations have also pioneered pre-competitive initiatives that are moving whole economic sectors toward safer chemical management. An industry-led consortium called the International Electronics Manufacturing Initiative has helped the electronics sector to comply with the European RoHS Directive and identified safer alternatives to the prohibited chemicals. In 2000, an automobile manufacturers consortium set up the International Materials Data System (IMDS) to help firms track chemicals among the thousands of automobile component suppliers and to develop a sector-wide RSL called the Global Auto Declarable Substance List (GADSL), which includes 111 chemicals that are either "prohibited" (e.g., hexavalent chromium) or must be "declared" (e.g., fluorotelemers) at the point of sale. The four largest construction companies in Sweden and the Swedish Construction Federation have developed a broad-ranging material screening tool called BASTA for helping building designers and contractors to avoid chemicals of high concern. Qualified construction material suppliers may sign up by registering their products on the BASTA list, which prohibits any ingredients recognized as CMRs or PBTs. The current database contains 13,000 registered products and materials.[13]

The clothing and apparel industry has developed one of the most comprehensive, sector-wide corporate measurement and rating systems. An initial collaboration between Patagonia and Walmart in 2008 led to the formation of a working group of twelve companies and several NGOs dedicated to creating an industry-wide measurement procedure for tracking the social and environmental impacts of apparel. The foundation for the new index built off of ongoing efforts by the Outdoor Industry Association to develop an "Eco-Index" and Nike's Apparel Environmental Design Tool. In 2010, the working group established the Sustainable Apparel Coalition and began recruiting members from the industry to work together to develop the Sustainable Apparel Index for evaluating materials, products, facilities, and processes in terms of energy consumption, greenhouse gas emissions, water use, waste generation, and chemical toxicity. The index was piloted among some sixty companies and finally released as the "Higg Index" in 2012. Future plans are to expand the index to include footwear and social and labor impacts. To improve the industry management of chemicals, a special Chemical Management Work Group has been convened, and it has released a chemical management framework to benchmark, maintain and improve chemical management "with the goal for all processes and products to ultimately use inherently safer chemicals and reduce or eliminate hazardous chemicals."[14]

These industry efforts have been spurred along by political advocacy. In 2010, Greenpeace organized a series of protests in China over hazardous waste water discharges. Some of the worst of these discharges came from local suppliers to the global footwear and apparel industry. Six leading brand firms—Adidas, C&A, H&M, Nike, Puma, and Li Ning—agreed to meet with the organization. The result was the Joint Roadmap for Zero Waste of Hazardous Chemicals, an ambitious plan to eliminate waste discharges from the supply chain, provide disclosure of production chemicals and chemicals in products, and observe the Precautionary Principle. The roadmap focuses on eleven hazardous chemicals common to the apparel and footwear industries, with plans to use RSLs, audit protocols, supplier training programs, and chemical information disclosure protocols to eliminate their use.[15]

Whereas major product manufacturers and whole industry sectors are focused on avoiding hazardous chemicals, these initiatives reveal a potent leverage point for changing the chemical production and consumption system. When a few firms refuse to accept chemical formulations containing known chemicals of concern, the chemical industry can often supply alternatives without changing production, but when many of these firms demand products free of these chemicals, the chemical industry must begin to remix its product lines. Bolder and less debatable than government regulations, changing demand from product manufacturers provides significant leverage in pressing the chemical industry toward safer chemicals.

8.3   New Directions for the Chemical Industry

The OECD and national chemical trade associations publish projections for the chemical manufacturing industry. It is expected in the years ahead that the demand for bulk and agricultural chemicals will track the growth in global gross domestic product, although growth in specialty chemicals and pharmaceuticals will significantly outpace that rate. Markets for all chemicals will remain rather flat in industrialized countries but grow more steadily in the developing regions.[16]

It is predicted that the future will see two kinds of new chemical plants: supersized, world-scale plants and smaller-scale, flexible, multipurpose plants. The large-scale production facilities primarily designed for manufacturing first-tier, basic chemicals will increase energy, resource, and production efficiencies, but they are unlikely to deviate far from the production of today's suite of high-production-volume chemicals. The newest ExxonMobil Chemical production facility in Singapore can produce 480,000 tons of polyethylene per year, and the new BASF fertilizer plant in Antwerp has a megaton per year capacity. These are likely to be scaled up "generation-isolation-purification" plants located in sprawling industrial complexes, such as those in the Rotterdam Harbor or the Houston Ship Canal that change little and demonstrate strong, structurally defined rigidities.[17]

However, small-scale production of custom-tailored specialty products opens opportunities for more environmentally sensitive, niche market chemicals that could be safer and safer to make. These highly versatile formulation facilities are to be built modular-style, involving an assembly of production units that can be combined and recombined to manufacture a diversity of chemical products. Networks of these smaller, more flexibly constructed plants could be linked up to make open and adaptive production systems of distributed chemical manufacturing.

To prepare for these new directions in chemical production, Europe has invested in a broad, multination modernization program called the European Technology Platform for Sustainable Chemistry (SusChem). Focused by a "Vision for 2025," SusChem provides incentives for industrial innovation and development, with particular attention to new material technologies, chemical reaction and process design, and "white biotechnology," which involves the use of biomass, bioprocessing, and biotechnology in industrial applications. In 2009, the European Commission released a "High Level" report on the future of the European chemicals industry that laid out pathways toward more economic sustainability. While the report presents a broad plan for the innovation, research, human resources, energy, feedstocks, and climate considerations, none of its thirty-one recommendations addresses safer chemicals or green chemistry.[18]

As ambitious as this well-planned "top-down" approach to restructuring the chemical industry is, a more likely driver for change is emerging in the less well-planned, "bottom-up" penetration of new science, processes, and technologies that will be presenting some of the most far-reaching innovations in the industry. These involve changes in scale, intensification, and chemistries.

Equipment miniaturization and a new generation of microreactors, micromixers, microseparators, micro heat-exchangers, and microanalyzers will greatly improve the flexibility and control of reaction conditions, leading to improved production yields, reduced wastes, improved process safety, and new routes for chemical synthesis and product production. Such process intensification involves new chemical engineering approaches that can result in "substantially smaller, cleaner and more efficient technologies." By employing new separation membranes, compact reactors, novel catalysts, and innovative solvents (e.g., ionic liquids, supercritical fluids), the result will be higher selectivity in safer and more simplified scale-up.[19]

Recognizing these opportunities, the European Commission has encouraged the development of multifirm collaborations to develop new ways to increase process intensity and production flexibility. Project "Impulse" aims at the integration of microstructured production components in basic chemical and pharmaceutical production, while the "Copiride" research consortium aims to develop standardized, microscaled equipment with enhanced functionality for modular plants. Another consortium linking universities and leading chemical companies is developing the "F³ Factory" (Flexible, Fast and Future) to combine flexibility and efficiency in smaller scale chemical production with fully containerized "process equipment assemblies," which can provide all of the common utilities, preparation services, and process management controls in one integrated system. These and other multiparty European initiatives are driving the technologies and management capacities of chemical and pharmaceutical manufacturing toward "versatile, continuous and modular processing infrastructure" for the manufacture of chemicals.[20]

The emerging chemistries of biotechnology and nanotechnologies are also likely to have increasing effects on the structure of the chemical industry, at least at the downstream level. What is central to these new technologies is their scale and specialized functionality, and this involves chemistry at the molecular level tailored to exacting functional requirements. The production of these materials does not require super-scaled, highly integrated complexes. At least at this stage, these new chemistries can be manufactured in batch plants by smaller firms or separate divisions of larger companies.

Leaders in the chemical industry tout these new directions as demonstrating the industry's commitment to advancing sustainability, and many of these initiatives will lead to more efficient production facilities that consume less energy and resources and generate less waste. However, these improvements all involve process changes and only secondarily changes in products. The environmental benefits are driven largely by the costs of energy, resources, and waste management, and the health benefits that may emerge come largely from improvements in process safety and reduced workplace exposures.

8.4   Safer Chemicals from the Chemical Industry

The chemical industry responded to the regulatory drivers of the 1970s environmental protection laws largely by installing pollution control and waste management equipment. However, safer chemicals have appeared. In responding to the Clean Air Act, alternatives were developed for replacing volatile solvents used in metal cleaning, and water-based or high-solid paints and powder coatings were developed to replace various aliphatic and aromatic compounds used in paints. Solid-state and radiation-cured coatings replaced volatile nitrocellulose lacquers used in paints and printing on plastic. The Clean Water Act drove the industry to develop replacements for phosphates in detergents and low-foaming agents in dish and clothes washing fluids. Growing concern inside the industry about liability and public relations has also meant that firms have not pursued certain chemicals or ceased their production. Monsanto did not pursue the production of cyclamine, a synthetic sweetener, when laboratory tests showed it to cause tumors, or the production of "Cyclesafe," a clear polymer resin, when it was found to leach acrylonitrile. 3M replaced the use of the perflurooctane sulfonate in its "Scotchgard" product once tests showed widespread presence of the compound and its breakdown products in human fluid samples.[21]

However, these were mostly chemical-by-chemical responses. In the future, the industry has many opportunities to develop more generic initiatives on safer chemicals. Reducing the amount of energy and raw materials inputs can improve overall environmental impacts of chemical production as can many green chemistry processing techniques. There are also opportunities for chemical recycling and reuse. Some production processes have been redesigned to take advantage of the reuse of process intermediates, such as solvents and reagents and the cleansing and reactivation of catalysts.

One particularly interesting approach to chemical reuse involves chemical leasing and chemical management services. Chemical leasing grew out of a cleaner production vision in which intermediary chemicals and chemicals used for services such as cleaning could be leased for use and returned for recycling and reuse. Typical programs involve contract chemical suppliers that are paid by the service performed (e.g., painting a car) rather than the chemicals consumed. The concept of chemical management services employs chemical leasing but takes it a few steps further to include employee training and regulatory compliance services. Chemical service arrangements present a new business model that shifts the objectives from selling volumes of chemicals to selling the functions that chemicals can perform and the management services that firms need. Because customers pay for the chemical services they want and avoid large chemical inventories and waste management expenses while chemical suppliers retain the chemicals for reuse, these services can provide incentives for reducing the use of hazardous chemicals. Chemical leasing has been heavily promoted by the Austrian government, with pilot projects in automobile painting, metal parts cleaning, paint stripping, powder coating, and electroplating baths. In the United States, these practices are promoted by the Chemicals Strategies Partnership, a nonprofit organization that has helped firms ranging from General Motors and Ford to Navistar and Delta Airlines in negotiating chemical management service contracts with chemical suppliers.[22]

These chemical leasing and chemical management services need to be carefully structured. At their best, they present a valuable model for strengthening the management capacities of the chemical industry; at their worst, they present a counterproductive form of contracting out dangerous work. Dow, Quaker, PPG, and BASF have all opened chemical management services in Europe to broaden their market share, secure customers, and offer information management and technical services. Although these initiatives remain small, they suggest the importance of the chemical supplier as a pivotal point of system changing leverage. Located between the chemical manufacturer and the product manufacturer, chemical suppliers provide important information linkages and potential feedback from chemical customers to chemical manufacturers. Enhanced roles for chemical suppliers and new chemical service business models offer important opportunities for innovation.

A third response already underway is the transition to renewable, plant matter and biomass as feedstock. Biorefineries making chemicals from biomass are touted as the alternative to petroleum refineries for producing both fuels and feedstocks for chemical manufacture. Originally promoted to produce ethanol as a fuel, biorefineries are now being developed as a source for chemicals. Current pilot programs show that biorefineries can successfully produce commercially competitive glycerol carbonate and succinic acid and promise a transition to other multifunctional chemicals.[23]

The first wave of biorefineries resulted in a series of facilities across the Midwest constructed as single product plants generating ethanol from corn. However, a second wave of biorefineries has begun to emerge in Europe and the United States, with facilities capable of producing multiple products from a single crop source. Under a French project called BioHub, Roquette is building biorefineries worldwide based on cereal crops with the goal of developing a portfolio of platform chemicals (e.g., isosorbide) for biopolymers and specialty chemicals.[24]

A third wave of anticipated biorefineries can make a variety of products using a variety of feedstocks. By employing multiple feedstock technologies, these new biorefineries could adjust to varying feedstock availability and prices and mix the composition of products to rapidly adjust to changing markets. Research is ongoing in Germany and Austria on "green biorefineries" that can convert grasses, clover, immature cereal crops, or algae into a wide range of useful products. Although no commercial third-wave facilities are in operation yet, development research is being carried out in both Europe and the United States.[25]

Just as cracking and distillation are the primary unit operations of a petroleum refinery, the basic operations in a biorefinery are fermentation, extraction and separation. The biorefineries of the future would generate raw source chemicals from which high-value chemicals such as fragrances, food-related products, and nuetracueticals would be extracted before processing the remaining polysaccharides and lignin to generate industrial chemical products. This would involve separating the green biomass into a nutrient-rich green slurry useful for producing amino acids, alcohols, carboxylic acids, and esters from a fiber-rich press cake that could be used to make biopolymers, insulation, or construction materials. As these processes are run to completion, there should remain enough waste lignin and carbohydrates to be used as fuels in the refinery or converted to syngas.[26]

However, the development of commercially competitive biorefineries confronts many challenges. Biomass typically exhibits low bulk density and high water content, making it costly to transport. Agricultural crops are also seasonal, meaning that off-season storage is necessary to maintain a steady supply, and the material is perishable and susceptible to molds and natural degradation, which requires sensitive storage conditions. Conversion processes are still relatively inefficient, with separation processes accounting for up to 60 to 80 percent of the production costs, and many of the conventional processes still relying on hazardous process chemicals.[27]

An ongoing slow transformation of the chemical manufacturing industry is occurring, and it is appearing on diverse fronts. Those sectors closest to consumer markets are being driven by consumer demands toward safer chemicals and safer chemical products. Firms in the pharmaceutical and specialty chemical sectors are adopting safer chemical processes. These are high-value sectors of the industry, with customers who are often sensitive to the health and environmental consequences of their purchases. These sectors have many firms producing many commercial products that use flexible, batch process technologies, and both sectors support reasonable levels of research and development on chemical products. The specialty chemical sector generated $127 billion in revenue in 2010; of that, an estimated 4 to 8 percent went into research, and 5 to 10 percent went into new plant and equipment.[28] From a strategic perspective, these sectors are important because a major share of both sectors is located in the United States.

However, bulk and basic chemical manufacturers far removed from consumer markets and rigidly bound to traditional chemical production technologies are not shifting to safer chemicals products. Where systems are this rigid, change often requires forces exogenous to the system. Although the chemicals industry has done little to encourage initiatives from outside the industry, some are emerging. Here we will examine three: private standard setting initiatives, nongovernment business dialogue networks, and investor activism.

8.5   Industry Standards Setting

Private industry standards in the chemical industry are not new, but recent standards are driving industry-wide attention to safer chemicals. Such standards include processes for evaluating product quality, standardizing products, and setting test methods and procedures for assessing product safety and environmental impacts. Many professional associations such as the American Conference on Government Industrial Hygienists, the Personal Care Products Council, the National Fire Protection Association, and the American Institute of Chemical Engineers publish standards on chemicals. The American National Standards Institute (ANSI), the American Society for Testing Materials (ASTM), and Underwriters Laboratory (UL) are among the oldest and most respected standards-setting bodies in the United States. ANSI has standards on chemical laboratory safety and scores of standards on the safe handling and storage of specific hazardous chemicals. UL has standards for electronic products, building products, polymeric materials, and wire and cable coverings and has recently established a new corporation, Underwriters Laboratory Environment, to provide tools for environmental product certifications.

Perhaps the most recognized and best studied example of an industry environmental standard is the standard for environmental management systems (EMS) that emerged during the 1990s. EMS include corporate policies, programs, assessments, and documentation systems that permit facilities to identify, monitor, and manage their impacts on the environment. The leading system is ISO 14001, developed as a standard by the International Standards Organization.[29] Today some 982,000 facilities worldwide are certified to the standard. To achieve ISO 14001 certification, a facility must implement a program composed of five parts—preparation of a corporate environmental policy, environmental planning (Plan), plan implementation and operation (Do), periodic monitoring (Check), and corrective action (Act)—and submit its program to a third-party audit. This standard provides a framework for environmental management procedures but does not specify performance outcomes. Therefore, an EMS may include a program to identify chemical hazards and reduce their use, but the standard does not specify this. Evidence on the effectiveness of ISO 14001 is equivocal. Some studies show a marked reduction in the environmental impacts of certified firms, whereas other studies find little evidence of improved environmental performance.[30]

One of the most successful environmental certification programs is the U.S. Green Building Council's (USGBC) Leadership in Energy and Environmental Design (LEED) standard. LEED is a voluntary, consensus-based certification program that provides a suite of rating systems for promoting sustainable building design and neighborhood development practices. For building construction, the program lays out various design categories (e.g., site development, materials and resources, water efficiency, indoor air quality) composed of preferable environmental quality attributes. By meeting each of these attributes, a building developer can collect points that, when accumulated, can earn the building a rating of one of three achievement levels: silver, gold, and platinum. Today more than 35,000 construction projects in ninety-two countries have been LEED certified, representing more than 4.5 billion square feet of building space. The LEED certification system is widely praised for driving the building construction industry toward more energy-efficient and environmentally sensitive building design.[31]

In 2013, the USGBC approved a fourth version of the LEED standard that includes two credits that address chemicals. One point is provided for selecting building products from suppliers that disclose the (non-trade secret) chemical ingredients of their products, and one point is provided for selecting building products from suppliers that have chemical management programs in place to address health and environmental impacts. Although two credits out of a necessary 100 credits does not make a major program around safe chemical management, it is at least a start, and it has not been adopted without controversy. While the LEED standard is today the internationally dominant building construction standard, it is not the only such standard. The Living Building Challenge (LBC) builds on the certification platform constructed by LEED and drives even higher performance standards, including a "Red List" of fourteen chemicals and classes of chemicals to avoid.[32]

Private industry standards can be powerful drivers or unmovable barriers to change. Although most private standards are voluntary, they may be so widely adopted by industry as to compel industry-wide conformance. Because these standards are drafted by practitioners in affected economic sectors, they are often seen by business leaders and professionals as better tailored and more acceptable than the more general standards set by government regulations. However, not all private standards are equal. Those initiated by a cross-section of firms in a specific economic sector without other potential stakeholders tend to set comfortably low standards so as to accommodate current practices, while those initiated by NGOs such as USGBC and LBC or collaborations that include them may provide more challenging standards. The difference between the aggressive standards set by the broadly based Forest Stewardship Council and the weaker industry-initiated Forest Stewardship Institute provides a telling illustration.

8.6 NGO Assistance

NGOs have also begun to play a role inside the chemical and product manufacturing sectors, at first as assessors but more recently as advisors and assistants. Following the major oil spill in Alaska's Prince William Sound in 1989, a group of national business, investment, and nongovernment leaders founded the Coalition for Environmentally Responsible Economics (Ceres) to promote environmentally and socially responsible business practices. Initially, Ceres established a ten-point code of corporate environmental conduct known as the Ceres Principles. In adopting the Ceres Principles, firms pledge to expand their environmental awareness and accountability and commit to continuous improvement and public dialogue on their environmental performance. In 2000, Ceres launched the Global Reporting Initiative (GRI) as an international template for sustainability reporting by firms, and today more than 1,300 corporate entities produce annual reports based on the GRI.

Neither Ceres nor the GRI has focused on safer chemicals; however, during the 2000s, new environmental and consumer advocacy organizations began to emerge focused on working with industries to promote safer chemical systems. Clean Production Action (CPA), a small advocacy organization, arose to promote cleaner production and provide trainings in the processes. Recognizing that more than advocacy was needed, CPA published a series of cases studies on "healthy business strategies" and developed two tools, the *Green Screen for Safer Chemicals* and the *Plastics Scorecard*, which can be used to compare and evaluate chemicals and plastics as to their desirability.[33]

The nonprofit, Healthy Building Network (HBN), was established by environmental advocates to promote safer chemicals in the building construction sector. HBN has developed a chemical ingredients database called *Pharos* to assist architects, contractors, and building developers in selecting safer materials in the building design and construction sectors. *Pharos* relies on an extensive chemical and material library to support a searchable Internet-based system for comparing more than 1,000 commercially available products ranging from doors and windows to paints and mastics. Currently, the library provides environmental, health, and social/community effects information on some 22,000 chemicals, polymers, wood species, and other substances. By searching in each product category, users can find lists of comparable products scored with color-coded and numeric ratings and deeper analyses that identify the attributes that underlie those scores.[34]

This new NGO focus on working with manufacturers has led to the founding of new business networks as well. The Lowell Center for Sustainable Production sponsors the Green Chemistry and Commerce Council (GC3), a network of some seventy U.S. companies convened to support company professionals working to promote green chemistry and design for environment programs. The GC3 has workgroups on advancing chemical information flow along the supply chain, engaging retailers in green chemistry and promoting green chemistry education. In addition, the GC3 provides support for the EPA's Design for Environment program and has worked extensively with congressional staff in seeking appropriations for a federal green chemistry research program.

A similar network, the Business/NGO Working Group on Safer Chemicals and Products (Biz/NGO), was organized by Clean Production Action to promote a dialogue between representatives from business and environmental advocacy organizations on chemicals of common concern and joint efforts to promote safer chemicals policies. This group has adopted four guiding principles for participating firms:

- Know and disclose product chemistry
- Assess and avoid hazards
- Commit to continuous improvement
- Support public policy and industry standards

These principles have been endorsed by some fifty businesses, health care organizations, and investor groups. In 2012, the Biz/NGO published the *Guide to Safer Chemicals*, which includes a scoring system with several performance tiers designed to assist companies in evaluating the degree to which they are implementing the principles.

These networks supported by NGOs suggests the potential opportunities for off-line, business-to-business dialogues by which corporate managers can share in identifying common chemicals of concern and promoting the adoption of safer alternatives. Mark Rossi, the Clean Production Action convener of the Biz/NGO, calls this the "most effective means of integrating progressive business and NGO interests in promoting safer chemicals. These dialogues show that businesses can work together to develop and support common principles for chemical management and promote alternatives to hazardous chemicals."[35]

8.7   Investors as Drivers

Private investors and socially responsible investor groups have also begun to press manufactures about the use of hazardous chemicals. Socially responsible investing began as a movement among investors interested in screening out potential investment opportunities tied to undesirable products (alcohol, tobacco, weapons) or objectionable labor or community relations practices, but it has expanded over the years to direct investments toward corporations that score highly on socially responsible investment indices. Beginning in the 1980s, trade unions such as the United

Mine Workers and the International Ladies Garment Workers Union and government pension funds such as California Public Employees Retirement System (CalPERS) began using social values or indices to direct pension fund investments. More recently, the Social Investment Forum has served as an information clearinghouse and convener for focusing the social and political impacts of the socially responsible investing movement. Today these funds are gigantic, with assets well over $3 trillion in so-called "socially directed" mutual funds and other portfolios.[36] Most of these use environmental screens among their many criteria to select preferred investment opportunities, with many focused on climate change issues. While none of these initiatives includes criteria on the hazards of chemical products or production processes, adding such values to the fund criteria could create another important driver for safer chemicals.

Investor activism has been a driver for corporate policy change since the 1970s global efforts to pressure firms to withhold investments from apartheid-bound South Africa. The establishment of the Interfaith Center for Corporate Accountability during that time helped to formalize and develop shareholder resolution campaigns. Such investor activism can take many forms, such as shareholder resolutions, proxy battles, publicity campaigns, litigation, and negotiations with management. One particularly interesting form is the development of scorecards for rating firms on corporate, social, and environmental factors. The Dow Jones Sustainability Index and the MSCI Global Environment Indices, which rate performance on energy sources, water protection, green buildings, clean technologies, and pollution, are two of the leading rating systems. Although there have been some campaigns by investors around environmental issues, such as environmental liability disclosure and climate change, few campaigns have centered around the risks of hazardous chemicals or production processes.

This has begun to change. The Swedish NGO, ChemSec, has created a "Chemicals Criteria Catalogue" that lists thirty-eight indicators that investors can use to evaluate the chemical safety of corporations.[37] Here in the United States, Clean Production Action and the Lowell Center for Sustainable Production are currently developing a scorecard for rating firms on their chemical management policies and practice. Two U.S. based organizations, the Investor Environmental Health Network (IEHN) and As You Sow, use investor strategies for pressing corporations on safer chemicals. Through the mobilization of investors, the IEHN was able to encourage Whole Foods to remove bisphenol A-based polycarbonate beverage bottles from its retail shelves and prompt Sears to embark on a PVC phase-out process in its products and packaging. As You Sow has used investor pressure to engage firms such as Heinz, Pepsi, Coca-Cola, Campbell Soup, and Hain Celestial on the use of bisphenol A-based polycarbonate in food containers, and it won an agreement with General Mills to phase out the chemical in some of its organic food can linings. Richard Liroff, director of IEHN and author of *A Fiduciary Guide to Toxic Chemical Risk*, sums up the experience by noting, "there is more potential here than the campaigns suggest. Sometimes the mere invitation to address corporate managers by major investors on issues of chemical risks is enough to begin some major rethinking."[38]

8.8   The Chemical Industry's Approach to Sustainability

The U.S. chemical industry, like U.S. industry as a whole, has been relatively cautious on the subject of sustainability.[39] Some large chemical manufacturers are members of the World Business Council on Sustainable Development and have participated in both the 1992 United Nations Conference on Environment and Development and the World Summit on Sustainable Development in 2002. Several of the top U.S. firms have a corporate sustainability statement on their website. Some industry leaders see sustainability as an emergent value. Bill Carroll, past president of the American Chemical Society, noted in a publication he co-authored in 2005 that, "By 2015, the chemistry enterprise will be judged under a new paradigm of sustainability. Sustainable operations will become both economically and ethically essential."[40]

Dow Chemical established a broad series of eco-efficiency goals as early as 1996. Working with an advisory board made up of international leaders, the company developed a sustainability index based on economic, environmental, and social metrics. DuPont also established a list of sustainable development goals and provides an annual progress report, and BASF has developed a life cycle-based eco-efficiency analysis that it has used to evaluate hundreds of its products and operations. Shell Chemical, Bayer, and AkzoNobel are among other chemical manufacturers that have launched similar sustainability programs. These programs primarily track energy conservation, greenhouse gas reduction, renewable energy, and waste reduction. Although Dow measures its performance on using renewable feedstocks, none of these programs addresses reduction in the production or use of chemicals of concern or the development of safer chemical alternatives.[41]

With so much public interest in environmental protection and global sustainability, it would be useful for the U.S. chemical industry as a whole to put forward a vision of a sustainable chemical industry. The capacity is there. In 1996, the Chemical Industry 2020 Technology Partnership published *Technology Vision 2020-The U.S. Chemical Industry*, which laid out a well-staged plan for the technological development of the industry. Around 2000, the American Chemical Society, the American Institute of Chemical Engineers, and the Council for Chemical Research all produced industry "roadmaps" for the future, although none of these addressed health, environment, or sustainability.[42]

Then in 2005, a pregnant moment appeared for doing so. The National Academy of Science convened a Committee on Grand Challenges for Sustainability in the Chemicals Industry as part of a broader series of "Grand Challenges" for industry and science. A work group of respected chemists and chemical industry leaders met nine times and held a well-attended two-day public workshop in Washington. The result was a thoughtful NAS report titled, attractively, *Sustainability in the Chemical Industry: Grand Challenges and Research Needs*.[43]

Sustainability in the Chemical Industry describes the important role of the chemical industry in the U.S economy, takes note of the major shift in production to other countries, and bemoans the declining corporate investments in chemical and process research and development. The report tackles the subject of sustainability ambitiously and offers eight grand challenges for promoting a sustainable industry future:

• discover ways to carry out fundamentally new chemical transformations through green and sustainable chemistry and engineering,

• develop life cycle tools to compare the total environmental effects of products...through the full life cycle,

• understand the toxicological fate and transport of all chemical inputs and outputs of chemical bond-forming steps and processes,

• promote renewable fuels,

• develop more energy-efficient technologies for current and future sources of energy for chemical processing,

• improve the separation, sequestration, and utilization of carbon dioxide, and

• support sustainability education.

These broad and compelling tasks could significantly advance the sustainability of the chemical industry. However, the response from the industry has been tepid and the follow-up limited to modest steps conducted by individual firms. There is an important opportunity here for the trade associations that represent the U.S. chemical industry, such as the American Chemistry Council, the Synthetic Organic Chemical Manufacturers Association, or the National Association of Chemical Distributors, but they have done little to follow up the report or promote sustainability. For the most part, these trade associations, like many U.S trade associations, are inhibited by their least progressive members, and this has forestalled bold discussions about industry-wide sustainability or the promotion of greener and safer chemicals.

8.9 Transforming the Chemical Industry

Restructuring the chemical industry and committing it to safer chemicals will not be easy. Much of the U.S. industry is mature and rigidly structured. The scale and efficiencies of production and the reluctance (outside of pharmaceuticals) to invest in big research has dampened the competitive drive for innovation and experimentation. The close ties between the chemical industry and the petroleum industry means the incentive to move away from oil and gas feedstocks is muted. While chemical industry leaders promote programs to cut emissions, wastes, and energy consumption, no effort has been made to fundamentally reconsider production so as to shift to a safer menu of chemicals.

The movement of the chemical industry out of the United States, particularly into China and developing countries, is not only a loss to the national economy, but it significantly complicates efforts in this country to transition the industry and its market to safer chemicals. Because the chemical industry is now global, much of its production and consumption is beyond the reach of the U.S. government. Therefore, it is important to focus on chemical suppliers and global supply chains that connect domestic drivers to foreign producers. Increasing the flow of chemical information in the supply chain is critical as are efforts to monitor chemical and product imports and press brand name product manufacturers to certify and audit suppliers. Safer chemical policies should also encourage investment in the innovative capacities of the domestic chemical industry, and a good place to start would be with the pharmaceutical, specialty chemical, and consumer product sectors.

Shifting to safer chemicals will be difficult for the production of first- and second-tier basic chemicals—the upstream building blocks of the chemical industry—which has been technologically and financially frozen by large historical investments. However, the more flexible and frequently changing systems of downstream specialty chemicals, pharmaceuticals, and consumer products provide an opportunity for a more innovative and decentralized chemical industry. More flexible forms of chemical production and use and new production technologies are being explored in Europe, and new recycled and plant-based sources of chemical feedstocks are emerging. Biorefineries offer promising opportunities for safer chemicals but only if health and environmental quality are over-riding design considerations.

Faced with complex and sensitive chemicals management issues and a public that is increasingly edgy about harmful chemicals, product manufacturing firms have taken divergent strategies. Some firms may be moving toward safer chemicals, but they disclose little. The large electrical appliance and pulp and paper production firms and many firms in the mining sector provide little public information. Others, particularly smaller manufacturers, tend to be reactive, waiting for regulations or significant media attention before moving to address hazardous chemicals. However, increasingly, leading brand manufacturers are taking proactive and visible steps to assess chemicals and, where necessary, substitute those chemicals likely to worry consumers, pose future liabilities or potentially cause serious health and environmental harm. These innovative efforts on chemicals are easy to praise. Going beyond regulatory compliance in addressing hazardous chemicals fits within the larger frame of corporate social responsibility. Both intentionally and less directly, these initiatives are shifting the responsibility for the next wave of chemical safety initiatives from government regulators to corporate managers. While Nike's John Frazier, HP's Helen Holder, and S.C. Johnson's John Long are certainly committed to improving their company's environmental performance, they are also ushering in a new role for brand producers that makes these firms increasingly agents of public health and safety. By pressing for full public disclosure of chemical ingredients and demanding safer chemical substitutes from suppliers, these corporate managers are using the purchasing and contracting power of their firms to reshape the existing chemical market.

However, such self-regulation has limits. Obviously, self-regulation goes no faster than the leading firms that promote it. Because customer demand is a primary driver, self-regulation works best where a sector is close to the end-use customers. Product manufacturers are more likely to push boundaries for safer chemicals than chemical manufacturers or firms in the oil and gas industries. Leading firms require effective champions, and if they are missing, corporate effort can be limited by economic concerns and lethargy. Finally, there is always the problem of the true laggards—those firms that will violate agreed-on standards to take advantage of gaps in regulations and available externalities.

Basically, self-regulation cannot achieve optimal outcomes without external pressures from customers, standards setting organizations, non-government advocates, investors, or the threat of government intervention. The market can be a driver for innovation, particularly where there are transparent measures and metrics for comparing chemicals, products, or corporations. However, without a background government presence to ensure a "level playing field," leaders will be threatened by laggards. But government can be more than a fair referee. There are simply things that governments do best, such as require reporting, set priorities, prohibit grievous behavior, and penalize delinquents. These functions not only correct "market failures," they also have the capacity to stimulate innovation and industrial development. Michael Porter, a Harvard Business School professor, has argued convincingly that carefully crafted government regulations can promote technology innovation and improved environmental performance that can offset the costs of regulatory compliance.[44]

Transforming the chemical and product manufacturing industries will take both external pressures on the chemical market and innovations from inside the chemical and product manufacturing industries. These efforts will work best where there is a basis for common dialogue and a broader movement that offers recognition, support, and some means of coordination. However, these strategic fronts cannot succeed without a third strategic front focused on engaging the innovative capacities of the science of chemistry. On this, much more work is needed, but success here could be far reaching and system changing.

9

Designing Greener Chemistry

Chelating agents are chemicals used in detergents, fertilizers, and household and industrial cleaners. Most applications involve dispersive uses that release the chemicals to the environment; however, conventional chelating agents are usually persistent and poorly degradable in the environment. Recognizing this, the Bayer Corporation developed a more environmentally friendly and biodegradable chelating agent incorporating sodium iminodisuccinate. Sodium iminiodisuccinate is a sodium salt that functions effectively in chelating iron, copper, and calcium, and it is both biodegradable and relatively benign in terms of its human and ecotoxicological characteristics. The development of this new agent demonstrates an innovation in chemical design. Instead of seeking a safer compound by progressively modifying conventional chemistries (a process called derivitization), Bayer shifted to an entirely new molecule. Sodium iminodisuccinate belongs to the class of aminocarboxylates, nearly all of which are derived from acetic acid. The production of acetic acid chemicals conventionally involves amines, formaldehyde, and hydrogen cyanide. To avoid such dangerous substances, Bayer developed a production process based on maleic anhydride, sodium hydroxide, ammonia, and water. This nontoxic, biodegradable chemical synthesized in a low-waste, low-energy process that avoids the use of formaldehyde and hydrogen cyanide earned Bayer Corporation a 2000 U.S. Presidential Green Chemistry Award.[1]

Changing the practice of chemistry, particularly the applied chemistry that serves as the foundation for the chemical industry, makes up a third strategic front in a chemical conversion strategy. For years chemists and chemical engineers have focused their research on questions of functional performance, processing efficiency, and cost, with little attention to the health or environmental effects of their chemicals. However, the increasing public concern about dangerous chemicals during the 1980s led some chemists to argue that there is sufficient knowledge about the mechanisms of toxicology and environmental contamination for designing chemicals and chemical processes that pose lower hazards.[2]

The concept of green chemistry has emerged from tentative explorations by a few pioneering chemists and firms to a worldwide movement with professional associations, journals, training programs, and supporting institutions. There are now green chemistry initiatives in more than thirty countries. For the leading figures and companies in this movement, these initiatives mark a significant break with conventional approaches to the field of chemistry and a commitment to a thorough overhaul of the chemical industry. For many others, green chemistry offers new opportunities to redirect current research, redesign and market new products, and find new sources of funding or sales.[3]

9.1 Developing Green Chemistry Conceptually

The current green chemistry movement appeared early in the 1990s when the Council for Chemical Research established a small program to support research on environmentally benign chemical synthesis and processing. Spurred by this early effort, the National Science Foundation (NSF) sponsored a conference on environment and chemistry in 1992 that led to the creation of a new NSF initiative called the Environmentally Benign Chemistry Program. In 1996, the EPA joined with the American Chemical Society (ACS) in convening a national environmentally benign chemistry conference that included a presidential awards program designed to recognize academic and corporate accomplishments in "green chemistry."[4]

Since then, green chemistry has been broadly institutionalized. In 1996, the prestigious Gordon Research Conferences convened a special Green Chemistry Conference, and the next year the ACS opened a Green Chemistry Institute to promote green chemistry. In 1999, the Royal Society of Chemistry began a new journal called *Green Chemistry*, and shortly thereafter, the Green Chemistry Network and Green Chemical Technology Awards were launched in the United Kingdom.

Today, the term green chemistry carries a lot of weight. Advocates note how green chemistry differs from conventional chemistry in terms of assumptions, tools, and goals. Those who seek fundamental changes in chemical processing—broad changes that include raw materials, intermediaries, processes, and products—use the term, as well as those who seek more pragmatic changes in synthetic pathways or chemical substitutions that lead to environmentally compatible processes or products.[5]

The most commonly referenced definition of green chemistry, and the definition most frequently used in the United States, is that put forth by two chemists, Paul Anastas and John Warner, in their book, *Green Chemistry: Theory and Practice*. Their definition is quite simple and straightforward:

Green chemistry is the utilization of a set of principles that reduces or eliminates the use or generation of hazardous substances in the design, manufacture and application of chemical products.[6]

This definition is then followed by a list of twelve "Principles of Green Chemistry" that lays out testable guidelines that can be used to guide those designing new chemicals or those seeking to differentiate environmentally benign chemicals from chemicals of greater concern. The widely recognized set of principles is presented in table 9.1.[7]

Green chemistry is hazard-based and directed at the intrinsic characteristics that make chemical processes toxic and hazardous. Green chemists argue that by carefully considering the molecular and physical properties of chemicals, it is possible to "design out" those properties that make them hazardous. Paul Anastas, director of the Green Chemistry and Engineering Center at Yale University, and Tracey Williamson, an EPA colleague, make this goal clear when they conclude, "(g)reen chemistry seeks to reduce or eliminate the risk associated with chemical activity by reducing or eliminating the hazard side of the risk equation thereby obviating the need for exposure controls and, more importantly, preventing environmental incidents from ever occurring through accident. If a substance poses no significant hazard, then it cannot pose a significant risk."[8]

Green chemistry has two different presentations. Within industry and government, it is another in a long line of policy reforms (such as pollution prevention and cleaner production) that requires a new approach to chemical selection, chemical processing, and waste avoidance. For science, it is a reform-oriented perspective that alters the ways of practicing chemistry and changes the mission of the field by infusing environmental consideration into the core of the discipline. The boundaries of the field are broadened, and the types of knowledge needed by the chemist (e.g., toxicology, environmental fate, environmental law) are expanded.[9]

9.2   Greener Chemical Products and Processing

Green chemistry involves some new and many conventional chemistry processes. Attention to atom efficiency, energy conservation, and hazard avoidance has been an element of good chemical processing for years. The

Table 9.1

Twelve Principles of Green Chemistry

---

1. It is better to prevent waste than to treat or clean up waste after it is formed.

2. Synthetic methods should be designed to maximize the incorporation of all materials used in the production process into the final product.

3. Wherever practicable, synthetic methodologies should be designed to use and generate substances that possess little or no toxicity to human health and the environment.

4. Chemical products should be designed to preserve efficacy of function while reducing toxicity.

5. The use of auxiliary substances (e.g., solvents, separation agents, etc.) should be made unnecessary wherever possible and innocuous when used.

6. Energy requirements should be recognized for their environmental and economic impacts and should be minimized. Synthetic methods should be conducted at ambient temperature and pressure.

7. A raw material of feedstock should be renewable rather than depleting wherever technically and economically practicable.

8. Unnecessary derivitization (blocking groups, protection/de-protection, and temporary modification of physical/chemical processes) should be avoided whenever possible.

9. Catalytic reagents (as selective as possible) are superior to stoichiometric reagents.

10. Chemical products should be designed so that at the end of their function, they do not persist in the environment and break down into innocuous degradation products.

11. Analytical methodologies need to be further developed to allow for real-time, in-process monitoring and control prior to the formation of hazardous substances.

12. Substances and the form of a substance used in a chemical process should be chosen so as to minimize the potential for chemical accidents, including releases, explosions, and fires.

---

novelty is in bundling together all of the principles and placing environ-mental and health protection alongside and equal to intellectual, scien-tific, and commercial values.

The technical literature and textbooks on green chemistry offer a wide array of examples.[10] These common conventions can be divided into those focused on chemical synthesis and those focused on chemical pro-cessing. Academics might suggest that this follows the professional dis-tinction between chemistry and chemical engineering. Indeed, an ongoing effort seeks to establish "green engineering" as an initiative similar to green chemistry; however, in practice, the term *green chemistry* is used for both environmentally benign chemical synthesis and chemical process-ing.[11] Here the conventions are simply noted.

Environmentally Benign Starting Materials

Green synthesis begins with environmentally benign feedstock chemicals such as biological and plant-based materials. The preferred chemicals are those derived from renewable sources that demonstrate low toxicity and appropriate rates of biodegradability. Employing polysaccharides as feed-stocks for polymers is an example of a renewable and nontoxic material for beginning a synthesis pathway. Likewise, glucose can be used as a raw material rather than benzene in the production of hydroquinone and adipic acid, both of which are important intermediaries in the production of commodity chemicals. Renewable agricultural crops such as corn, potatoes, sugar, grain, or grasses offer a source for carbohydrates. Glu-cose and polysaccharides generally provide nontoxic, biodegradable feed-stocks. Indeed, relatively nontoxic silicon has been suggested as a useful replacement for carbon as a starting base for the synthesis of some organic chemicals.[12]

Efficient and Low-Waste Reactions and Processing

Atom economy is a simple way of describing improvements in yield—getting the maximum amount of chemical product out of each reaction. In application, it is defined as the molecular weight of the desired prod-ucts divided by the molecular weight of all the reactants. As convention-ally used, atom economy does not consider wastes, although the "E-factor," a similar concept, incorporates wastes, auxiliaries, solvents, and excess reagents into the calculation of yield efficiencies. The E-factor is defined as the weight of all waste produced per weight of product. Atom economy and the E-factor provide useful calculations at the process design stage in selecting the most efficient reactions. The objective is to improve reaction efficiency and minimize the number of process steps. For instance, addition reactions are preferred over subtraction reactions because they incorporate much of the starting materials and are less likely to produce large amounts of waste.[13]

Green Reagents and Green Catalysts
In many chemical reactions, catalysts are used to increase speed and efficiency. Greener processing involves identifying catalysts that function in chemical transformations with minimal environmental harm (e.g., minimize energy inputs, reduce waste outputs, minimize occupational exposures, and eliminate accidents). Alternatives are sought for highly toxic heavy-metal catalysts. Solid acid and heterogeneous oxidation catalysts can be used to replace hydrogen fluoride and other highly corrosive catalysts. In addition, phase transfer and photosensitive catalysts offer high functionality and low toxicity.[14] The use of liquid oxidation reactors replaces metal oxide catalysts with pure oxygen and permits lower temperature and lower pressure reactions with higher selectivity and no metal-contaminated wastes. New catalytic techniques that rely on enzymes, microwaves, ultrasound, or visible light obviate the need for harsh chemical catalysts.[15]

Alternative Solvents
The single largest source of wastes from many reactions involves solvents. In addition, many solvents generate volatile organic compounds, and several halogenated solvents (methylene chloride, perchloroethylene, carbon tetrachloride) and aromatic hydrocarbons (benzene) are recognized carcinogens. Investigations on alternative solvents have demonstrated potentially wide applications of ethyl acetate, ethanol and various ionic liquids, immobilized solvents, and supercritical fluids. Supercritical fluids, which are typically gases (CO_2) liquefied under pressure, are commonly used in coffee decaffeination and hops extraction. Supercritical CO_2 can be used as a replacement for organic solvents in surface preparations, cleaning, polymerization reactions, and surfactant production. Indeed, there are many applications where solvents can be eliminated altogether by working in solid state chemistries or where the feedstock provides its own solvency. Future work may involve solventless or "neat" reactions, such as molten-state reactions, dry grind reactions, plasma-supported reactions, or solid materials-based reactions that use clay or zeolites as carriers.[16]

Aqueous Phase Reactions

Water-based processing provides many benefits because it is low cost, safe, and nontoxic and because water is simple to store and use. The use of water in organic reactions offers unique properties because of low volatility and noncovalent hydrogen bonds. Water has been shown to be an effective solvent in chemical reactions such as free radical bromination. Both oxidations and reduction of organic molecules can occur in aqueous media. Recent successes in conducting reactions thought not to be possible in water suggest that these reactions actually take place "on water" because the reactants are hydrophobic and insoluble in water. In addition, water proves to be an effective agent in separating homogenous catalysts from a reaction mixture, allowing the catalysts to be recycled and reused.[17]

Biocatalysts

Biological enzymes and cells of organisms used as catalysts offer many environmental advantages. They are typically biodegradable, nontoxic, and water soluble, and they perform comfortably in aqueous environments at ambient temperature and pressure. They work in small quantities, are fairly rapid, and require comparatively low energy. The reactions are typically one-step processes that avoid protection and de-protection of functional groups. The use of enzymes, such as hydrolases, lyases, transferases, and ligases, have all found applications in environmentally sensitive bioprocessing. However, biocatalysts have limits because they are difficult and expensive to make, are unstable under extreme conditions, and require dilute solutions.

Energy-Conserving Processes

Many conventional routes of chemical synthesis require large amounts of thermal energy for heating, cooling, purifying, and separating. Designing processes that avoid exothermic conditions reduces the amount of separations and limits the need for filtration, or distillation and can reduce the consumption of energy. Microwave, ultrasound, and sonic energy used as catalysts require relatively little energy. Microwave heating has developed rapidly as a low-energy component of conventional chemical transformations, particularly in condensations, oxidations, and some carbon-carbon or carbon-nitrogen bonds. Localizing production steps and reducing material transport can also reduce energy consumption.[18]

Inherently Safer Process Design

Inherently safer design (ISD) seeks production process safety through fundamental process design rather than the application of ancillary physical, mechanical, and management processes. The concept arises from early work by Trevor Kletz, a safety advisor for ICI Petrochemicals. In a landmark presentation in 1977 titled "What You Don't Have, Can't Leak," Kletz laid out the basic argument for reducing chemical plant hazards by redesigning chemical processes and replacing hazardous chemicals, rather than creating barriers, buffers, and control redundancies. Since then, efforts to infuse inherent safety into chemical process design has proceeded, although slowly. Green chemistry offers a new driver for designing safer and less accident-prone chemical processing plants.[19]

Alternative Products

Greener chemical processes are intended to produce greener chemical products. There is an inherent contradiction in striving for a green synthetic pathway for production if the product itself is highly hazardous. Green chemistry avoids making chemical compounds that are well-recognized carcinogens, reproductive toxins, or substances that are persistent, bioaccumulative, and toxic. Indeed, green chemistry avoids making chemicals that appear on lists of high-hazard chemicals or government lists of severely restricted or banned chemicals. Not surprisingly, green chemistry has not found much value in pesticide production.

9.3   Chemicals from Renewable Feedstocks

Shifting from fossil fuels to renewable feedstocks is one of the principles of green chemistry. This means renewable resources such as biological and agricultural materials, organic wastes from forestry, food and fiber industries, and municipal sewage. There is a fine history here. During the 1930s, a small group of well-respected chemists and corporate leaders, including William Hale, Wheeler McMillan, Thomas Edison, and Irenee du Pont, launched a movement to promote agricultural crops as a feedstock for manufacturing chemical products and fuels. Henry Ford, an enthusiastic advocate, was convinced that he could manufacture industrial organic chemicals from the oil of soybeans. Over the years, he supported research into crop-based polymers, and in 1941, a "farm grown automobile" was displayed at the Dearborn, Michigan, Ford factory. This so-called "chemurgy movement" grew in size and legitimacy up through World War II, but it collapsed when petroleum prices fell below the price of the required farm crops. While the movement was largely driven by a desire to promote the economics of impoverished farmers, it did demonstrate the viability of a chemical industry based on agricultural feedstocks.[20]

Plenty of chemicals and products on the market today are made from agricultural crops. More than 15 percent of the dyes and 16 percent of inks are made from plant matter, as are many shampoos, cosmetics, soaps, and detergents. Biomass provides the basis for nearly 6 percent of pigments, 35 percent of surfactants, and 40 percent of adhesives. Soy-based inks are the standard for color reproduction in newspapers. Linseed oil is a key feedstock for many surface coatings and linoleum flooring. Coconut and palm oils are used extensively in soaps and detergents. Epoxidized oils from soy beans are used as plasticizers. Ethylene is being produced from sugar-based bioethanol dehydration in Brazil. There is a slowly growing market for biopolymers-plastics made from corn, potatoes, sugar beets, sugar cane, or cellulosic feedstocks. Several biobased resins are available today for use in various applications such as fibers, films, and extruded and thermoformed containers for packaging.[21]

Carbohydrates make up the largest volume of plant material. This includes starches, sugars, and the cellulose and hemi-cellulose that compose the structural elements of plant tissues and calls. Each of these has feedstock potential for the production of polymers of glucan, coniferyl, and coumaryl. During the 1990s, Cargill and Dow Chemical formed a joint venture to manufacture polylactic acid-based polymers from Midwestern corn. A heavy investment in production capacity in Nebraska led to the development of a commercial grade product called *NatureWorks*. Although Dow withdrew from the enterprise, the market for the product has continued to expand, although slowly. Other chemical companies have pursued plant-based polymers as well. DuPont has an active market in *Sonora*, a polymer made from corn-based 1,3 propanedial and petroleum-derived terephthalic acid, which is used in the production of carpets. Metabolix, Bayer, and BASF all have active research programs in the area. Researchers at the University of Maine are exploring the use of the state's cellulosic waste from the forest products industry as a source material for biobased polymers. Other production facilities are online or in development to create millions of pounds of additional biobased resins.[22]

In 2000, the U.S. Congress passed the Biomass Research and Development Act to provide federal support for research on biobased fuels and chemical products. Primarily viewed as a means of expanding markets for agricultural products and reducing export dependence, this initiative, nonetheless, has spurred the search for "new uses" of biobased chemicals. A biomass research and development initiative was launched in 2002 that prepared a *Roadmap for Bioenergy and Biobased Product in the United States*. The roadmap identifies the many challenges that lie ahead for bio-based production, including improving the conversion and separation technologies needed to process oils, sugars, and proteins and lowering the unit cost so as to effectively compete with petroleum feedstocks.[23]

Many economic, environmental, and social factors are driving a renewed interest in biobased chemicals:

• increasing costs of fossil fuels
• concern over predicted limits to fossil fuels
• pressure to reduce greenhouse gas emissions
• pressures to reduce nondegradable wastes
• consumer demand for safer and "greener" products
• corporate desire to enhance sustainability and social responsibility[24]

In terms of green chemistry, plant-based chemicals also have technical benefits. Chemical products derived from renewable resources tend to be less hazardous and inherently more biodegradable. The feedstocks are generally safe and nontoxic. The chemical processing takes place in aqueous environments under ambient thermal and pressure conditions.

Still a green chemistry-driven shift toward renewable feedstocks presents concerns. Simply being biobased does not mean that a chemical is safer. It is quite possible to make many current chemicals of concern, such as formaldehyde and epichlorohydrin, from biomass sources. The production processes for making chemicals from crops can also be hazardous. For instance, the conventional extraction of many oils from vegetable crops uses hexane, a hazardous air pollutant. The production of agricultural crops in the United States has become highly industrialized, and the amounts of hazardous pesticides, synthetic fertilizers, and processing chemicals involved in modern farming compromise the possibility that agricultural feedstocks provide net benefits over petroleum feedstocks. Intensive agriculture often leads to decreases in soil biota, nutrient cycling, and soil quality. The tillage of cornfields is particularly intensive and results in substantial amounts of soil loss (ranging as high as two bushels of topsoil for every bushel of harvested corn).[25]

The energy and resource consumption of industrial agriculture is significant. Pesticide and fertilizer production is energy-demanding, and farm management from tilling to harvesting involves large amounts of fuel-driven machinery. During the past century, the energy intensity in corn production rose eightfold. It is estimated today that U.S. food production requires two units of fuel energy for every unit of food energy produced. Intensive agriculture also consumes large volumes of water. Although irrigation from surface water supplies a large amount of farmland, excessive pumping of ground water for agricultural uses is progressively depleting some of the nation's largest aquifers, such as the Ogallala Aquifer that underlies seven Midwestern and supplies water for 20 percent of U.S. irrigated farmland. The huge petroleum-based fuel inputs, the water use, and synthetic chemical pesticides and fertilizers used in industrial agriculture chasten the easy assumption that crop-based chemicals are safer than petroleum-based chemicals.[26]

The scale of agriculture also raises concerns. It is estimated that the land needed for generating enough conventional crops to replace the fossil fuels base of the current petrochemicals market would take up a land area equal to the entire state of Connecticut. Because of this, it is becoming more widely recognized that waste from the pulp, marine, dairy, and food industries; organic sewage; and lignocellulose materials such as wood, straw, and grasses offer better raw materials because they do not require more farmland or compete with the food supply.[27]

In an attempt to promote biobased materials that present net-positive environmental profiles, a group of environmental and small farm advocates has collaborated to form the Sustainable Biomaterials Collaborative. In 2009, the collaborative released a performance standard called "Biospecs" for bioplastic food service ware that meets environmental and sustainable agriculture values. Because sustainably produced crops cannot be traced though the complex processing of the agricultural industry, Biospecs uses a sustainability certificate as a marketable offset for farmers selling sustainably grown crops (corn, soy, sugar beets) that meet the Biospecs standard.[28]

9.4 Chemistry Modeled on Natural Systems

The energy-intensive, waste-generating, and hazardous production systems of the chemical industry are polar opposites of the low-energy, nearly wasteless, and relatively benign production processes of biological systems. An examination of these processes suggests new models for safer chemical synthesis. Several early polymers were designed as derivatives of natural polymeric compounds, and today many pharmaceuticals are based on compounds found in living plants and animals. Recognizing this, the emergence of a research specialty, called "biomimetics" or "bio-mimicry," has drawn together a group of natural scientists to identify processes that might serve as models for environmentally sustainable industrial production. These studies explore how organisms make and use materials to compose physical structures, communication systems, habitats, and tools.[29]

A broad exploration of biological systems reveals processes for making hundreds of different chemicals and materials ranging from pliable polymers, rigid membranes, crack-resistant ceramics, and durable surface coatings, to adhesives, gels, lubricants, inks, dyes, and disinfectants. Natural processes involve fats, carbohydrates, amino acids, and the vitamins and nutrients necessary for life. These can be used to make coatings, structures, insulators, electrical conductors, and data processors. Considering how organisms make tough, polymeric materials such as skin, hair, or shells reveals a wide variety of natural chemistries. Indeed, for many common synthetic chemical products, similar natural products are made through relatively benign and renewable processes. This involves digestible substances that are safely metabolized and biodegradable and readily compostable materials that fit comfortably into ecological nutrient streams.

Studying nature provides many clues on what chemicals are safe for living things.[30] Over thousands of years, organisms have used a carefully selected menu of chemicals to build up their bodies. Identifying materials and processes that are safe and ecologically sound could begin with a list of those chemical compounds that make up the human body. Unlike the relatively simple reactions of synthetic chemistry that employ much of the periodic table, the elaborate reactions in nature's chemistry rely on only a few common chemicals. The many compounds and chemical structures that are used for common human foods could be added to this list. People have been exposed to the carbohydrates, starches, proteins, minerals, and vitamins that make up the human diet for centuries. In many ways, humans and their diets have co-evolved to sustain one another. The human food source, rich in organics, and the processes by which seeds, grains, legumes, and other food plants develop could provide a host of chemicals that are likely to be compatible with human and ecological health. However, this is not the same as saying that all naturally produced chemicals are likely to be safe. Nature's chemistry focuses on both the chemical product and its production process. Industrial processes that mimic nature need to be as exacting in their design.

9.5   Green Chemistry and the Chemical Industry

Green chemistry has found uneven acceptance within the chemical industry. The leading examples of green chemistry typically involve high-value/complex chemical compounds such as pharmaceuticals and specialty chemicals where there are big opportunities for processing changes or polymers and materials where feedstocks can be shifted from fossil fuels to renewable resources.

Firms within the pharmaceutical industry, such as Merck, Pfizer, Bayer, and GlaxoSmithKline, have been leaders in promoting green chemistry. Because drug development is so research intensive and the health care industry is so sensitive to health objectives, these firms have found competitive benefits in promoting green chemistry initiatives that generate safer products. "If a pharmaceutical company is truly committed to health, that mission is incomplete without a similar commitment to a safe environment," says Berkeley (Buzz) Cue, a retired Vice President at Pfizer, who has been an outspoken champion for green chemistry. Dr. Cue goes on, "Green chemistry makes sense because it lowers drug production costs, communicates to the public a sense of corporate responsibility and inspires students who one day will be Pfizer employees."[31]

There are also promising green chemistry innovations among the consumer product formulators and in the specialty chemical sector. The Green Chemistry Formulators Roundtable sponsored by the ACS lists a series of innovative processes and chemical substitutions that have reduced the use of hazardous antimicrobials, boron compounds, amines, solvents, alkanolamides, UV absorbers, surfactants, and fragrances. A good example of a successful green chemistry initiative in specialty chemicals involves photoresists used in semiconductor production. Legacy Systems of Fremont, California, developed a chilled ozone process that uses only oxygen and water in a cold process vessel to replace hazardous "piranha" solutions (sulfuric acid, hydrogen peroxide, or ashers). The process digests and oxidizes the photoresists into soluble chemistries, thereby eliminating the hazardous solutions and thousands of gallons of wastewater at each silicon wafer production station. Another example involves pigments. The Engelhard Corporation, a "minerals-to-chemicals" company, won a 2004 Green Chemistry Challenge Award for the development of azo pigments based on calcium, strontium, and barium as substitutes for its own heavy-metal based pigments that contain lead, hexavalent chromium, and cadmium. These new pigments have been deemed safe enough to be approved by the FDA for indirect applications with food, and the production process takes place in an aqueous medium that eliminates the polychlorinated intermediates and organic solvents associated with the production of conventional pigments.[32]

However, further up the chemical synthesis chain, there are fewer and fewer green chemistry examples. The closer a production step is to the refinery or smelter where platform chemicals are dominant, the less flexibility for innovation there is. Consider benzene, ethylene oxide, and phthalates again. There are many green chemistry alternatives to phthalates in cosmetics, personal care products, toys, cables, and other plastic products. The large uses of ethylene oxide as a sterilizing agent in hospitals can be replaced with hydrogen peroxide and peracetic acid-based products. Benzene presents a more complex problem. Benzene, chlorine, fluorine, bromine, formaldehyde, styrene—these are all highly hazardous chemicals, and they are also fundamental platform chemicals critical to the production of many other substances.

These are commodity chemicals whose production processes are technologically fixed and mature. The basic process technologies were developed decades ago, and the path of development since then has largely been in optimizing those processes. Years of seeking yield efficiencies, energy and resource conservation, and waste minimization for cost reduction (not environmental protection) objectives have left little room for significant process innovations. Since the development of linear low-density polyethylene during the 1980s, there have been no real big "breakthrough" product innovations in commodity chemicals. The large-volume, platform chemical production facilities built over the past twenty years in Asia and the Near East tend to be the highly efficient and technologically advanced. However, the production of bulk chemicals in the United States takes place in long-established, capital-intensive plants that are generally run until technological exhaustion, with little pause for technological upgrades. Therefore, green process improvements for these conventional technologies are unlikely.[33]

The chlor-alkali industry provides an example. Chlorine and sodium hydroxide are critical platform chemicals. There are three types of production processes based on how the chlorine gas and sodium hydroxide once separated are prevented from remixing. The oldest production technology is based on the mercury-cell process that included many hazardous intermediates including mercury. Most of the U.S. mercury cell facilities were driven to retirement, and today mercury cell production makes up less than 10 percent of national production. Production based on diaphragm technologies that employ an asbestos fiber diaphragm appeared during the 1930s and became the preferred alternative to the mercury cell technology. However, concern over asbestos drove several countries (e.g., Brazil, Saudi Arabia) to ban the diaphragm technology and led Japan to phase out all diaphragm technologies. During the 1970s, a new process for sodium hydroxide production based on ion-exchange membranes developed, and today the newest facilities in Asia and the Near East use such technologies. Each of these steps in the technological development of the industry has led to cost reductions, yield improvements, lower energy consumption, and reduced environmental impact. However, no major chlor-alkali facility has been built in the United States since the 1980s, and 70 percent of U.S. production remains based on the diaphragm technology even though the ion exchange technology is more efficient. Energy and yield improvements have been made in U.S. diaphragm production facilities; however, the fixed capital investments and "sunk" investments made in these facilities have delayed the shift to even more energy- and resource-efficient technologies.[34]

Finding alternatives to these platform chemicals requires exploring substantively new chemistries. Historically, the shift from stoichiometric to catalytic chemistry brought significant advances in synthetic efficiencies, and laid the foundation for the tightly integrated systems of the modern chemical industry. Green chemistry could have similar transformative effects today. Starting chemical synthesis with the hydrocarbon derivatives of petroleum requires additive steps such as chlorination to support reactions and then subtractive steps to generate pure products. Starting chemical synthesis with biobased feedstocks such as lactic acid, citrates, methyl esters, cellulose, glycerol, sorbitol, and polysaccharides offers hydrocarbon compounds that are rich in oxygen, nitrogen, and diverse functional groups. Instead of stripping off functional groups to fit these substrates into conventional processes, new chemistries could be built directly from these complex structures The high degree of chirality in amino acids, sugars, and enzymes could be exploited as well as the biodegradability that keeps carbon in the lithospheric carbon cycle. This is where the new biorefineries become important: not only do they employ new starting materials, they offer opportunities for new synthetic routes to platform chemicals. However, this will require adjustments. Conventional chemistries use and generate high-purity substances, while biobased processes are inherently "messy," generating fairly non-homogeneous products. Either separation/purification processes need to be greatly improved or the market needs to adjust to less than pure substances.[35]

The Department of Energy has completed a study designed to identify promising organic platform chemicals that could be derived from biobased feedstocks. The study identified twelve compounds with multiple

Table 9.2
12 Platform Chemicals from Renewable Resources

1,4 succinic, fumaric and malic acids
2,5-furan dicarboxylic acid
3-hydroxy propionic acid
aspartic acid
glucaric acid
glutamic acid
itaconic acid
levulinic acid
3-hydroxybutyrolactone
glycerol
sorbitol
xylitol/arabinitol

functional groups that could be transformed into new families of molecules with a wide potential for useful applications (table 9.2).[36]

In addition to biobased synthesis, chemists could explore different approaches in underutilized chemistries. By manipulating the spatial geometry and timing of molecular interactions, chemists could impart reactivity with high selectivity and low energy even with quite unreactive substrates. John Warner has long noted that chemists have relied too heavily on covalent bonds in building up molecules and ignored the non-covalent structures common in nature. More could be done respecting weaker bonds, such as the hydrogen bonds of aqueous systems. Paul Anastas recognizes these possibilities: "Green chemistry innovation is not just about less hazardous processes and products, it can open up opportunities for changing the fundamental way in which we make molecules and how we transform them once they are available."[37]

9.6   Redesigning Chemistry

The introduction of green chemistry might be further advanced if it were more seriously embraced by mainstream chemistry. The concern for safety and sustainability in the future of chemistry could be as potent a driver for research and innovation as performance and cost have been in the past. Academic departments of chemistry could be churning out young chemists inspired by chemical safety and sustainability who could provide the energy and intellectual capital for changing the chemical industry. However, changing mainstream chemistry is a significant challenge.

The academic field of chemistry has long focused on the composition, structure, behavior, and properties of chemicals. Little attention has been given to public health or the environment. The conventional practice of chemistry and the material sciences tends to distance scientists from information on the potential health and environmental consequences of their inventions. Green chemistry challenges this separation by broadening the disciplinary boundaries to embrace the effects of chemicals on health, environment, and, by implication, society. John Warner has often noted that even at the leading university departments of chemistry, there is no requirement for coursework in toxicology, ecology, environmental fate of chemicals, or the laws and regulations that affect chemical processes or products. Without a broader approach to chemistry education, chemists are limited in their green chemistry capabilities.

All this may be changing. Today, a growing number of textbooks are available for teaching green chemistry, and classes in green chemistry have been introduced in scores of undergraduate chemistry programs. Green chemistry principles are penetrating into conventional graduate chemistry education. The University of Toledo has a new School of Green Chemistry and Engineering, and the University of Massachusetts Boston offers a doctoral concentration in green chemistry. Green chemistry research centers have been established at several universities, including Carnegie-Mellon, Yale, the University of Oregon, and the University of California Berkeley. A chemistry that is not only about making new molecules but about making them safely might be just what college students are looking for today. The popularity of the green chemistry classes currently offered by some colleges suggests that green chemistry could be a key to reversing the long-term decline in students interested in chemistry degrees.

However, the fundamental curriculum for teaching college chemistry is fairly rigidly defined, and the teaching staff comes largely from the traditional approach to the science of chemistry. Changing a science is not easy. Thomas Kuhn, a leading figure in the philosophy of science, noted the staged way that science develops through periods of turmoil, consolidation, and resistance with fixed paradigms pushing fundamental innovations to the margins. Green chemistry may be a paradigm buster, but it has not been promoted as such. Rather, green chemists have sought to slide the new perspective unobtrusively and ubiquitously into conventional science practice with the hope that, eventually, all chemistry will be green.[38]

This strategy requires that green chemistry be integrated throughout organic, inorganic, and analytical chemistry courses. Coursework and laboratory exercises need to integrate toxicology, environmental science, law, and policy. Green chemistry principles need to be integrated into conventional graduate and undergraduate chemistry curriculum, and safer chemistry examples and illustrations need to be incorporated into high school chemistry classes and elementary science programs. This requires a shift in curriculum. Traditional engineering education has been pushed to include principles of sustainability by the national accrediting board for engineering schools. Although the ACS's accrediting program encourages a green chemistry concentration, it does not require that green chemistry be integrated into the core curriculum. This has led Amy Cannon, director of Beyond Benign, a center promoting green chemistry education, to promote a "Green Chemistry Commitment" similar to the nationally recognized American College and University Presidents' Climate Commitment. Launched in 2013, the commitment is to be signed by university administrators who are actively integrating green chemistry into their educational programs. It may be a small step, but over time, it could broaden the awareness and commitment to green chemistry education in the nation's schools.[39]

9.7   The Challenge of Green Chemistry

The greening of chemistry has not been easy. Like any professional reform, green chemistry faces serious challenges and, in some cases, outright resistance. The slow pace of adoption of the green chemistry principles can be explained by many factors, some institutional, many economic, and some simply personal. Many large chemical companies have no green chemistry programs, and many others see green chemistry as merely a niche enterprise. Chemical trade associations have been slow to embrace green chemistry, and many large academic chemistry programs remain skeptical of green chemistry. Indeed, government agencies have been hesitant to promote green chemistry or link green chemistry to existing regulatory or technical assistance programs, and the only national legislation written to directly support green chemistry research has languished in the U.S. Congress for years.

John Warner understands these challenges but sees many opportunities. "Green chemistry is the mechanics of doing sustainable chemistry," he observes. "Focusing on green chemistry puts us in a different innovation space. It is an approach to science that is ripe with new opportunities.

It takes our current knowledge about chemicals and health and the environment and builds a new platform for the materials of the future. It is not a revolution; it is just the next step in a long journey, and, best of all, it is really inevitable."[40]

There is a growing network among secondary and college-level educators around green chemistry education, and these could be better fostered and organized with conferences, associations, and websites. Although there is an increasing amount of green chemistry education in the United States, the wave of green chemistry training is more visibly cresting in Asia. Green chemistry is increasingly integrated into Chinese college chemistry curriculum, and the Indian Department of Higher Education has mandated that green chemistry principles be introduced into all undergraduate chemistry education in the colleges of Delhi.

Chemistry is about innovation and the design of new chemicals and synthesis routes for making chemicals is critical to a chemical conversion strategy. Where safer chemical policies create a demand for safer chemicals, green chemistry provides the means for generating the supply. Substituting toxic and hazardous substances, optimizing yields, eliminating wastes, and minimizing energy consumption generates a broad set of objectives for encouraging innovation and opening opportunities for corporate leadership. It introduces a renewed interest in biobased processes and renewable feedstocks for chemicals and encourages an emerging interest in how nature makes materials and solves chemical problems. Green chemistry broadens the field of chemistry and offers the promise of a new generation of chemists who could be promoting a safer chemical economy.

So these are three strategic fronts that are driving the search for safer chemicals. Although they are not currently coordinated or mutually supportive, they could be, and this possibility opens the opportunity for a broader chemical conversion strategy. Such a conversion strategy addresses the systems of the chemical economy and reveals the places where changes could be made, but it does not suggest how these varied initiatives could be incorporated into a broad policy shift. This is where a safer chemicals policy framework is needed. Such a policy framework requires new and better developed methods and tools for characterizing, classifying, and prioritizing chemicals; developing and making accessible chemical information; selecting safer alternatives; and developing safer chemicals. These are the building blocks of such a framework, and we turn to these next.

IV

Safer Chemical Policies

Adopt policies and regulatory and nonregulatory measures to identify and minimize exposures to toxic chemicals by replacing them with less toxic substances and ultimately phasing out the chemicals that pose unreasonable and otherwise unmanageable risks to human health and the environment and those that are toxic, persistent, and bioaccumulative and whose use cannot be adequately controlled.

—*Agenda 21*, United Nations Conference on Environment and Development (1992)

10

Characterizing and Prioritizing Chemicals

In 1979, the Chemical Industry Institute of Toxicology (CIIT) released findings showing that formaldehyde caused cancer in rats. The EPA convened a federal panel on formaldehyde that reported in 1980 with findings similar to the CIIT as did similar studies by the National Toxicology Program. On this basis, the agency determined that there "may be a reasonable basis to conclude that formaldehyde poses a significant cancer risk" and proceeded to prepare a priority review for regulatory action. However, the Formaldehyde Institute challenged the findings, and in 1982, a new EPA administrator halted the process and initiated a risk assessment.[1]

The risk assessment continued for seven years. In 1989, the EPA released its final risk assessment on formaldehyde, but the study came under such heavy attack that the agency withdrew the study for further consideration. In 1998, the EPA began an update of the formaldehyde risk assessment, but it was continually delayed awaiting further research. Although the EPA's risk assessment was due to be released in 2004, it was delayed again in anticipation of new findings from the National Cancer Institute (NCI). In 2009, the NCI released results of an extended study demonstrating strong evidence of a link between formaldehyde exposure and blood and lymphatic cancers and leukemia. However, the risk assessment was then further delayed when a powerful Senator demanded a review by the National Academy of Science (NAS). In 2011, the NAS released a review noting both strengths and weaknesses in the EPA assessment, and the agency has begun revisions again. Some 30 years after the first experimental evidence of formaldehyde's carcinogenicity and a "known human carcinogen" designation by the National Toxicology Program and the International Agency for Research on Cancer (IARC), the EPA has yet to complete its formaldehyde assessment and instead continues to pursue a risk-based approach that is decades late.[2]

The chemical control policies of the twentieth century laid the basis for addressing the hazards of the chemicals that could present unreasonable risks. However, those laws were narrowly focused, risk-based, and largely dependent on the authority of government regulations. They provided important standards for exposure to some chemicals and drove the substitution of others. However, they tended to chase one chemical after another rather than address the broader system that created chemicals of concern. Efforts to phase out the use of DDT resulted in the substitution of aldrin and dieldrin, both highly toxic and persistent and, now, globally banned under the Stockholm Convention. The regulations that phased out the use of lead in paints and gasoline left lead in many other commercial applications, such as flashing, tire weights, solders, pewters, bullets, and lead-acid batteries. Efforts to phase out well-recognized carcinogens such as arsenic, benzidine, and benzene in commercial products have left those chemicals still in use as common production intermediates that threaten workers.

The safer chemical policy framework proposed here is broader and more focused on systems change. We begin by considering the building blocks that comprised the older chemical control policies and then consider how they could be reworked and rewoven into a safer chemical policy framework. We start here with identifying, characterizing, and prioritizing chemicals.

10.1   Characterizing Chemicals

The first step in a safer chemical policy framework involves identifying and characterizing chemicals. For a comprehensive policy, it is critical that every chemical on the market has a unique identification. This is not necessarily easy. Formaldehyde, for example, has some twenty common, scientific, and commercial names. For well over a hundred years, the Chemical Abstract Service (CAS) of the American Chemical Society (ACS) has maintained a registry of all chemicals that provides a unique CAS number for each individual substance. Today, the registry contains identities for some 91 million organic and inorganic substances.[3] While such a detailed registry can be useful for the intricate needs of research scientists, its meticulousness (e.g., there are 42 CAS numbers for titanium dioxide) creates problems for chemical users. There are other identification systems, such as the Royal Society of Chemistry's ChemSpider, the federal FDA's Unique Ingredient Identifier (UNII), the Registry of Toxic Effects of Chemical Substances (RTECS) number, and the Simplified Molecular

Input Line Entry System (SMILES) string notation. All of these are useful, but ultimately there is a need here for a single international standard for chemical identification appropriate for chemical policy.[4]

Once a chemical (an element, a singular substance, or a molecular compound) has a unique identity, it needs a robust file of information—a chemical profile. Such a chemical profile would be more than the Material Safety Data Sheets (MSDSs) commonly available today with many commercially sold compounds and mixtures. Like an MSDS, a chemical profile would include the chemical identification (common names, molecular formula), the physical and chemical properties (mass, density, melting point, boiling point, flash point, water solubility, viscosity, and physical appearance) that make up the immutable characteristics of a substance. However, in addition, a chemical profile would include common uses of the chemical and likely exposures and the chemicals that are conventionally associated with the substance along its life cycle, such as the production intermediaries and the expected breakdown, decomposition, and metabolic compounds.

A complete chemical profile should also include the known and reasonably predicted hazard traits of a chemical substance. The hazards or hazard traits of a chemical are derived from the physical and chemical properties of the substance. There are physical hazards such as flammability and reactivity, there are environmental hazards such as ozone-depleting potential and greenhouse gas generation, there are wildlife hazards such as aquatic toxicity and high biological oxygen demand, and there are acute and chronic human health hazards based on toxicological properties. Some of the chemicals of highest concern are persistent, bioaccumulative, and toxic (PBTs), a few are very persistent and very bioaccumulative (vPvBs), and others are carcinogens, mutagens, and reproductive toxins (CMRs). [5]

The California Office of Environmental and Health Hazard Assessment (OEHHA) has prepared a preliminary typology of hazard traits in response to legislation to establish a state Toxics Information Clearinghouse. In all, OEHHA identified forty-one hazard traits that fall into four broad categories and more than a hundred potential adverse endpoints. Endpoints are specific adverse outcomes, such as dermatological systemic reactions, allergic sensitization, allergic reactions, acute or subacute irritation, and photosensitivity. Table 10.1 lists OEHHA's hazard traits.[6]

There are several different reasons that chemicals commonly used in the economy are hazardous. Some chemicals, such as those used in insecticides, have "intended hazard characteristics"—they are designed to be

Table 10.1

Chemical Hazard Traits

Toxicological Hazard Traits	Wildlife developmental impairment
Carcinogenicity	Wildlife growth impairment
Cardiovascular toxicity	Non-target phytotoxicity
Dermatotoxicity	Loss of genetic diversity
Developmental toxicity	Eutrophication
Endocrine toxicity	Exposure Potential Hazard Traits
Epigenetic toxicity	Ambient ozone formation
Genotoxicity	Bioaccumulation
Immunotoxicity	Environmental persistence
Hematotoxicity	Global warming potential
Heptotoxicity	Lactation or transplacental transfer
Musculoskeletal toxicity	Mobility in environmental media
Nephrotoxicity	Particle size or fiber dimension
Neurotoxicity	Persistence in biota
Ocular toxicity	Stratospheric ozone depletion
Ototoxicity	Toxic environmental transformation
Reactivity in biological systems	Physical Hazard Traits
Reproductive toxicity	Explosivity
Respiratory toxicity	Flammability
Environmental Hazard Traits	Nanomaterial hazards
Wildlife survival impairment	Oxidation
Wildlife reproductive impairment	Self-reactive substances or mixtures
	Radioactivity

poisonous. Other chemicals like the halogens are highly reactive, and their powerful chemical energy makes them quite hazardous when they enter an organism and react with the life-supporting chemicals of cells and organs. These are "derived hazard characteristics." There are also "unintended hazard characteristics," such as those that appear where heavy metals, such as mercury, lead, cadmium, and chromium, are used. The hazardous properties of these metals are not required in meeting the desired function but are intrinsic to the substance and therefore appear as undesirable side effects.

Some chemical compounds are not hazardous when used in a product or function, but their production and disposal involve hazardous intermediaries. For instance, the polymer, polyvinyl chloride (PVC), as a compound in a product is not an immediate hazard during its use, but the vinyl chloride monomer that it is derived from is a recognized carcinogen. In addition, PVC has a tendency to generate hydrochloric acid and dioxin during fires, adding to its hazardous profile. Because such compounds cannot be manufactured without hazardous chemicals and/or are likely to generate hazards when disposed, the properties of these chemicals should be considered "embedded hazard characteristics," and the intermediary and decomposition chemicals should be considered an integral part of their chemical profile. Thus, a chemical profile on bisphenol A should acknowledge the acetone and phenol from which it is derived, the phosgene used in its preparation, and the propanic acid that it decomposes into.

Under a safer chemical policy framework, a government agency (EPA, FDA) could create a common template for all chemical profiles, although this could also be done by a private standards-making body (e.g., American National Standards Institute) or a respected nongovernmental organization (e.g., ACS). However, only a chemical profile that presents physical, toxicological, and ecological impacts of a chemical and acknowledges the embedded production and disposal decomposition chemicals associated with it can provide a complete and comprehensive portrait of the substance and its "health and ecological footprint."

Assembling the information necessary to develop a complete chemical profile has been made easier by the information requirements of the European Union's REACH regulation. The physical, chemical, and hazardous properties of every chemical registered under REACH (potentially 143,000) and produced over ten tons per year in Europe must be identified in a "chemical safety dossier." The dossiers are required to include a minimum data set including:

- Identity of the substance
- Environmental fate properties
- Human health hazard assessment
- Physiochemical hazards (explosivity, flammability, oxidizing potential)
- Environmental hazard assessment
- Persistent, bioaccumulative toxic, and very persistent/very bioaccumulative assessment
- Exposure assessment
- Risk characterization (human and environmental)
- Overall exposure assessment[7]

These dossiers can provide a large part of a complete chemical profile. All of the dossiers have now been completed for the first and second phase of registrations in Europe—those covering the PBTs, vPvBs, and the

CMRs and some 3,500 of the largest production volume chemicals—and they already appear to offer a rich information source for assembling chemical profiles. Even if these dossiers are not fully comprehensive, they have become the global standard. Firms that prepare such dossiers for the European Union should be able to expand on them to include the embedded hazard characteristics of a chemical profile. The requirement for generating such chemical information would only weigh heavy on smaller, domestic chemical producers and those manufacturing chemicals not sold in Europe. A temporally phased process for generating the chemical profiles could be used to take account of the special needs of these domestic producers.

It is important that the responsibility for identifying a chemical and developing its chemical profile be shouldered by the manufacturer, supplier, or importer of the chemical. No other stakeholder has access to as much information about the chemical. The REACH regulation in Europe made this responsibility clear. By making the continued access of a chemical to the market dependent on the presentation of a chemical dossier, the European Commission created an effective incentive for firms to provide the minimum data set required. It is important that there is an equally effective process for firms to provide notification about each chemical that they place on the U.S. market and, in so doing, to properly present a complete chemical profile.

10.3   Classifying Chemicals

Characterizing chemicals to develop chemical profiles can benefit many users, including consumer safety advocates trying to inform consumers; retailers and procurement officers trying to identify safer products; product manufacturers trying to squeeze chemical information from suppliers; and green chemists trying to synthesize safer chemicals; however, the information becomes even more valuable where chemical profiles can be compared with each other. This could be accomplished by creating a chemical classification scheme, where chemicals could be ranked as to their degree of danger. Such a chemical classification scheme would provide a visual landscape within which chemicals could be differentiated across a gradient of values ranging from highest to lowest concern.

Such a large effort to classify chemicals is not infeasible. In amending the Canadian Environmental Protection Act in 1999, the Canadian Parliament called on Environment Canada and Health Canada to categorize chemicals commonly manufactured or used in Canada and then, where necessary, screen the substances to identify those of highest concern. Working together, Health Canada and Environment Canada completed a monumental effort to screen all of the chemicals manufactured and used in Canada and classify them in order to set chemical priorities. The screening was conducted on Canada's national Domestic Substances List (DSL), which includes some 23,000 existing chemicals that were reported to be on the Canadian chemical market between 1984 and 1986.[8]

To carry out the classification, the government reviewed publicly available data sources, sought data voluntarily submitted by industry, conducted specific chemical tests, and relied on modeling and chemical structure assessments. Using two specific criteria—substances posing the most significant potential for human exposure and substances identified to be persistent, bioaccumulative, and toxic (PBTs) the government identified 4,300 substances as high hazard (850 substances posed significant potential for human exposure, and 3,450 were identified as PBTs). To conduct the screening, Health Canada developed simple and complex exposure and hazard screening tools. Using the two criteria—greatest potential for human exposure and PBTs—the high hazard list of substances was screened to identify 1,200 low-priority substances, 2,600 medium-priority substances, and 500 high-priority substances.

Based on this tiered list of priority chemicals, the government prepared a national Chemical Management Plan (CMP) that included a "Chemical Challenge" program on the highest priority 200 chemicals and used a mandatory "call in" to acquire any outstanding information on these chemicals that had not been previously discovered. The CMP has then been used to integrate the government's various chemicals policies from across its health and environmental protection agencies, including those involving industrial chemicals, pesticides, cosmetics, and chemicals in products and focus them for concerted action on these priority substances.

The Canadian initiative demonstrates how, with substantial effort, a government could characterize and classify chemicals. The task for a safer chemical policy would be to promote a similar screening but shift the burden and require that chemical manufacturers and suppliers characterize and initiate the classification. The EPA or some respected private body such as the ACS or ANSI could establish criteria for characterizing chemicals and set up a tiered classification framework. Chemical manufacturers, distributors, or users could then fill in the classification categories by placing chemicals into the appropriate tiers of concern. If the classifica-

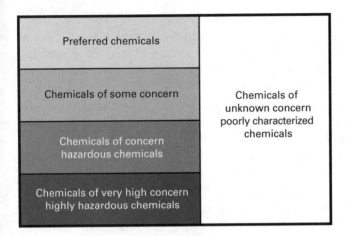

Figure 10.1

Illustrative Example of a Tiered Chemical Classification Scheme

tion scheme were posted on the Internet, its veracity could be managed by comments and critiques from the hundreds of contributors it most certainly would attract. Figure 10.1 suggests a possible chemical classification scheme with such a tiered approach.

This classification structure provides four tiers arrayed across a hierarchy of concern. The lowest tier would be for chemicals of very high concern (CVHC) and include chemicals such as CMRs and PBTs. The second tier would include chemicals that appear on authoritative lists of hazardous chemicals but are not of such very high concern. The third tier would include chemicals of some concern, such as common solvents, surfactants, and acids and those presenting moderate acute hazards, while the upper most tier would include those substances "generally regarded as safe."[9] By including a category for substances that are poorly characterized, this scheme creates a default for unstudied chemicals or chemicals with conflicting research results. This category would also be useful in identifying areas where new health and environmental effects research would be most effective.

The actual number of tiers is important: too few provides too little differentiation, whereas too many makes categorization complex and controversial. Criteria for determining each tier need to be based on solid, physiochemical, and hazard characteristics that can be identified by carefully defined indicators that have either well-recognized metrics or easily employed qualitative ratings (high, medium, low). Today, there are several comparative chemical hazardous assessment tools that could be used to help in these classifications. The Cradle to Cradle Products Innovation Institute, and GreenWERCs provide proprietary tools. The EPA has a Design for Environment Program Alternative Assessment Framework, the Healthy Building Network has *Pharos* and Clean Production Action provides the *Green Screen for Safer Chemicals.*

The *Green Screen* is particularly well suited for such classification because it separates chemicals into four tiers. A *Green Screen* user seeking to rate a chemical as to its level of concern assembles a broad range of health and environmental effects information as input data to calculate scores that are displayed as four benchmarks ranked in terms of chemical safety. Each benchmark includes a set of criteria that a chemical, along with its breakdown products and metabolites, must pass. The criteria are based on hazardous endpoints and "levels of concern" as defined by categories (H=high, M=moderate, L=low) of threshold values. The threshold values are based on those developed by various government authorities, including the EPA, the European Commission (REACH), and the Stockholm Convention for determining whether a chemical poses a specific type of hazard. The *Green Screen* considers seventeen factors related to human health, eco-toxicity, environmental protection, and physical hazards in making its determinations (see figure 10.2).[10]

Such a tiered classification scheme would provide obvious utilities. It could clarify the relative hazards of chemicals, reveal what is known and not known about chemicals, and establish the basis for setting priorities for government or industrial action. Working within a tiered chemical classification scheme, governments could design policies, programs, and regulations to promote conversions toward safer chemicals. Chemical users could use the tiers to assist in converting from more to less concerning chemicals and chemical manufacturers could use the tiers in developing safer chemicals for the market. In addition, the framework could be

	Group I human					Group II human							Ecotox		Fate		Physic	
	C	M	R	D	E	AT	ST	N	SnS	SnR	IrS	IrE	AA	CA	P	B	Rx	F
Chemical 1	M	L	L	M	M	L	L	M	L	DG	L	H	L	L	vH	M	L	L
Chemical 2	L	L	M	M	H	L	H	M	L	L	M	M	H	H	vH	vH	L	L
Chemical 3	L	L	M	H	DG	L	H	DG	L	DG	L	L	L	DG	M	M		M

Figure 10.2

Green Screen Hazard Categories with Three Chemical Ratings

used to identify where science and research should be targeted to either characterize understudied chemicals or develop safer alternatives.

The keys to classifying chemicals into such a scheme are a series of graded hazard criteria and a set of decision rules for locating chemicals with multiple hazard traits. The Swedes and Danes have developed such screening protocols for classifying chemicals based on degree of concern. However, the Globally Harmonized System of Classification and Labeling of Hazardous Substances (GHS) promises the most widely accepted framework for classifying chemicals.

The GHS was designed to define health, physical, and environmental hazards; classify processes for comparison with defined hazard criteria; and communicate hazard information and protective measures. The classification scheme of the GHS is inclusive, covering all chemicals, and hazard-based, classifying chemicals based on their inherent properties. The program is actively managed under the United Nations Economic and Social Council by a committee of experts. It is designed to promote "self-classification" such that anyone supplying a chemical can properly classify the substance. The *Green Screen*, *Pharos*, and several of the new proprietary chemical screening tools use the GHS criteria.[11]

Criteria are the key to the classification scheme. The GHS criteria are divided into three categories: physical hazards, human health hazards, and environmental hazards (see table 10.2).

The physical hazard criteria are largely quantitative or semi-quantitative, with multiple hazard levels within each endpoint. For instance, there are four categories of flammable liquids ranging from category 1 with a flashpoint of <23°C and an initial boiling point of <35°C to category 4 with a flashpoint of 60°C and an initial boiling point of <93°C. The human health criteria are quantitative where possible but also qualitative. For instance, the carcinogenicity criteria adopt the IARC classification of categories 1A, 1B, and 2. The reproductive toxicity criteria include category 1A (based on human evidence), category 1B (based on experimental animals), category 2 (human or animal evidence with other information), and an additional category for effects "on or via lactation." The environmental criteria are divided into three categories based on toxicity, bioaccumulative potential, and degradability.

The GHS includes a catalog of numerically coded hazard and precautionary statements that when applied to a chemical indicate the type of hazard presented by that chemical and the recommended safety procedures. These statements are modeled on and replace earlier European Risk Phrases (R-phrases) and Safety Phases (S-phrases) developed under

Table 10.2

GHS Classification Categories

Physical Criteria	Human Health Criteria	Environmental Criteria
• Explosives	• Acute Toxicity	• Hazardous to the
• Flammable Gases	• Skin Corrosion/	Aquatic Environment—
• Oxidizing Gases	Irritation	Acute Aquatic Toxicity—
• Gases under Pressure	• Serious Eye Damage/	Chronic Aquatic Toxicity
• Flammable Liquids	Eye Irritation	Bioaccumulative
• Flammable Solids	• Respiratory or Skin	Rapid Degradability
• Self-Reactive	Sensitization	
Substances	• Germ Cell	
• Pyrophoric Liquids	Mutagenicity	
• Pyrophoric Solids	• Carcinogenicity	
• Self-Heating	• Reproductive Toxicity	
Substances	• Target Organ Systemic	
• Substances which in	Toxicity—Single	
water emit Flammable	Exposure	
Gases	• Target Organ Systemic	
• Oxidizing Liquids	Toxicity—Repeated	
• Oxidizing Solids	Exposure	
• Organic Peroxides	• Aspiration Toxicity	
• Corrosive to Metal		

the Dangerous Substances Directive. Each hazard statement is designated by a code starting with the letter H and followed by three digits. Altogether there are seventy hazard statements separated into physical, human health, and environmental hazards. For instance, there is a physical hazard statement ("H225—highly flammable liquid and vapor"), a human health hazard statement ("H340—may cause genetic defects"), and an environmental hazard statement ("H401—toxic to aquatic life"). Using one or a combination of these hazard statements (e.g., H225, H340, H401), a chemical's hazard profile can be indicated. The precautionary statements are numerically coded phrases (beginning with the letter P followed by three digits) that indicate how a chemical should be handled to minimize danger to the user.[12]

The GHS is a voluntary standard; however, many nations are now working to adopt it. The U.S. Occupational Health and Safety Administration has updated the federal Hazard Communication Standard to align parts of it with the GHS. The World Health Organization has restructured its pesticide classification system to harmonize with the GHS's Acute Toxicity Hazard Categories.[13] In 2008, the European Commission adopted a Regulation on Classification, Labeling and Packaging of Substances and Mixtures (CLP) to introduce the GHS into the European Union. This is a self-classification process relying on firms to classify their own chemicals. The European Chemicals Agency (ECHA) recommended a four-step protocol to assist firms in making these classifications:

1. Collect available information,

2. Evaluate the adequacy and reliability of the information,

3. Review the information against the classification criteria, and

4. Decide on classification.[14]

Under the regulation, chemical manufacturers and importers were required to notify the ECHA by January 2011 on the classification and labeling of all hazardous chemicals and chemicals required to be registered under REACH. The ECHA received more than 3.1 million notifications concerning 24,529 substances, and it is now working to put all of the notifications into a publically accessible inventory. By presenting the information in a publically available inventory in the form that it was received, ECHA hopes that it will encourage a process of refinement and adjustment. It is anticipated that the multiple submitters on the same substance will note points of agreement and disagreement on the chemical classification and, through an interactive dialogue, approach consensus on a harmonized classification for each chemical.[15]

The ECHA effort plus contributions from around the world will drive the GHS into an international standard, even without an international treaty. Relying on the power of a common classification framework and many participants, the GHS may generate internationally accepted definitions of the hazards of thousands of chemicals. Although the GHS does not take the next step of slotting each chemical into the tiers of a comprehensive chemical classification scheme, it lays the basis for doing so. With or without the federal government, such a classification framework could be launched here in the United States; however, it would be more effective to have one global classification framework. Experience with REACH suggests that were the European Union to create such a comprehensive chemical classification scheme, it would likely become the internationally accepted framework by default.

10.4   Locating Chemicals

Europe's ECHA's experience in characterizing chemicals under the CLP demonstrates whether and how a general consensus can be achieved for many chemicals by simply displaying the multiple classifications of a chemical and waiting for an Internet dialogue to shake out the agreement. A similar multiparticipant, open access process might also achieve consensus for classifying a majority of chemicals into a universal chemical classification scheme like the one presented above.

Creating the initial list of Chemicals of Very High Concern (CVHC) for the lowest tier of the classification framework might be fairly noncontroversial. Many of the heavy metals, halogens, and aromatic hydrocarbons would likely be located in this lowest tier. Most of the organochloride and organophosphate pesticides also would fall into this tier, but naturally occurring pesticides such as neem (azadirachtin) and ryania (antranilic diamide) might be located in higher tiers. On the basis of their inherent hazards alone, a general consensus would be likely regarding those substances that are recognized carcinogens, mutagens, and substances toxic for reproduction (CMRs). The International Agency on Research on Cancer (IARC) currently lists twenty-eight (group 1) confirmed human carcinogens, twenty-seven (group 2A) probable human carcinogens, and 113 (group 2B) possible human carcinogens. An additional 122 substances are classified as "may" cause or "suspected" of causing reproductive damage.[16]

The ECHA has compiled data from the initial 25,000 REACH registrations dossiers and the 3 million CLP notifications it has received to identify those chemical substances identified by their manufacturers or importers as carcinogens, mutagens, or reproductive toxins and developed a list of just over 2,000 such substances. The Oslo and Paris Convention for Protection of the Marine Environment of the Northeast Atlantic (OSPAR) publishes a "List of Substances of Possible Concern" that includes 315 substances listed as PBTs. This list plus the two substances (short chain chlorinated paraffins and musk xylene) recognized in the European Union as vPvB could be added to the ECHA list to generate a starting baseline of some 2,000 to 2,300-plus chemicals in the lowest tier (CVHC).[17]

This would be similar to other lists of chemicals of highest concern. For instance, California's Office of Environmental Health Hazard Assessment currently lists 840 chemicals on its Proposition 65 list of chemical carcinogens and reproductive toxins. Currently, the European Chemicals Agency has identified 155 candidate Substances of Very High Concern (SVHC) under REACH; however, the International Chemicals Secretariat (ChemSec—an NGO based in Sweden) has issued a list of 626 substances that are likely candidates for the REACH SVHC list.[18]

Let's consider again the three chemical compounds—benzene, ethylene oxide, and phthalates—described in chapter 6. Benzene is a recognized reproductive toxin and an IARC group 1 human carcinogen causing both acute myeloid and non-lymphocytic leukemia. Long-term exposure to benzene is associated with harmful effects on bone marrow and a reduction in red blood cells leading to anemia. In use, it is a highly flammable liquid. The inherent characteristics of benzene would suggest that it be located in the lowest tier as a CVHC.

Although ethylene oxide is commonly used in the production of many low-hazard chemicals and products, it is also a very hazardous substance. At room temperature, it is an explosive and flammable gas. It has acute irritating, sensitizing, and narcotic effects, and it is both mutagenic and carcinogenic (an IARC class 1 carcinogen). On the basis of such inherent hazardous characteristics, ethylene oxide would be located in the lowest tier as a CVHC.

Phthalates present a more complex task. There are many phthalates, and some have been studied whereas others have not. The unstudied compounds would be located in the side bar of chemicals of unknown concern. Those with research histories present divergent findings. The European Commission has identified three phthalates—di(2-ethylhexyl) phthalate (DEHP), benzyl butyl phthalate (BBP), and dibutyl phthalate (DBP)—as among the first substances to be listed as SVHC for Authorization. These chemicals would fall into the first tier of the chemical classification framework. DEHP would be harder to classify. It is listed on some carcinogen and reproductive hazard lists, and it is suspected of disrupting endocrine, although there is strong disagreement over the current research. DINH, diisonyl phthalate, which is largely thought to have lower hazard properties, might be slotted into the second tier, Chemicals of Some Concern.

10.5 Setting Priorities

The classification of chemicals by hazard requires a single universal chemical classification scheme for all chemicals. Prioritizing those chemicals for assessment, regulation, or substitution need not be so centralized. Firms might prioritize the chemicals they make or use, economic sectors might prioritize chemicals for substitution, and governments might prioritize chemicals for regulation. Some science-informed, transparent, and rapid means of sorting chemicals to set priorities is needed. Although the categorization of chemicals should be based on the inherent hazards of chemicals, prioritization should consider both hazards and potential exposures. The Scandinavian countries use a two-stage approach based on hazard traits that involves a broad list of chemicals of concern and a more select sublist of priority chemicals of high concern. Similarly, Minnesota, Maine, and Washington State have created staged methodologies for prioritizing chemicals.

The Washington State program provides a good example. In 2008, the state legislature enacted a law, the Children's Safe Product Act, which directed the Department of Ecology (DOE) to reduce the risks of chemicals that had the potential for exposure to children. The law required that the DOE set priorities for those hazardous chemicals most likely to significantly affect children and then take regulatory action to reduce those exposures. To set those priorities, the DOE developed a three-part methodology that generated a list of high-priority chemicals (HPCs), a list of chemicals likely to be in children or in their immediate environment, and a combined list of high-priority chemicals that are in children or their immediate environment (chemicals of high concern to children [CHCCs]). The initial list of 2,160 HPCs was developed through an extensive search of authoritative scientific and regulatory sources. A separate list of 2,600 chemicals likely to expose children was developed from a search of biomonitoring studies and studies of residential exposure media (e.g., indoor air, household dust, drinking water, toys, domestic products). By comparing the lists and noting those substances that appeared on both lists, the DOE developed a consolidated list of 476 CHCCs.[19]

The REACH regulation requires a "prioritization process" for identifying SVHCs. The Member States have the responsibility for posting chemicals to a SVHC Candidate List, which the ECHA then prioritizes using one of two approaches: a qualitative (verbal-argumentative) approach or a semi-quantitative (scoring) approach. Both approaches involve two steps: documenting the hazards, volume, and wide dispersive use, and considering the "regulatory effectiveness and coherence." The verbal-argumentative approach has been used for PBTs and vPvBs. The scoring approach uses several indicators with numeric scores. Four numbers are used to rank intrinsic hazards, six numbers represent product volumes, and three numbers represent the number of sites using the chemical and the potential for worker or consumer exposure. The scores are then weighted and added to provide an aggregate indicator that is used to assist but not determine a selection decision.[20]

Over the years, the EPA has developed several schemes for prioritizing chemicals of concern, including a Use Cluster Scoring System, a two-tiered risk management ranking system, and, most recently, Chemical Work Plans based on hazard, exposure, and chemical use. The recently developed method for identifying TSCA Work Plan Chemicals provides a useful example.[21] Like other methods, this is a two-step process. Step One is designed to create a candidate list and includes chemicals meeting one or more of the following criteria:

- Chemicals identified with reproductive or developmental effects.
- Chemicals identified as PBTs.
- Chemicals identified as probable or known carcinogens.
- Chemicals used in children's products.
- Chemicals used in consumer products.
- Chemicals detected in biomonitoring programs.

Using these criteria to screen chemicals, the agency searched some twenty authoritative databases and developed a list of 1,234 chemicals that were then cut down to 345 by excluding several categories of substances because they were either not under TSCA authority (pesticides, drugs) or were not of immediate concern (polymers, gases, physical hazards). Step Two involves a screening methodology focused on three characteristics: hazard, exposure, and persistence/bioaccumulation. Numeric scores were assigned based on available data, and where data were insufficient, chemicals were placed into a separate category for further study. Figure 10.3 suggests a similar process adapted to the chemical classification scheme noted above.

The first step creates a candidate list. An initial list could be developed by a government agency or a collaboration among firms in a particular region or economic sector; alternatively, a list could be created by nominations from governments, businesses, NGOs, or the public at large. The hazard scoring would be based on the chemical classification tiers. Scores for human exposure could be based on size of population, vulnerability of populations, function and use of a chemical, volume and types of releases, and types of occupational, domestic, and consumer exposures. Scores for environmental exposures could be based on persistence, bioaccumulation, types of environmental releases, fate, and transport and ecological significance. The scores would be added up to determine the level of priority. If there were no data available for one or several of the categories, a chemical would still receive a sore for uncertainty and be referred for further information generation.

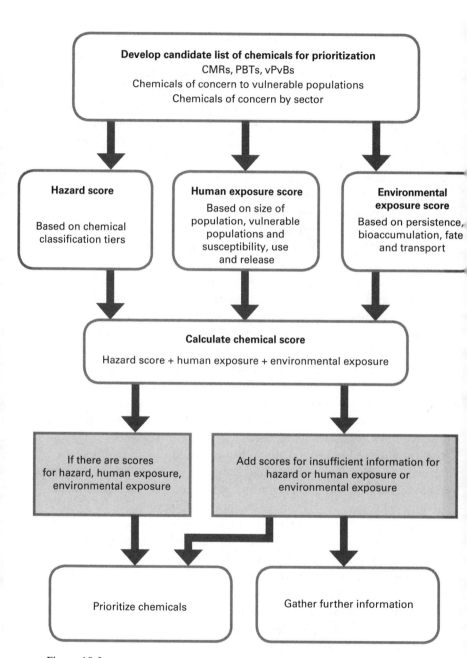

Figure 10.3

A Safer Chemical Prioritization Method

Using a prioritization method like this, priorities for corporate or sector decisions or government regulatory actions could be set using the tiers in the chemical classification scheme and various indicators of chemical exposure.

A hazard-based approach does not replace the need for risk assessment. Risk assessment and its attendant focus on potential human and environmental exposures could be valuable in setting priorities for government or business attention. However, the current practice in conducting risk assessments presents several long-recognized methodological problems that have eroded public confidence. The National Academy of Science has published two recent reports that present recommendations on how risk assessment methodology could be improved. Recommendations from the first report, *Science and Decisions*, suggest that future risk assessments should justify assumptions, such as a safe threshold for exposures, as well as take special consideration of vulnerable populations, such as fetuses and children, and aggregate the possible multiple exposures to a chemical. A second report recommends that risk assessments group chemicals with similar biological mechanisms and consider the effects of repeated and cumulative exposures.[22]

Prioritizing chemicals for government or corporate attention need not start with the most hazardous; the range of exposures, the types of exposure, and the condition of those exposed might all act as criteria for prioritization. Special attention should be paid to the persistence and bioaccumulation in ecological systems as well as the possibility of exposure to sensitive biotic populations. Prioritization should be guided by a formal protocol; however, it should also be open to unforeseen new chemical information and heightened public concern. Additional attention should be given to the unique conditions of the young and the elderly and inequalities of exposure based on social conditions such as race, poverty, or geographic location. This has long been called for by environmental justice organizations such as West Harlem Environmental Action, the Deep South Center for Environmental Justice, and the Indigenous Environmental Network.[23]

Prioritizing chemicals for government response also requires thoughtful analyses of the complex, integrated systems in which those chemicals are manufactured, used, and disposed. In addition, the function of a chemical, how it is used, and the amount used may affect its prioritization. For instance, the hazard assessments and information on the functional uses of a chemical could be used together to rank use functions for policy prioritization. A decision-making protocol using this information could be devised as a series of screens starting with hazard and use assessments and then moving progressively through various exposure screens. For each of the chemicals with a hazard assessment that displays the highest levels of concern (across hazard traits), the top functional uses could be identified. Then, for each functional use category, the volume and specific conditions of use could be ranked by scale and type of exposure in that function to rank the most "hazardous functions."

Although this information might be useful just in terms of its codes, it could also be loaded into software programs that could display the information graphically with chemicals and their uses displayed in landscapes of colors and tones that could visually display patterns and trends. Such graphic representations could be useful in presenting a "big picture" of chemicals displayed by hazard and use that could be drilled down to indicate specific hazard traits, or most common types, volumes, or conditions of use. Such a prioritization based on chemical hazards and uses would be useful at the level of a production facility or an industrial sector; however, a truly ambitious government or NGO might conduct such an exercise to determine the most hazardous functions in an economy as a whole in order to effectively prioritize policies.

10.6 Creating an Orderly Chemical Landscape

Because a safer chemical policy framework considers all chemicals, it must develop procedures for identifying, characterizing, sorting, and prioritizing chemicals. A hazard-based approach offers significant benefits for such a policy framework. Because it focuses on the intrinsic structural determinants of a chemical, a hazard-based approach offers opportunities to compare one chemical with another and note the differential preferences. To make this opportunity real, every chemical, from the most hazardous pesticide to the least hazardous vegetable oil, needs to have a chemical profile that presents its characteristics. But a fully developed chemical profile is not enough. A safer chemical system requires that chemicals be comparatively evaluated among other chemicals and classified into a comprehensive chemical classification scheme, with a name like a "Universal Classification of Chemical Substances."

For some well-studied chemicals, characterizing and classifying them should be fairly straightforward. They already appear on the lists of governments or other authoritative bodies. There is another collection of chemicals for which the European REACH dossiers should provide enough information, to develop reasonable chemical profiles and classify the chemicals into tiers. A third collection of chemicals would exist where there is some but insufficient information. The GHS classification criteria could be used where information does exist, and analogues and similarities to other chemicals might provide for tentative conclusions. A fourth group of chemicals would be so understudied that the chemicals can only be classified in the category of unknown concern. There will remain an assortment of chemicals where there are ongoing debates about test methods, data sources, missing data, and contested uncertainties, and these will not be easily or comfortably located into the classification scheme. However, recognizing that some chemicals will not be easy to classify should not deflect attention from the many chemicals that can be classified and the more general value of a classification scheme that creates a visible map of the universe of chemicals.

The open and participatory process for the classification of chemicals is further extended to the setting of priorities. Here the issues of exposure and risk become salient. The exposure of workers throughout the supply chain (including those in foreign countries) and vulnerable and sensitive populations, children, pregnant woman, the elderly, lower income people, and those with limited means can all be considered in determining the degree of exposure and risk. A weighting and calculation system is suggested above; however, it should not be considered determinant. Instead, like the classification scheme, it is a tool for assisting in decision making.

Priority setting could be advanced in stages, relying on current knowledge now and revisiting it later as more information becomes available. Graphic representations of chemicals coded by currently known or computationally predicted hazards and commonly identified types, volumes, and conditions of use may be enough for priority setting. Scaling this up to cover groups of chemicals or, more ambitiously, the chemical market as a whole could provide a broad framework for visualizing where the highest hazard chemicals provide the most concerning exposures.

Generating effective chemical profiles, classifying chemicals into tiers, and setting priorities all require good chemical information. It could be argued that the significant lack of chemical information limits such an ambitious classification and prioritization process. The next chapter explores the availability of chemical information today and how more can be generated. However, characterizing and classifying chemicals and setting priorities need not wait for complete and final information. Tools for chemical hazard ranking are available today from Clean

Production Action, SciVera, and the Cradle to Cradle Product Innovation Institute. Screening a chemical for known hazard traits and well-recognized exposures may be all that is necessary to initially classify a chemical and set a priority, and for a large number of high-production chemicals, that process could begin today starting with what is currently known. Decision making with imperfect knowledge is normal, whether in business or government. Acting on such imperfect knowledge defines a precautionary approach. Shifting the burden of characterizing chemicals and acting in the presence of uncertainty are hallmarks of the Precautionary Principle.

11

Generating Chemical Information

Let's start with a chemical such as tert-butyl acetate (TBAC). TBAC is a solvent with potentially wide commercial applications. However, little is known about the hazards of this chemical. TBAC has not been tested for carcinogenic, mutagenic, teratogenic, or other chronic health effects, and its eco-toxicity has not been determined. It decomposes in water, but its decomposition products have not been identified or tested. In 2004 the EPA exempted TBAC from its volatile organic chemical (VOC) regulations because of its low volatility. In 2009, the California Air Resources Board followed the EPA in exempting TBAC in several applications, and TBAC is now steadily replacing higher VOC chemicals in many commercial products, such as coatings, inks, adhesives, and industrial cleaners. In making their decisions, both the EPA and the California Board acknowledged the incomplete record of environmental and human health effects information. When the EPA exempted TBAC, the agency noted that its metabolite, tert-butyl alcohol, was a recognized carcinogen, but without enough information on the biotransformation of TBAC, neither the EPA nor the California Board have moved to further regulate it.[1]

This is not an unusual case; policymaking is often inhibited by the lack of good chemical information. Information on chemicals is a fundamental requirement for an effective chemical policy. However, large numbers of chemical substances and mixtures are on the market with little or no production, use, or health and environmental effects information. The EPA's High Production Volume Chemical Program has generated screening information on some of the health and environmental effects of large-volume chemicals, but for most small- and moderate-level production chemicals, even such screening data are absent.

The generation of high-quality and readily accessible chemical information makes up the second building block of a safer chemical policy framework. Although it is possible to roughly characterize and locate chemicals into a tiered chemical classification framework without complete information, such classification is more accurate and convincing when it is based on good chemical information. Good chemical information is also needed by product manufacturers when selecting potentially safer chemicals for production, retailers trying to identify and market safer products, chemists trying to make safer chemicals, and consumers seeking to protect themselves and their families. Although some of this information may be locked away by corporations concerned about trade secrets, there is an increasing supply of good chemical information, and there are new, efficient methods for predicting chemical hazards and exposures and new means to make such information available to those who need it.

11.1   Types of Chemical Information

Some types of chemical information are available in easily accessible databases but not for all chemicals, and many types of information are substantially deficient. It is useful here to consider how much information we have, how we might get more, and what level of information is sufficient.[2]

To fully support safer chemical policies, four broad types of chemical information are needed:

• *Information on the inherent characteristics of chemicals.* To properly classify, prioritize, and regulate chemicals, a minimum set of chemical health and environmental hazard information is needed for every chemical on the market. Such a base set of data is fundamental to a sufficient chemical profile.

• *Information on the production and use of chemicals.* To understand and account for how chemicals appear, change, and disperse in the economy, information is needed on the manufacture, importation, distribution, use, and disposal of chemicals across their life cycle.

• *Information on release and exposure of chemicals.* To predict threats from chemicals and prioritize attention and interventions, information is needed on the releases of chemicals as wastes and emission, the actual and potential human exposures to chemicals, and the presence of chemicals in environmental media, humans, and other organisms.

• *Information on alternatives to hazardous chemicals.* To shift from chemicals of concern to safer alternatives, information is needed on chemical substitutes and nonchemical solutions that could replace chemical of concern or make them unnecessary.

11.2   Chemical Hazard Information

In constructing a chemical profile, information on the intrinsic hazards of chemicals is fundamental. Basic information on chemical hazards can be drawn from the information already assembled in existing national and international data sources. The U.S. National Library of Medicine (NLM) provides an international portal to a wide range of chemical property databases. These include the NLM Toxicology and Environmental Health Information Program (TEHIP) and the web-based TOXNET (Toxicology Data Network) system, including a series of databases such as the EPA's Hazardous Substances Data Bank (HSDB), the Integrated Risk Information System (IRIS), the Toxics Release Inventory, the National Cancer Institute's Chemical Carcinogenesis Research Information System (CCRIS), and the TOXLINE bibliographic file. The Hazardous Substances Data Bank provides peer-reviewed data of 5,000 potentially hazardous chemicals, the IRIS contains data from hazard and dose-response assessments on some 550 chemicals or groups of chemicals, and TOXLINE contains scientific references to the world's toxicology literature.[3]

The European Chemicals Agency (ECHA) maintains a searchable data system that provides chemical hazard and classification information on chemicals manufactured or imported in Europe. It includes information from the European Inventory of Existing Commercial Chemical Substances (EINECS), the European List of Notified Chemical Substances (ELINCS) and the European Classification and Labeling Inventory. The Organization of Economic Cooperation and Development (OECD) maintains an Internet-based Global Chemicals Portal ("e-Chem Portal") where twenty-eight chemical hazard databases are grouped to provide direct links to retrievable data, allowing queries based on chemical name, synonym, or CAS registry number.[4]

The newest of these grand-scale searchable databases is the EPA's Aggregated Computational Toxicology Resource (ACToR), a product of the agency's Computational Toxicology Research Program. ACToR aggregates data from more than 500 public sources on more than 500,000 chemicals searchable by chemical name, CAS number, and chemical structure. The data cover chemical structure, physiochemical values, in vitro assay data and in vivo toxicology data on high- and medium-production volume industrial chemicals, pesticides (active and inert ingredients), and potential ground and drinking water contaminants.[5]

These hazard-based databases provide data on the inherent hazards of a chemical. Although the databases are impressive to use, they only include information from government or professional databases. A large amount of additional information on chemicals exists in journal articles, unpublished studies, lab reports, and the propriety files of corporations. Add to this the many chemicals that have never been studied, and this leaves many "data gaps" on various hazard traits and little information on the "embedded" chemicals required to manufacture chemicals or (with some exceptions) the decomposition products when chemicals are released to the environment.

11.3   Chemical Hazard Screening

Given these data gaps, various chemical hazard screening tools are available from government or private sources to assist in characterizing chemicals. These screening tools typically use analogs among compounds with similar structures that are grouped into categories. The most common device is a structure-activity relationship (SAR) or a quantitative structure-activity relationship (QSAR). Widely used in the development of drugs, SARs are based on the assumption that similarly structured molecules have similar activity in terms of reactivity, solubility, and biological effect. By knowing the behavior and effects of a chemical with a specific molecular structure, such as one benzidine dye, the behavior and effects of similar benzidine dyes can be predicted. The EPA has developed several chemical screening tools based on QSARs. For instance, the Analog Identification Method (AIM) relies on a database containing more than 31,000 potential analogs derived from publicly available data sources, which can be searched for similarities based on each molecular fragment of a chemical compound.[6]

The EPA's "PBT Profiler" is used to predict the persistence, bioaccumulative potential, and aquatic toxicity (PBT) of organic chemicals in the absence of measured data. Included in this protocol are several computer models that predict the physical/chemical and fate properties of a chemical based on structure-activity relationships and default scenarios and provide easy-to-read, color-coded comparisons of predicted PBT values.[7]

EPA's OncoLogic and ECOSAR are more specific hazard screening tools. "OncoLogic" is a chemical screening tool designed to predict the carcinogenicity of a chemical based on SARs. It is designed as an "expert system" and works by asking the user to input answers to a series of questions, which it then compares to an internal database in order to provide an estimate of carcinogenic potential. ECOSAR provides a library of QSARs for predicting aquatic toxicity and an expert system for selecting the most appropriate QSAR model.[8]

11.4   Chemical Testing

Sceening tools can be valuable, but efforts to build robust chemical profiles are limited where there are insufficient sources of data or the data are of questionable quality. When the Health Canada and Environment Canada collaborated to screen the Domestic Substances List of chemicals on the Canadian market, they conducted a broad search of the literature and sent requests to most of the largest chemical suppliers. For more than 11,300 organic chemicals, they found experimental nonhuman toxicity data on only 1,051 substances, of which 25 percent was of questionable quality; experimental persistence data on 850 substances, of which two-thirds was questionable; and experimental bioaccumulation data on 410 substances, of which three-quarters was questionable. Filling such data gaps requires either laboratory testing or computational modeling.[9]

The EPA maintains testing requirements for pesticides and industrial chemicals whereas the CPSC and FDA have their own testing requirements. These test guidelines cover tests for acute toxicity, subchronic toxicity, chronic toxicity, genetic toxicity, and neurotoxicity and include specifications for many aspects of experimental design, including animal selection, dose levels, administration methods, dosing periods, and methods of data collection, recording, and evaluation. However, interpreting test results with confidence can be challenging. There are often several chemical test methods available, and it can be difficult to generate consensus on the appropriate test method for each end point, particularly for chemicals with high market value. After some twenty years of research, it is still not possible to find agreement on the appropriate tests and test interpretations for endocrine disruption.[10]

In vivo, animal-based testing is typically complex and costly. In vitro biological models are of particular importance because, for many purposes, they can provide a sufficient alternative to the more labor-intensive and costly in vivo experimentation. This rapidly developing area of work has important developments in research on subjects ranging from discrete initiating events to the biological mechanisms that lead to adverse outcomes. Further research focuses on the metabolic pathways of chemicals within biological systems and the degradation processes in environmental ecosystems. Highly sensitive assays can now be performed to identify biotransformational patterns that provide valuable kinetic information. Testing of chemical compounds that focus on the synergistic, promoting, inhibiting, and antagonistic effects and screens that indicate pre-hazardous end points such as oxidative stress provide valuable information for characterizing and comparing chemical compounds and mixtures.

However, there are many criticisms of the current state of chemical testing. It is expensive, slow, open to bias, prone to methodological problems, and sorely out of scale with the large number of untested chemicals. In addition, there is growing public and professional sentiment against the use of laboratory animals in in vivo chemical testing. Negotiations over REACH in Europe generated significant debate over the need for the extensive chemical testing anticipated under the proposed regulation. Whereas industry was concerned about the projected costs, animal welfare advocates were alarmed over the potential increase in animal testing. REACH, therefore, takes special precautions to reduce the need for new tests and encourage alternatives to animal testing. For example, REACH allows firms to submit existing data, historic human data, weight-of-evidence studies, QSARs, computational analyses, and various non-animal, in vitro studies. In addition, REACH encourages firms to cooperate together in so-called Substance Information Exchange Forums (SIEFs) to collectively conduct chemical studies or use the studies of one firm as evidence for all firms producing similar chemicals.

The growing recognition of the problems of current chemical testing has led to a search for a new methodological paradigm. In 2007, the National Research Council (NRC) released a report that attempted to chart such a path.[11] The NRC recognized that current toxicity testing procedures face several competing objectives. To meet these objectives, the NRC suggested using computers to predict toxicity using "automated high-throughput" in vitro and in silico (computer-simulated) studies involving ranges of doses based on perturbations of critical cellular responses and chemical screening using computational approaches. Moving away from the conventional animal exposure-response approach toward this more direct molecular level approach assumes that cellular changes are sufficient to indicate toxic pathways.

The EPA has begun moving in this direction with the establishment of the National Center for Computational Toxicology and the development of the Toxicity Forcaster (ToxCast), a rapid, computation-based program for screening and assessing chemical exposure, hazard, and risk. ToxCast is designed to determine how chemical exposures impact human biological processes and how chemicals most likely lead to health effects. ToxCast draws on data and screening tools from the FDA, National Toxicology Program, and National Institute for Environmental Health Science. The program compares the results of the new high-throughput tests with well-characterized chemicals to assess the tests' reliability. To date, more than 1,000 chemicals have been screened through ToxCast. The move from conventional animal-based tests to new human cellular tests could provide earlier and less costly evidence; however, the transition will not be easy. Some predict that the transition could take a decade or two with costs up to $1 to $2 billion.[12]

11.5   Chemical Production and Use

In general, there is little reliable data on where and how chemicals are used in the world's economies.[13] The periodic reporting on chemical production required under TSCA's newly reformed Chemical Data Reporting Rule (CDR) or the European EINECS provide a rough survey of the production volumes of the largest volume industrial chemicals in the United States and European Union, respectively. For instance, the CDR requires manufacturing site and production volume data on chemicals manufactured or imported at or above 25,000 pounds during a single year at a single site (see table 11.1). For chemicals manufactured or imported in annual quantities of 300,000 pounds or more per site, additional "readily available" information must be reported on downstream users, types of commercial and consumer uses, and maximum concentrations in commercial products. The new CDR Rule requires chemical manufacturers, suppliers, and importers to report every fourth year, including reports on the three intervening years. In 2012, 1,528 companies reported on 7,674 chemicals.[14]

Although these chemical inventories provide a broad picture on chemical production, little effort has been made to gather data on what chemicals in what volumes go into what product production processes or what commercial products. Chemical manufacturers and distributors may have reasonable data on immediate chemical sales, but they quickly lose sight of those chemicals as they penetrate further into the market. This is a problem for governments trying to set chemical policy; however, it is particularly a problem for product manufacturers, retailers, and other

Table 11.1
TSCA Chemical Data Reporting Reportable Elements

- Parent company and site identification
- Specific chemical name
- Chemical Abstracts Service (CAS) Registry Number (or other identifying number)
- Domestically manufactured production volume
- Imported production volume
- Site-limited status
- Maximum concentration
- Number of workers reasonably likely to be exposed during the manufacture of the chemical substance
- Physical form and the percentage of the chemical substance in each physical form

---

downstream chemical users who are trying to identify the chemical ingredients in their supply chains and the products they use.

Although the federal government has not moved far on chemical use reporting, there has been more activity among the states. Since 1990, California has required pest control operators to report annually on all pesticide products used by product name, operator name, location of application, amount of product used (by weight), purpose of application, and application method. All agricultural pesticide applications must be reported on a monthly basis to the county agriculture commission, and this includes pesticide field application, post-harvest application, and all pesticide applications in livestock, poultry, and fish production operations. Although there have been some problems in harmonizing the methods for reporting across the counties, the state program is now making all of its data available through a special California Pesticide Information Portal.[15]

Massachusetts collects data on chemicals used in industrial production under its Toxics Use Reduction Program. These data are collected annually from some 550 facilities and covers 192 chemical substances. The data are posted by facility on a public database maintained by the Toxics Use Reduction Institute. New Jersey collects similar chemical use data by facility. Recent legislation in Maine and Washington provides authority for state agencies to collect chemical use information on selected priority chemicals of concern to the health of children.[16]

Chemical use information can be particularly valuable to product manufacturers seeking to identify the ingredients in chemical formulations purchased from chemical suppliers. Such information can also be valuable to product retailers who increasingly feel customer pressure to identify and disclose the presence of hazardous chemicals in the products they sell. However, chemical or component suppliers are often reluctant to reveal chemical ingredients for fear of losing the competitive advantage of proprietary information or because they do not actually know the ingredients supplied to them.

11.6   Chemicals in Supply Chains

The many gaps in communications among different parts of the chemical production and consumption system were noted in chapter 5. When chemical suppliers do not receive information from product manufacturers, retailers, or product consumers about chemical safety concerns, chemical manufacturers have less incentive to shift toward safer chemicals. Improving the business-to-business flow of chemical information is important in advancing the adoption of safer chemicals.

Conventional Material Data Safety Sheets (MSDSs) do not provide sufficient information for effective dialogue within most supply chains. MSDSs rarely provide full information on all the chemical ingredients in a mixture, and what little information is provided is often in categories and ranges. Generally, the information needed in supply chain communication should provide:

- Chemical identification (e.g., trade name, chemical name, CAS number),
- Function of a chemical ingredient in a product,
- Amounts and concentrations of a chemical ingredient,
- Human, environmental, and physical hazards of chemical ingredients, and
- Potential for human exposure to chemical ingredients.[17]

To provide chemical information, chemical suppliers must first acquire the information. When the chemical supplier is a chemical manufacturer, this may be straightforward; however, when the supplier has several tiers of suppliers to search through, the search may be complicated and resisted.

Because of the scale and costs, collecting and managing chemical information in supply chains can be better implemented at the level of an economic sector, and chapter 8 described several industry sector-based chemical information exchanges. Some involve the simple transfer of information among suppliers, whereas others involve the disclosure of such information to the public. Government policies have been important in driving these programs. REACH and the European product directives all require that product manufacturers know some of the chemical ingredients in their products. For instance, the European Union's End-of-Life Vehicle Directive drove the automobile manufacturing sector to set up the International Materials Data System (IMDS) that today tracks some 9,000 substances used by industry suppliers in making vehicle components, subcomponents, and materials. Similarly, the European RoHS Directive encouraged the electronics sector to set up the Joint Industry Guide for Material Composition Declaration for Electronics Products (JIG). The JIG provides a "standardized list of materials [that] suppliers must disclose when present in products and components provided to electrical and electronic equipment manufacturers." Table 11.2 documents several of these industry sector chemical databases.[18]

Table 11.2
Economic Sector Chemicals in Products Information Exchange Systems

Program	Sector	Information
International Material Data System (IMDS)	Automobile	Chemicals used in automobile assembly
IPC 1752—Joint Industry Guide (JIG)	Electronics	Chemicals used in electronics
BOMCheck	Electronics	Chemicals covered by REACH and the EU RoHS, Battery and Packaging Directives
Outdoor Industry Association, Chemical Management Framework	Apparel and Footwear	Chemicals used in footwear and clothing production
Apparel and Footwear International RSL Management Group	Apparel and Footwear	Clothing manufacturers' RSL chemical information
Cleangredients	Cleaning Products	Chemicals used in formulated cleaning products
BASTA	Building Materials	Chemicals used in Swedish building construction industry
Pharos	Building Materials	Chemicals used in building materials and equipment

The formulated cleaning products sector provides a good example of a sector-wide chemical information initiative, but in an industry particularly sensitive to chemical information disclosure. *Cleangredients* is a chemical constituent database designed to help consumer cleaning product formulators select safer chemical ingredients. It was established as a partnership among several formulators, the EPA, and GreenBlue, an environmental technical services organization, to help formulators identify safer and more environmentally preferred ingredients and to help chemical suppliers showcase cleaner and safer chemicals. Through an annual subscription, chemical suppliers can list an unlimited number of chemical ingredients providing standard chemical characteristics that are verified by an independent third-party reviewer. Similarly, by paying an annual subscription fee, chemical product formulators can search the provided data and compare the potential ingredients across the designated characteristics. At present, *Cleangredients* includes listings for surfactants and solvents, with plans to expand to cover fragrances and chelating agents in the near future.[19]

11.7   Chemicals in Products

Low-level chemical exposures can occur from the chemical ingredients in formulated chemical products (adhesives, paints, shampoos) or assembled articles (toys, apparel, furniture). However, information on chemicals in products is not often publically available, and it is difficult for product manufacturers to acquire such information. Some nongovernment organizations (NGOs) have set up programs to test products to determine their chemical constituents. Both the Ecology Center in Michigan and the Silent Spring Institute in Massachusetts maintain programs for testing products and revealing the constituents on their Internet websites.

There are several chemical testing methods for revealing the chemicals in formulated products. A good laboratory with a mass spectrometer can analyze chemical mixtures and identify the compounds in formulated chemical products. Identifying chemicals in articles and the components of assembled products presents greater challenges. Laboratory tests are expensive and typically require trained personnel. Handheld X-ray fluorescent (XRF) analyzers now on the market at reasonable costs provide an inexpensive alternative to laboratory testing and can be used to identify the chemical constituents of products; however, these analyzers only detect elements and so cannot identify congeners or compounds such as

Table 11.3
Chemical in Product Information Disclosure Systems

Program	Sponsor	Information	Method of Disclosure
Scandinavian Product Registries	Sweden, Finland, Norway, Denmark	Chemicals in formulated products	Internet Site
Federal Hazardous Substances Act	U.S. CPSC	Some formulated chemical consumer products	Product label
Federal Pesticide Law (FIFRA)	U.S. EPA	Active pesticide ingredients	Product label
Material Safety Data Sheets	U.S. OSHA	Chemicals in Products	Safety sheet
California "Prop 65"	California EPA	Certification of products containing carcinogens or reproductive toxins	Product/shelf labels
Mercury-Added Product Database	Interstate Mercury Education and Research Clearinghouse	Products containing mercury	Internet site
Cleaning Products Chemical Ingredient Disclosure	New York State	Chemicals in Cleaning Products	To the state
Safe Cosmetics Program	California	Chemicals in cosmetics	To the state

types of brominated flame retardants. Such product testing has limits. Although the identity of chemical constituents can be identified through testing, the volume of chemicals and uses in a product are more difficult to determine.

Several government initiatives have been launched to track and catalog information on chemicals in products. Table 11.3 provides some examples.

The Scandinavian countries have taken a lead in identifying chemicals in products by requiring that product suppliers disclose (sometimes through customer-friendly "product declarations") the chemical ingredients of some products on the market (particularly chemical formulations). Since 1982, Norway has required the reporting of all products containing substances classified as dangerous under the European Union Dangerous Substances Directive. All nonconfidential data are made available to the public on a Product Information Bank. The Swedish Product Registry contains chemical use information on some 78,000 chemical products on the Swedish market. Information is posted annually by manufacturers and distributors on all products manufactured or imported into Sweden in quantities greater than 100 kilograms per year. Along with Denmark and Finland, these countries have taken steps to harmonize their product ingredient reporting systems; however, the differing statutory authorities have made this difficult.

Some U.S. states require product manufacturers to disclose the identities of chemical ingredients in products. New York requires manufacturers of household cleaning products to report each ingredient contained in each product by weight, and the California Safe Cosmetics Act of 2005 requires that any manufacturer, packer, or distributor of a cosmetic product that contains any ingredient known or suspected to cause cancer, birth defects, or other reproductive harms to report the ingredients to the state.[20] Fourteen states require manufacturers and distributors of products containing mercury (with the exception of fluorescent lamps) to identify and report those products to a common product database called the Interstate Mercury Education and Reduction Clearinghouse (IMERC). The IMERC was established in 2001 to create a searchable database of information submitted to the IMERC member states on the amount and purpose of mercury in products. The database is intended to inform consumers, recyclers, policymakers, and others about products that contain intentionally added mercury, the amount of mercury in specific products, and the amount of mercury in specific product lines.[21]

Some product manufacturers have initiated programs to fully disclose the chemicals used in products and in their production. S.C. Johnson, Clorox, and Seventh Generation have posted the chemical ingredients of many of their household and personal care products on their corporate websites. S.C. Johnson has an interactive site that allows consumers to search by product and by ingredient. The Clorox website is more limited, only listing the chemical substances for a range of products in descending order of concentration. Seventh Generation has the most comprehensive listing of chemical ingredients by product and ingredient, including substances that are not intentionally introduced or appear only as trace amounts.[22]

The problem of accessing information about chemicals in products becomes more acute where product production, sales, use, and recycling or disposal take place in different countries. The United Nations Environment Program has launched a special "Chemicals in Products" project under SAICM that is focused on supporting the development of internationally harmonized systems for transferring chemical information among companies in supply chains and disclosing chemical information to customers, governments, civil society organizations, and waste managers. This project is only in an early stage, but its focus on an international information exchange for chemicals in products appears appropriate for a globalized product market.[23]

11.8   Chemical Releases and Exposures

Although safer chemical policies focus on the inherent properties of a chemical in generating chemical profiles and classifying chemicals, assessing chemical releases and exposures is important in setting priorities for chemical substitution or regulation, identifying vulnerable populations or ecosystems, and conducting risk assessments for chemical risk management.

The EPA collects and discloses information on chemical releases and waste generation. Statutory requirements for waste, effluent, or emission reporting under the environmental protection laws combined with permit data provide a rough guide to the disposal of some regulated chemicals. The Toxics Release Inventory (TRI) currently provides information on emissions and wastes covering 593 individual substances and thirty chemical categories released from 20,800 U.S. generators. Similar national Pollutant Release and Transfer Registries (PRTRs) exist in eleven other nations around the world. These registries provide data on chemical emissions and wastes; however, they cover only a small number of chemicals from a limited universe of sources.[24]

Exposure information is more difficult to collect. Chemicals reported in the ambient condition of air and water document environmental exposure. The EPA's Drinking Water Program periodically surveys drinking water supplies for the presence of synthetic chemicals, and the Clean Air Program conducts surveys of regional air districts to identify chemical contaminants. Some of the best data on the presence of synthetic chemicals in air come from surveys conducted by the California Air Resources Board.[25]

Starting with the 2006 survey, some human exposure information is now collected under the EPA's Chemical Data Reporting Rule. Use- and exposure-related data are reported for chemicals manufactured or imported in amounts of 300,000 pounds during a single year at a single site. This includes information on the physical form of the chemical, the number of workers potentially exposed, and the maximum concentrations (in ranges).

Biomonitoring surveys provide evidence of chemicals or their metabolites in human tissue or fluids. The Centers for Disease Control's National Center for Environmental Health has maintained a National Biomonitoring Program since 1999 that relies on surveys conducted as part of the biennial National Health and Nutrition Examination Study (NHANES). Blood and urine samples from some 2,500 individuals are analyzed for environmental chemicals. The programs' *Fourth National Report on Human Exposures to Environmental Chemicals* was released in 2009 and covered 219 chemicals ranging from lead, cadmium, arsenic, and mercury to perchlorates, phthalates, and polycyclic aromatic hydrocarbons.[26] In addition, several states administer biomonitoring surveys focused on chemical contaminants, including the California Environmental Contaminant Biomonitoring Program established by law in 2006.

Biomonitoring data provide direct evidence of population exposure, but in the absence of such direct data, models can provide reasonable estimates. The EPA has developed several models for this purpose, among them the EPI Suite™, ChemSTEER, and E-FAST. The EPI Suite™ (EPI stands for Estimation Programs Interface) provides an integrated computer program for running chemical property and fate and transport models sequentially. It includes fate information on some 40,000 chemicals and estimation modules for chemical partitioning among air, soil, sediment, and water, as well as more complex modules on rates of volatilization from water bodies and degradation in sewage treatment. ChemSTEER is a chemical exposure tool for estimating workplace exposures and releases resulting from the manufacture and use of industrial chemicals. E-FAST is used to convert chemical release information into population exposure estimates.[27]

11.9   Safer Alternatives Information

Converting to safer chemicals could be better supported if there were better information sources on safer alternatives. Throughout government and industry, there are hundreds of ready examples of chemical substitutions or changes in the design of products and production processes that have led to safer chemicals. Much of this information is described in published case studies, held in corporate records, or available in government manuals and reports. However, there is no common catalog of potentially safer alternatives.

Such a catalog could be organized in sections around specific chemicals of concern (lead, mercury, bisphenol A, etc.) or around certain chemical-using functions (dry cleaning, painting, printing). Setting the catalog up as a database would allow it to be searchable by chemical, chemical alternative, or functional use of a chemical, such as "formaldehyde used in adhesives." The catalog could be developed in an open and broadly participatory process. In the same way that chemical suppliers could populate the tiered Universal Classification of Chemical Substances, with chemicals classified by their level of concern, government agencies and chemical users could fill in a catalog of successful safer chemical assessments. Making the catalog available to the public in a progressively developing "wiki" fashion could tap the powers of many reviewers to catch errors, enlarge upon analyses, and provide for an evolving addition of new alternatives.

In 2008, several NGOs in Europe collaborated to launch a special Internet site called SubsPort to promote the substitution of hazardous chemicals.[28] SubsPort has been designed as an information resource for European firms seeking to comply with REACH, and it has recently opened a catalog of analyses of alternatives to the chemicals identified under REACH as substances of very high concern. Because SubsPort's information is available on the Internet, it can be of assistance to any company seeking to identify safer alternatives to chemicals of concern. However, besides this one effort, little has been done to address this important information need.

11.10   Confidential and Proprietary Information

Historically, trade secret and confidential business information (CBI) protections have presented a confounding problem for transparency. Chemical and product manufacturers often need to protect some proprietary information as confidential business information (CBI); however, such protections can be misused. Studies of new chemical submissions under TSCA have found that some submissions claim CBI protections for a wide range of noncompetitive information ranging from health and safety information to the chemical identity of the compound. Such CBI protections appear to be driven more by liability concerns than truly technical secrets. Chemical suppliers can be reluctant to tell suppliers that their products contain carcinogens, mutagens, or reproductive hazards, even when the concentrations are well below standards. Such restrictions tend to block the free flow of chemical information even for those firms along a chemical's supply chain.

Some CBI protection is necessary. Proprietary information, particularly process manufacturing information and information on mixtures (like fragrances), can be of critical competitive value and needs protection. However, chemical content information is difficult to protect even with CBI protections. Industrial competitors can gather confidential information by reverse engineering products, hiring away knowledgeable technical staff, reviewing marketing data and even more clever forms of espionage. Although a mass spectrometer can reveal the chemicals in a formulated product, it cannot reveal the exact quantities (the "formula"), the process steps, or the nonchemical contributions such as mixing times, temperatures, and pressures.

Legitimate confidential business information could be protected in chemical profiles with clear criteria for defining such information and a substantial means to protect its disclosure. Common criteria include:

• reasonable measures can be taken to protect the confidentiality of the chemical information,

• the chemical identity is not required to be disclosed by any government law or regulation,

• the disclosure of the chemical information is likely to cause substantial harm to the competitive position of the company, and

• the chemical information cannot be discovered through literature searches, reverse engineering, or other means.

A common means for protecting such information in public information exchanges involves accepting both a protected and a redacted version of a chemical profile, although other procedures might be appropriate given certain conditions (such as national security). Patenting and copywriting have long been used in other areas where intellectual property needed protection and patenting has been used on various chemical production processes. Pharmaceuticals and certain pesticide formulas are routinely patented. In a similar manner, the chemical formula of novel chemical products could be patented if it were not that some foreign governments have been lax in enforcing such patent protections.

There is growing international agreement that some forms of chemical information, such as the basic identity of chemicals and their health and safety information, should not be protected information.[29] Governments in Europe are developing initiatives to review and "sunset" CBI protections after a certain number of years. As part of a broader transparency initiative, the EPA has recently completed a review of some 22,000 CBI claims and found that more than half of its CBI protections had not been asked for. The agency is planning a requirement that such claims be periodically resubstantiated.[30] There are now various third-party testing and certification programs that can provide assurances to product manufacturers about chemical ingredients without compromising confidentiality. Each of these strategies bares consideration in setting safer chemical policies.

11.11   Generating, Distributing, and Interpreting Chemical Information

The market works best when it is information rich. However, generating and distributing chemical information is complex and costly. Therefore, it is important that new chemical policies shift the burden for developing and publically presenting chemical information to those who manufacture chemicals and know or should know their effects. Shifting this responsibility opens up a new set of accountability relations. Government agencies set information standards, chemical and product manufacturers present their dossiers, and the role of determining conformance falls to government reviews, third-party audits, or the critical judgment that results when information is fully displayed on public databases or Internet websites.

The EPA has set up various voluntary programs to collect chemical hazard and exposure data from private corporations. The Voluntary Children's Chemical Exposure Program established in 1998 identified twenty-three chemicals of concern to children's health and asked manufacturers of those chemicals to voluntarily provide information on toxicity and risk. The agency's High Production Volume (HPV) Challenge provided a similar effort to coax manufacturers and importers of large market volume chemicals to voluntarily submit health and environmental effects information to a publically accessible information system.

Chapter 7 described the value of "radical transparency" in improving corporate accountability and driving change. The EPA's experience with emission reporting under the TRI demonstrates the potential effects of public chemical information disclosure. The TRI was initially intended in the Emergency Response and Community Right to Know Act as a vehicle by which the agency could acquire more accurate data on chemical releases to the environment. However, the mandatory annual reporting and its public release with the associated names and addresses of the generators of emissions and wastes proved a significant incentive for many generators to reduce the volume of their releases. This same reliance on public scrutiny can be valuable in a safer chemical policy framework. All that may be necessary for similar effects is the uploading onto a public database of the chemical profiles for each of the chemicals classified under the Universal Classification of Chemical Substances in an easily accessible manner.

However, transparency requires more than raw information disclosure—information requires interpretation. As the previous sections document, many chemical information databases and catalogs are available on the Internet; however, there are too much data for some chemicals, too little for many others, and not enough of the interpretation that makes knowledge out of raw data. Converting these data into information that supports and encourages the transition to safer chemicals requires sorting the data using appropriate metrics and interpreting them in ways that lead to public and professional understanding. Much can be learned from recent efforts to improve popular chemical literacy.

During the 1990s, the Environmental Defense Fund developed a Web-based chemical information site, called the *Chemical Scorecard*, which provided a means of interpreting the chemical information on the EPA's TRI such that citizens who used the TRI could understand the health and environmental significance of the chemicals released in their communities. The *Chemical Scorecard* provides information on human health hazards, use and exposure characteristics, and relative hazard rankings based on several human health and exposure and ecological ranking systems.[31]

Additional efforts have been made to assist workers in understanding more about chemical hazards. The right to know campaigns of the 1980s and the OSHA Hazard Communication Standard provided workers access to the names and MSDSs of worksite chemicals. The Spanish Trade Unions' ISTAS commissioned by the European Union Trade Federation has developed an Internet-based site for European workers to find information on chemicals that they might be exposed to at work. The site provides information on the health hazards, environmental risks, environmental and health related regulations, and classification and labeling of chemicals according to the European Union CLP regulation.[32]

A more recent information resource for workers can be found in *ChemHat*, the Internet-based chemical information system developed by Charlotte Brody at the BlueGreen Alliance. Charlotte explains, "We started out to help workers better connect to on-line chemical hazard databases, but soon realized how inadequate Websites filled with terms like IARC and STELs were. Seeing chemical information through the eyes of workers led us to create a Website that was understandable to people like them." *ChemHat*, which is an acronym for Chemical Hazard and Alternatives Toolbox, is designed to assist workers by providing chemical information labeled with a range of icons indicating various hazard traits (ten chronic hazard traits, three acute hazard traits, two physical hazard traits, and five environmental impact traits). In addition, the system describes common uses for each chemical, regulatory requirements (if any), and whether there are known alternatives that could be adopted as substitutes. "We train on *ChemHat*," Charlotte explains. "With *ChemHat* workers get chemical information in the form they need and they become better prepared to promote safer chemicals in their workplaces and the products that they make."[33]

There are many avenues by which chemical information can be distributed, including fact sheets, numeric or color-coded indicators, ranking algorithms, scorecards, and globally harmonized chemical safety data sheets (SDSs). A chemical ingredient disclosure box similar to the nutritional box label on prepared foods or a barcode-reading cell phone app similar to the *GoodGuide* app could provide consumers with more chemical information at the point of product purchase. A chain of custody file similar to a trucking manifest could be used for product manufacturers to track the chemicals introduced into components through complex supply chains. The common template developed for chemical profiles might use colors, numbers, and icons to indicate hazard traits and chemical classification. Once the chemical data are made publically available, civil society advocacy organizations, business associations, and academic centers could sort and interpret the data so as to use them effectively to call attention to the most common chemicals, the most persistent chemicals, or the "safest" chemicals. Today, there is significant capacity for chemical information interpretation and distribution in the thousands of consumers, citizen activists, and consumer-facing firms that are eager to share what they know through popular tools such as Google, Wikipedia, and Toxipedia.

11.12   Solutions-Oriented Science

Some hundred years of developing toxicology, pharmacology, and environmental science have expanded scientific understanding of the biological functions and effects of chemicals in the environment and living organisms. With 80,000 chemicals to characterize in the United States and some thirty or more hazard traits to consider, the scientific task remains substantial, even where the responsibility for generating chemical information lies with the chemical manufacturers and suppliers. Chemical screening and computer modeling offer valuable tools for initially predicting chemical behavior and effects, and newer, faster computational analyses offer significant promise for filling the many current gaps in chemical hazard and effects characteristics.

The rapid advances in information technology and informatics are opening opportunities for multifunctional information portals that can link massive amounts of information together into searchable and interactive information machines. Generating and validating such information could be aided by tapping the same multiparticipant, open access process that was suggested for classifying chemicals. For instance, college and university classes could be equipped with the new computational toxicology software tools and challenged to validate chemical hazard data. If there were chemical information services ready to post those data, they could be reviewed publically and corrected as needed, and over time an interactive and reasonably trustworthy global chemical information resource could be built. Such public information disclosure and transparency can promote accuracy, authenticity, and accountability in a manner that is currently described as "crowd sourcing." The Internet and social networking are creating a whole new information environment for public transparency and participation in managing chemicals.

To further that understanding, scientific information on chemicals and data from many sources need to be compiled and interpreted. There is a need for chemical information management services either within the government or as independent bodies to manage and interpret the data and make it available to industrial supply chains and public media. The EPA's recently opened "ChemView," an information portal that covers all of the agency's nonconfidential TSCA chemical information, provides a useful new model for such a service.[34]

However, more information on chemicals needs to be generated. There is a need for more chemical testing, predictive modeling, and basic health and environmental effects research, but it is important that this research is directed toward adding knowledge to the chemical landscape rather than simply deepening existing furrows. Too much research on chemical properties has been expended on studies of a small set of chemicals or on the methodologies for chemical testing. This is a defensive approach to research that paralyzes action by burying science in endless analysis.

Although good science can inform and support policy decisions, it should not determine them. A critical literature suggests that science is not independent of economic and political contexts, and that more and more research may reinforce, rather than settle, disagreement and controversy.[35] Scientific uncertainty is endemic to basic research on causality, and waiting for enough good science to fully characterize chemicals before appropriate action can be taken places an undue burden on science and usurps the role of public opinion and good professional judgment.

Daniel Sarewitz and David Kriebel criticize this "knowledge-first" tendency in science and call instead for a "solutions-oriented" science. As an example, they note methylene chloride, a toxic industrial solvent whose mechanism of biological action is not well understood. The standard knowledge-first approach would focus more and more research on better understanding how methylene chloride behaves in biological organisms. Instead, they argue that research could be better focused on identifying or designing safer alternatives to methylene chloride.[36]

Like the health sciences that are motivated by a desire to treat or prevent disease, research on chemicals needs to be "solution-oriented" and driven by the clear mission of shifting the chemical economy toward safer alternatives. Scientific studies remain important. However, the objective is recast by focusing on the development of alternative solutions rather than the degree of safety of current technologies. The strategy is technical and pragmatic, working within the bounds of what is known at the current time and adjusting research directions as new information arises.[37]

More chemical information made widely accessible provides the foundation for a safer chemical policy framework. Just as the Apollo Project required a national research plan, the transition to safer chemicals requires a well-conceived and implemented research agenda. Like the Apollo Project, the investments necessary to develop chemical information need to be strategically directed toward a transformative mission. The research should be change-focused and solutions-oriented. This is science on a mission, and it is a mission critical to characterizing, classifying, and prioritizing chemicals in order to promote safer chemicals.

12

Substituting Safer Chemicals

The STD Gear and Instrument Company of West Bridgewater, Massachusetts, is a 30-year-old manufacturer of high-performance gears and mechanical transmission components. For years the company cleaned oils from its metal products in dip tanks of trichloroethylene (TCE), an inexpensive and effective cleaning and degreasing agent. During the 1930s, TCE replaced the flammable petroleum distillates commonly used in metal parts degreasing, but in the 1950s, concern over the toxicity of TCE led many shops to switch to trichloroethane (TCA). When the production of TCA was phased out during the 1990s under the Montreal Protocol, STD Gear and many other metal component manufacturers switched back to TCE. However, in the late 1990s, the EPA listed TCE as a "Hazardous Air Pollutant," and the International Agency for Research on Cancer classified TCE as a probable human carcinogen. Recognizing the potential hazards of TCE, the managers at STD Gear searched for an alternative degreasing chemical and replaced TCE with n-proypl bromide (nPB), a nonflammable chemical that was somewhat more costly but performed to high standards as a degreaser. However, a growing body of research now suggests that nPB might have adverse effects on human reproductive and nervous systems, and in the early 2000s, California and North Carolina issued occupational hazard warnings. Concern over worker safety has now led the STD Gear managers to replace nPB with an alkaline aqueous cleaning bath.[1]

This short story of STD Gear's long struggle to find a safer metal degreasing agent focuses on the concept of substitution. The substitution of safer chemicals makes up the third building block of a safer chemical policy framework. The substitution of hazardous chemicals at the firm level is important in ensuring a safer workplace or product; however, the substitution of those same chemicals at the broader levels of economic sectors or entire supply chains can have system-changing effects.

Conceptually, there are three strategies to promote a shift toward safer chemicals: prohibiting the use of a chemical, avoiding a chemical of concern, or seeking safer substitutes. A "prohibition strategy" typically involves a government regulation or a corporate policy that bans the continued use of a chemical. Where government agencies or private sector chemical users voluntarily seek to avoid chemicals that appear to present unacceptable risks, they are engaged in an "aversion strategy." The third strategy is a more prospective strategy looking ahead at what alternatives could be used as substitutes for a chemical of concern and evaluating the health and environmental impacts, the costs and the technical consequences of a potential substitution. Because it focuses on substitute chemicals, this strategy is here called a "safer chemical strategy."

12.1   The Prohibition Strategy (Banning Chemicals)

Many national governments have banned the production, importation, use, or disposal of dangerous chemicals. In 1991, the Swedish government published a list of eight chemicals and chemical groups that it planned to "sunset."[2] The U.S. EPA has used its authority under the federal pesticide laws to prohibit the registration and use of highly hazard pesticides such as aldrin and dieldrin (1974), heptachlor (1976), chlordane (1976), and chlordecone (1978). The European Union's Restriction on Hazardous Substances directive prohibits the use of lead, mercury, cadmium, hexavalent chromium, and brominated flame retardants in electrical and electronic products. During the 1980s, the United Nations compiled a list of some 600 chemicals that had been banned or severely restricted in at least some countries around the world.[3]

Ideally, the outright banning of a chemical can be an immediate and effective response to a chemical of high concern. However, because chemicals are always components of systems, a prohibition strategy presents many challenges. These include:

• Vacated markets. The sudden absence of a chemical may not only adversely affect immediate consumers but also downstream users who previously depended on products made with the banned substance.

• Stranded inventories. A ban on further use of a chemical can leave suppliers holding large inventories that no longer have market value and encourage the abandonment of those supplies in poorly regulated stockpiles.

• Illegal markets. Where the legal market for the use of a chemical is closed but the demand persists, there is an opportunity for illegal manufacture and use.

• Rushed Substitutions. Immediate bans may lead chemical users to adopt substitutes with minimum research and testing, leading to compromised products or unanticipated risks.

The most successful global example of a prohibition strategy has been the international effort to phase out the use of ozone-depleting chemicals. During the 1980s, a broad international consensus supported the United Nations in negotiating a global phase out of the chlorinated and fluorinated compounds recognized under the Montreal Protocol as ozone-depleting substances. This same prohibition strategy underlies the international Stockholm Convention on Persistent Organic Substances, which is proceeding to phase out a short list of highly hazardous pesticides and industrial chemicals.

However, single chemical bans are often compromised. Some of the substitutes adopted under the Montreal Protocol turned out to have milder but still adverse effects on the ozone layer. The phase out of the chlorofluorocarbon, Freon, in the United States led to the illegal importation from Mexico of millions of pounds of replacement Freon to supply automobile air conditioners.[4] The phase out of DDT, although justified due to its effects on wildlife, has raised controversy in Africa because no more cost-effective insecticide for killing mosquitoes in combating malaria has been developed.

The problems generated by a prohibition strategy are evident in the case of daminozide, a growth hormone at one time commonly used in the ripening of orchard fruits. Daminozide (2,2-dimethyl hydrazide) was initially registered under FIFRA in 1963 by Uniroyal Chemical Company under the trade name "Alar." During the 1970s, Alar was widely marketed for use in apple orchards, where it was applied in mid-summer to decrease fruit cracking, increase fruit firmness, and delay pre-harvest fruit drop. During the 1980s, the EPA began studies of daminozide's potential carcinogenicity. In 1989, the Natural Resources Defense Council released a report on Alar's presence in children's applesauce, and a news segment appeared on the popular television newsmagazine "Sixty Minutes" linking Alar to cancer. Applesauce sales plummeted, and within six months, Uniroyal voluntarily withdrew the registration for Alar. The loss of Alar caused changes in the orchards. Without Alar to prevent apple "drop," the picking season was shortened, and orchard workers were pressed to work faster and extend their working days. The loss of Alar also increased the use of insecticides in the orchards and meant that workers were more likely to be in the orchards earlier when residuals of the insecticides were higher and their exposure was more extensive. The sudden withdrawal of Alar was never accompanied with an investigation of the systemic consequences of the loss of the chemical, and no effort was made to identify or develop safer alternatives.[5]

Anticipating such problems, governments often avoid sudden bans in favor of staged phase outs. Such phase outs require establishing an optimum timetable: too short a schedule leads to the problems noted above, whereas too long a schedule leads to procrastination, price gouging, and inventory hoarding. In implementing such phase outs, industry resistances can be high, and governments often respond with special use exemptions, compliance extensions, and tolerance for missed deadlines.

12.2   The Aversion Strategy (Avoiding Chemicals)

An aversion strategy focuses on identifying chemicals of high concern and seeking to avoid them.

Government agencies employ an aversion strategy when they draw up lists of chemicals to avoid as do private firms that create Restricted Substance Lists (RSLs). Such lists often include chemicals that are already restricted by other governments or included on lists of dangerous chemicals. Many governments, international agencies, and other "authoritative bodies" maintain lists of these hazardous chemicals. The Oslo and Paris Convention for the Protection of the Marine Environment of the Northeast Atlantic (OSPAR) publishes a "List of Substances of Possible Concern" in the marine environment. IARC publishes an internationally recognized list of human carcinogens, and the European Commission maintains a list of chemical substances known or suspected to disrupt the endocrine system. Table 12.1 identifies a selection of such lists.

Some European governments publish lists of dangerous chemicals that, although not prohibited, are to be avoided. In 2000, the Danish Environmental Agency published a list of 1,400 undesirable substances. During the 1990s, the Swedish government developed a hierarchy of lists that ranged from a short list of substances that were to be phased out of use to a longer list of substances that should be voluntarily avoided.[6] The EPA has developed several lists that have been used as avoidance lists. The agency maintains a list of suspended, canceled, and restricted use pesticides. The 630 substances and chemical categories on the agency's Toxics

Table 12.1

Lists of Hazardous Chemicals Compiled by Authoritative Bodies

Carcinogens	• International Agency for Research on Cancer • U.S. National Toxicology Program, *Report on Carcinogens* • State of California, *Proposition 65*
Reproductive Hazards	• U.S. National Toxicology Program, Center for Evaluation of Risks to Human Reproduction • State of California, *Proposition 65* • Japan, International Center for Occupational Safety and Health
Endocrine Disruptors	• European Commission, *List of Category 1 and 2 Endocrine Disruptors*
Persistent, Bioaccumulative, Toxics	• OSPAR Convention, *List of Substances of Possible Concern* • U.S. Toxics Release Inventory, *List of Persistent, Bioaccumulative Toxics* • State of Washington, *List of Persistent, Bioaccumulative Toxics* • Environment Canada, *List of Persistent, Bioaccumulative and Inherently Toxic Chemicals*
Eco-Toxins	• OSPAR Convention, *List of Substances of Possible Concern to the Marine Environment*
Ozone Depletion	• U.S. Environmental Protection Agency, *Ozone Depleting Substances—Class 1 and Class 2* • European Commission, *Regulation (EC) No. 1005/2009—Substances that Deplete the Ozone*

Release Inventory (TRI) are viewed by some businesses as chemicals to avoid. Other businesses use the agency's Waste Minimization Program list of some thirty-one persistent, bioaccumulative, and toxic (PBT) substances as an avoidance list.

Once such hazardous substance lists have been accepted as lists of undesirable substances (such as RSLs), a production manager or chemical supplier needs only to review the lists to identify chemicals that should be avoided. Likewise, the synthetic chemist designing a new substance or synthetic process needs only to consult the lists to develop chemistries that avoid the use of these listed chemicals. However, consulting lists and avoiding substances identified as dangerous present several challenges:

• Inconsistent lists. Although there are many cross-listings among lists, lists have diverse criteria such that chemicals classified by one hazard trait on one list may be differently classified on another list.

• List changes. Keeping such lists current requires frequently adding and deleting chemicals. New information can lead to unanticipated additions or deletions from lists

• Unlisted advantages. The prospect of moving from a listed chemical of concern to an unlisted chemical favors potentially hazardous but unstudied chemicals.

Avoiding chemicals of concern is a reactionary and reductionist approach. It can successfully remove an undesirable chemical; however, it may neglect a thorough assessment of future consequences. Consider the search for a safe wood preservative. Waxes and tars have long been used for preserving wood exposed to environmental forces. During the nineteenth century, Europeans began experimenting with solutions of mercury chloride to preserve wood, but the chemicals leached from the wood and proved dangerous to apply. In 1838, John Bethell received a patent for injecting creosote under heavy pressure into wood posts. By mid-century, coal tar creosote had become the dominate chemical for treating wooden railroad ties and telegraph poles. However, early studies of the health effects of creosote suggested its probable carcinogenicity, and during the 1980s, both IARC and the EPA classified coal tar creosote as a carcinogen. Although creosote remains widely used in commercial and industrial applications, chrominated copper arsenate (CCA) began to replace creosote in consumer-related products during the 1990s. However, in 2003, the CPSC released a study linking arsenic treated lumber to increased cancer rates in children. Under a voluntary agreement with the EPA, in 2004, the wood products industry began substituting alkaline copper quat (ACQ), cooper azole, and sodium borate for CCA in treated timber. The sequential steps transitioning away from mercury chloride to creosote to CCA and then to ACQ demonstrates how an aversion strategy can lead to a long series of regrettable substitutes. The lesson here is that it is important not only to avoid hazardous chemicals, but to be assured that the substitutions that follow lead to safer alternatives.[7]

12.3   The Safer Chemical Strategy (Seeking Safer Substitutes)

Instead of focusing narrowly on eliminating undesirable chemicals, a safer chemical policy framework focuses on identifying safer alternatives to chemicals of concern and, where possible, substituting them. Focusing on the process of substitution shifts the hazardous chemicals problem from a problem-oriented to a solution-oriented activity that can be anticipatory and forward looking and open to opportunities for creativity and innovation.

Since the 1970s, the Swedish government has promoted the idea of substitution of dangerous chemicals through its "Substitution Principle" first in industrial processes and, more recently, in the chemistry of products.

Box 12.1
The Substitution Principle

> Substitution means the replacement of a substance by an alternative that delivers the same or similar performance or functionality. The Stockholm Convention calls such alternatives "locally available, safe, effective and affordable."[8] To tailor this concept to chemical policy, in 1973, the Swedish government developed the "Substitution Principle":
>
> > If risks to the environment and human health and safety can be reduced by replacing a chemical substance by another substance or by some non-chemical technology, then this replacement should take place.[9]
>
> The principle is tied closely to the idea of prevention and inherent safety. In considering the Substitution Principle, exposure conditions and risk may be addressed, but the Swedish policy is clear that such consideration should not significantly cause delay of appropriate action. Like the European Polluter Pays Principle and Precautionary Principle, the Substitution Principle has served as both an underlying legal principle of environmental regulations and an informal guidance for protective practice.

There are many successful examples of safer chemical substitutions in industry. Substituting aqueous and semi-aqueous (terpines and alcohols) solvents for chlorinated solvents in industrial parts cleaning and degreasing provides a common example. Converting from mineral-based inks to soy-based inks offers an example that swept the newspaper business during the 1980s and 1990s. During this same period, many large mills in the pulp and paper industry moved from hazardous chlorine to more benign chlorine dioxide and peroxide for bleaching and delignification. Many conventional hydrocarbon-based paints and coatings have been reformulated into less volatile water-based mixtures that have eliminated ingredients such as toluene, methyl ethyl ketone, formaldehyde, and various isocyanides.[10]

The substitution of chemicals of concern can occur at the chemical, material, product, or process level of a production system and can result in a replacement chemical or a nonchemical engineering or process management change. At the chemical level, the easiest substitution involves a simple "drop-in" chemical-for-chemical replacement. Here, safer chemical lists may help. The EPA's Office of Wastewater Management publishes a list of Environmentally Acceptable Lubricants. The agency's Design for Environment Program (DFE) has published a Safer Ingredient List of safer chemical that can be used in formulated household products. To be listed on the DFE list, the chemical must not be recognized as a CMR, PBT, systemic or internal organ toxicant, asthmagen, sensitizer, or endocrine disruptor.[11]

Where a simple chemical-for-chemical substitution is not available, a chemical of concern can still be replaced by shifting the material that requires it. The substitution of brominated compounds used as flame retardants in polymer resins provides an example. Beginning in the mid-2000s, public and government pressure led many product manufacturers to seek substitutes for decabromodiphenyl ethers (decaBDE). The decaBDE used in the high-impact polystyrene (HIP) casings of electronic products was commonly replaced with phosphorous compounds (such as resorcinol diphenyl phosphate). This was a chemical-for-chemical substitution. However, for the HIP used in television housings, there were no functionally equivalent flame retardants, so instead the polymer resin was replaced with another material—polycarbonate/acrylonitrile-butadiene-styrene (ABS)—that could use the phosphorus-based fire retardants. For similar reasons, Apple shifted from plastic to aluminum clad housing for its laptops. These shifts are examples of material-for-material substitutions.[12]

At the product or process level, the replacement of a chemical can occur through redesign. Here it is important to consider the function of a chemical of concern in a product or the function of the product containing a chemical of concern. Is the hazardous chemical necessary? Is there another way to achieve the function? Is the product necessary? Is there another way to satisfy the need without the product? The brominated flame retardants conventionally used in foam mattress applications were eliminated by redesigning mattresses to include a barrier material to keep a flame (e.g., from a cigarette) from reaching the foam. Packaging materials have been redesigned to avoid the need for adhesives, printing inks, and plastic labels that contain hazardous chemicals.

Consider phthalates again. Table 12.2 suggests a range of alternatives to phthalates used as plasticizers in PVC-based children's toys.[13]

Substitution is best seen as a staged process that may involve "bridging chemicals"—chemicals that are less hazardous than those they replace, but

Table 12.2
Alternatives to Phthalate Plasticizer in Children's Toys

Chemical Alternatives in PVC	acetyl tributyl citrate (ATBC) di-isononyl-cyclohexane-1,2dicarboxylate (DINCH) di(2-ethyl hexyl) adipate (DEHA) butylated hydrxytolluene (DEHA) 2,2,4-trimethyl,1,3-pentanediol diisobutyrate (TXIB)
Material Alternatives to PVC	ethylene vinyl acetate high density polyethylene polymers from polylactic acid polymers from polyhydroxyalkanoate
Product Redesign Alternatives	wooden toys woven fiber toys

still present some hazards. For instance, in the chemical classification scheme presented in chapter 10, a lower tier substance of very high concern such as lead as a paint pigment might be replaced with a higher tier substance of some concern such as zinc or titanium oxide before a fully preferred chemical such as a calcium- or clay-based pigment could be adopted. Solvents used as intermediaries in product manufacturing provide a common example. Because process solvents are not intended to end up in the product, there is often latitude on what substances are used. The transition from hazardous solvents to safer alternatives may progress through a series of increasingly safer steps. Table 12.3 provides a simplified model of preference categories including a list of "usable" bridging chemicals that could provide interim steps toward "preferred" chemicals.[14]

The substitution process can be quite straightforward when alternative chemicals or technologies are already on the market. Substitutions can also be quite complex and costly, where the substitution requires significant process changes, awaits new chemical development, or faces conflicting objectives. Substitution often involves trade-offs and the balancing among several values. Some now regrettable hazardous chemicals introduced during the last half century, such as brominated flame retardants, chlorofluorocarbons, chlorinated solvents, and PCBs, were adopted to lower acute occupational hazards or the risks of fires and explosions. It is important in seeking safer substitutes to a chemical of concern that safety risks are not inadvertently shifted to workers or environmental burdens are not created at other points in a chemical's life cycle.

In industrial or commercial applications, identifying safer alternatives to hazardous chemicals and adopting them involves two steps: chemical

Table 12.3
Alternative Solvents Ranked by Preference

Undesirable	Usable	Preferred
pentane	cyclohexane	water
hexane (s)	methycyclohexane	acetone
di-isopropyl ether	toluene	ethanol
diethyl ether	heptanes	2-propaol
dichlorometane	methyl t-butyl ether	1-propanol
dichlorethane	isooctane	ethyl acetate
chloroform	acetonitrile	isopropyl acetate
dimethyl formamide	2-methylietrahydrofuran	methanol
N-methylpyrrolidinone	tetrahydrofuran	methyl ethyl ketone
pyridine	xylenes	1-butanol
dimethyl acetamide	dimethyl sulfoxide	t-butanol
dioxane	acetic acid	
dimethoxyethane	ethylene glychol	
benzene		
carbon tetrachloride		

action planning and alternatives assessment. A chemical action plan lays out the purpose and process for the replacement of a chemical of concern with a safer alternative. Alternatives assessment provides the methodological basis for selecting the preferred alternative. Figure 12.1 outlines the relationships.

12.4   Chemical Action Planning

The successful conversion of a manufacturing process or consumer product to safer chemicals requires a planning process, which is here referred to as chemical action planning. This planning requires setting goals and boundaries, allocating resources, selecting participants, conducting alternatives assessments, persuading decision makers, and setting timetables for implementation. Substitution is a central objective of chemical action planning. Because the substitution of a safer alternative requires a reasonable amount of informed analysis and professional judgment, the EPA coined the term "informed substitution" to identify chemical replacements for which the alternatives and consequences have been carefully considered. Informed substitutions of chemicals are made every day in industry. Instead of adding an entirely new procedure to these practices, it is more effective to slip health and environmental factors into those conventional decision-making processes.

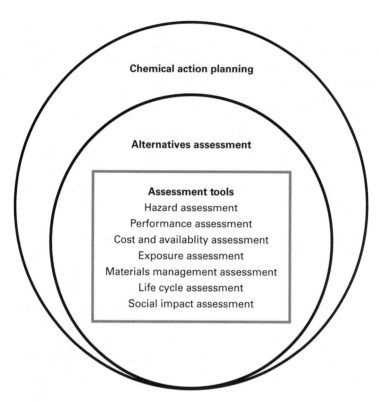

Figure 12.1

Chemical Action Planning

There is some experience with chemicals action plans in U.S. chemical policy. The State of Washington prepared chemical action plans to guide its regulatory initiatives on a list of persistent, bioaccumulative, and toxic chemicals. Several states have prepared chemical action plans as part of their mercury elimination programs. In 2009, the EPA posted chemical action plans for four chemicals of concern—polybrominated diphenyl ethers, long-chain perfluorinated compounds, certain phthalates, and short-chain chlorinated paraffins. These plans identified the range of substances covered (some of these substances have many congeners), described the functional uses, listed a set of proposed government responses, and identified specific government responses ranging from initiating chemical data "call-ins" to issuing significant new use rules, adding the chemical to the TRI reporting list, encouraging voluntary phase outs, and limiting or banning continued use. Since 2009, another seven chemical action plans have been posted.[15]

Less common have been chemical action plans developed by industry. Industry-prepared plans for chemical management lie at the heart of the Massachusetts Toxics Use Reduction (TUR) program. The law identified a list of priority chemicals and then required firms to prepare facility-level "TUR plans" that must be updated every other year. The planning process is organized as a series of steps that can be pursued through several iterations to identify safer alternative chemicals, technologies, or processes. The process starts with goal setting and an assessment of who should be involved. Some firms rely on externally contracted toxics use reduction planners to prepare the plan while others set up internal teams of production managers, environmental, health and safety professionals, procurement managers, and shop floor workers.

The Massachusetts program does not require that the plan be implemented (although studies have shown a high level of substitute adoption), but a good planning process needs to follow through on the substitution and evaluate the results. Although the plans are proprietary, they need to be certified by a specially trained and state-licensed "TUR planner" and are available for state inspectors when state inspectors are conducting routine inspections. Altogether more than a thousand firms have completed facility-level "TUR plans," with more than two-thirds of firms surveyed reporting that they used the plans to assist in reducing the use of toxic chemicals.[16]

12.5   Alternatives Assessment

An alternatives assessment is the central component of a chemical action plan. Alternatives assessment is a conventional decision-making process recently updated to meet the needs of businesses, governments, and scientists in selecting or designing alternatives to hazardous chemicals. It provides a framework and a set of tools for assessing the comparative advantage of chemical and nonchemical alternatives, evaluating the functional and performance characteristics of alternatives, and assessing the cost consequences.[17]

Procedures for assessing alternatives are taught in business and industrial design schools. Formally, alternatives assessments are practical examples of a business tool called multicriteria decision analysis (MCDA), which was developed to assist in making decisions among options displaying numerous and conflicting attributes. These indicators may involve measurable attributes such as costs, attributes that can be ranked such as performance, and sometimes nonquantifiable attributes where subjective judgments are necessary.[18] Product designers typically weigh a broad range of factors in the design of products, and design students are taught criteria and metrics for determining a careful balance among performance, marketability, and costs. Until recently, few industrial designers or design schools included considerations of health, environment, or sustainability factors in making such design decisions. Today, designers can use several protocols for selecting among alternative materials and designs that include software programs with built-in databases for sorting through relevant environmental and health impact information.[19]

Various issues make alternatives assessment challenging. The absence of data on health and environment, as well as data on the technical performance and availability of alternatives can lead to problematic uncertainties. Determining what factors to include and exclude and determining the boundaries of impacts can significantly affect the weights and preferences rankings given to the various factors included in the comparisons. If there are not strong decision rules and open and transparent procedures, the subjective judgments made throughout the process may preclude the replication of an alternatives assessment and leave the results open to question. Recognizing these limitations, alternatives assessments tools can still provide significant value in seeking safer chemicals by providing logical decision-making steps, formal procedures for handling large amounts of incompatible data, transparent processes for ensuring confidence, and the opportunity to reduce the probability of regrettable decisions.

The Toxics Use Reduction Institute and the Lowell Center for Sustainable Production at the University of Massachusetts Lowell have been leading centers in the development of alternatives assessment. In 2004, the Lowell Center convened an international workshop on alternatives assessment and, from this workshop, prepared a generic framework for alternatives assessment processes (see figure 12.2). The "Lowell Framework for Alternatives Assessment" provides an "open source" framework that can be used either for making selections among alternatives or as a guide in designing safer chemicals.[20]

The framework is divided into three sections. The foundation section provides an opportunity to clarify the purpose of the assessment and to make transparent the underlying assumptions, decision rules and weighting criteria. The assessment section is divided into two tracks: one for chemists designing new chemicals and the other for chemical users making chemical selection decisions. The final section, the evaluation module, presents a menu of assessment tools such as life cycle assessment,

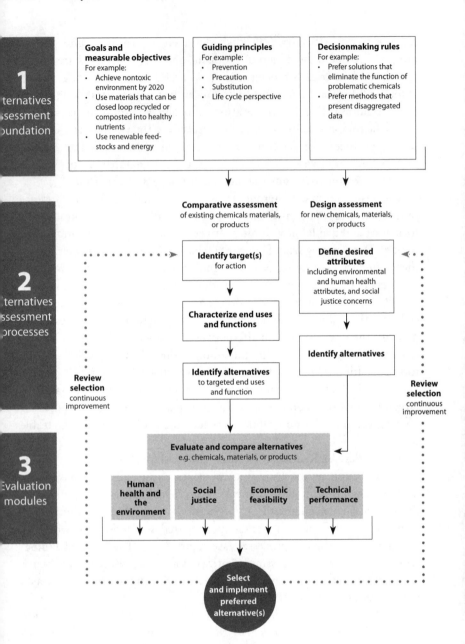

Figure 12.2

Lowell Center Alternative Assessment Framework

comparative hazard assessment, performance assessment, and social impact assessment that can be used to assist in comparing alternatives.

In 2005, the Massachusetts legislature funded the Toxics Use Reduction Institute to complete a one-year alternatives assessment on five chemicals of concern to determine whether there were functionally equivalent and cost comparative safer alternatives available. The legislature identified formaldehyde, lead, perchloroethylene, hexavalent chrome, and di (2 ethylhexyl) phthalate (DEHP) as the five chemicals but left the design of the alternatives assessment up to the Institute.[21]

The Institute refined the alternatives assessment method used in preparing toxics use reduction plans to create a participatory process that would engage a wide range of potential stakeholders, including chemical manufacturers, product manufacturers, suppliers of alternative chemicals, scientists, and health and environmental advocates. Because the targeted chemicals have multiple uses, priority use categories were selected for each substance. Sixteen different use categories were identified for the five chemicals. The range of alternatives included drop in chemicals, changes in materials, changes in production processes, product redesigns, and alternative procedures for achieving the chemical's function. The identified alternatives were then screened to delete carcinogens and other substances that presented serious hazard traits. The remaining alternatives were evaluated in terms of technical performance, financial equivalency, and health and environmental effects. The results were displayed in matrix form with symbols representing positive, negative, or equal attributes. Michael Ellenbecker, the director of the Institute, notes, "Completing five assessments in parallel was a big challenge. Overall, we found at least one functionally equivalent alternative that displayed lower hazard traits for each of the selected uses for each of the chemicals, but not all were cost competitive. For most uses there were several safer alternatives. We made no effort to rank or compare them, although that would have been the next step."[22]

The Toxics Use Reduction Institute procedures for alternatives assessment are organized around a series of steps.[23] These steps can be summarized into a procedure involving seven steps (see figure 12.3).

Step One: Define Goal and Scope of Assessment. The alternatives assessment may be initiated by a firm seeking a safer substitute to a chemical of concern or by a government seeking to identify safer alternatives to a chemical targeted for regulatory attention. The objectives of the initiator need to determine the goals, system boundaries, range of alternatives,

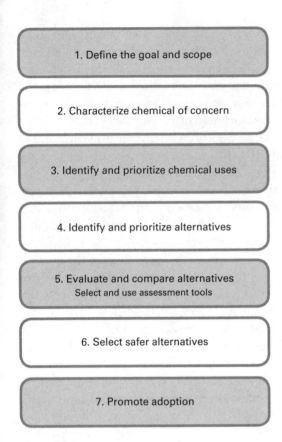

1. Define the goal and scope

2. Characterize chemical of concern

3. Identify and prioritize chemical uses

4. Identify and prioritize alternatives

5. Evaluate and compare alternatives
Select and use assessment tools

6. Select safer alternatives

7. Promote adoption

Figure 12.3

Steps in an Alternative Assessment

decision rules, and value weightings to be used in conducting the assessment.

Step Two: Identify and Characterize Chemical of Concern. The targeted chemical of concern needs to be sufficiently characterized in terms of its physical and chemical characteristics, hazard traits, functional use, performance values, costs, and availability.

Step Three: Identify and Prioritize Functional Uses. If the assessment is initiated by a firm, the functional use may be clearly defined; however, if the assessment is initiated by a government agency seeking alternatives to a hazardous chemical, the uses under consideration may need to be prioritized because there may be many and varied uses of the targeted chemical.

Step Four: Identify and Characterize Alternatives. Alternatives can be discovered by Internet searches, market surveys, interviews with chemical distributors, or reviews of professional and scientific literatures. The potential (chemical and nonchemical) alternatives need to be characterized using attributes similar to the chemical of concern. It is usually useful to prescreen alternatives to narrow the array by rejecting chemicals that are clearly more hazardous than the targeted chemical or technically unacceptable.

Step Five: Evaluate and Compare Alternatives. The evaluation always involves some kind of comparative hazard assessment and a technical performance and cost assessment and may involve a life cycle analysis, an exposure assessment, a materials management assessment, and, possibly, a social impact assessment. Typically the assessment will use both quantitative and qualitative measures.

Step Six: Select Safer Alternatives. The assessment may lead to a single preferred alternative or provide a range of alternatives displayed in a spread sheet, "spider graph," or range of color-scored values to provide information for assisting decision makers in making a final choice.

Step Seven: Promote the Adoption of Safer Alternatives. Once preferred alternatives have been identified, the chemical action planning process can continue.

12.6   Alternatives Assessment Tools

Today, there are several alternatives assessment frameworks. Besides the Lowell Center Framework, there are procedures set out by EPA's Design for Environment Branch, the Biz-NGO Working Group for Safer Chemicals, and the Interstate Chemicals Clearinghouse.[24] All of these include a series of iterative steps that constitute the alternatives assessment process. For each of these steps, there are several assessment tools that can be employed to differentiate the alternatives. The Lowell Center Framework refers to these as modules that can be plugged into an alternatives assessment depending on the goals and context of those conducting the assessment. The guidance on alternatives assessment put out by the Interstate Chemicals Clearinghouse identifies seven such tools. It is not necessary to use all of the tools; the goals and interests of the user will determine which tools to use and how intensely to use them.

Hazard Assessment

All chemical alternatives assessments rely on some form of comparative chemical hazard assessments.[25] The EPA's Design for Environment (DfE) program has an assessment tool that uses a functional use approach to screen among alternatives and to narrow a list down to those most appropriate for substitution. The hazard assessment process involves assigning a value of high, moderate, or low for a suite of hazard end points (hazard traits). The assessment is based on hazard thresholds derived from the Globally Harmonized System of Classification and Labeling of Chemicals (GHS) to identify those points that display significant variation in human and environmental toxicity among alternatives. These end points are considered the "distinguishing" factors for differentiating the "safer" from the "less safe" alternatives.[26]

The *Green Screen* described in chapter 10 is also useful here in comparing various alternatives in terms of their hazard traits. An abbreviated form of the *Green Screen*, the *Green Screen List Translator*, can be used to screen out chemicals of concern and identify those chemicals that are best suited for a full *Green Screen* assessment. Using the full *Green Screen* protocol, the calculated scores for each alternative can be used to rank alternatives in terms of their preference and the rankings can then be used to compare the alternatives. The *Green Screen* has found favorable use in both government and industry, including HP, which has used the *Green Screen* to assess scores of alternatives to chemicals banned by the European Union's RoHS directive and Washington State, which has used the tool to screen alternatives to persistent, bioaccumulative, and toxic (PBTs) chemicals that it has targeted as chemicals of concern.[27]

Performance Assessment

Performance qualities can be readily determined where there are formally established technical standards. However, the depth of technical analysis will depend on the needs of the decision maker. In some cases, a chemical supplier's recommendations or simply recognizing that competitors have already adopted an alternative may be sufficient for a firm to accept the performance value of an alternative. In more sensitive cases, a technical assessment may require a market assessment, a review of technical reports, and laboratory testing. Larger firms may have onsite testing laboratories, whereas smaller firms may seek testing services from university and commercial laboratories. For government policy, a performance assessment may be satisfied by identifying that a safer alternative to a chemical of concern is on the market, being used for the same functional use, and readily available.

There are many conventional tools available for predicting technical performance. At the chemical level, performance variables such as sensitivity, selectivity, robustness, accuracy, precision, and speed may involve tests such as gravimetric or volumetric analysis or mass spectroscopy. At the product level, technical performance tests may involve stress and durability analysis, aging assessment, failure analysis, energy assessments, manufacturability analysis, process safety assessments, regulatory analysis, flammability tests and various corrosion, reactivity, and microbiological tests. The EPA's Cleaner Technologies Substitutes Assessment includes a section on technical assessment that involves a seventeen-step process protocol for evaluating product technical performance and formalizing procedures for testing and comparing results.[28]

Cost and Availability Assessment

Analysis of the initial costs associated with an alternative may include the costs of the chemical and any capital costs for new equipment or production changes. For simple comparisons among alternatives, chemical suppliers can often provide sufficient information. Such costs can be factored out over time using conventional net present value (NPV) analyses based on a discount rate and period of investment. Recognizing that a chemical is available and currently used for a comparable function may provide a sufficient cost and availability assessment for a government policymaker.

However, direct price comparison among chemical substitutes will miss many "hidden costs." Conventional cost accounting procedures that cover the costs of raw materials, energy, and labor often neglect "indirect costs" that arise from waste and environmental management, regulatory compliance, special handling, storage, liability, worker training, and worker health impacts. Therefore, for more fully predicting the costs of a chemical or process change, it is useful to use comprehensive capital budgeting tools. Full Cost Accounting (FCA) expands conventional capital budgeting to include indirect and hidden costs, whereas Life Cycle Costing (LCC) adds a time component by including projected costs of a project from initial research and development to final retirement and disposal.[29]

Exposure Assessment

Exposure assessments evaluate potential exposure scenarios and determine whether alternatives pose significant or greater concerns than the chemical of concern. There are many methods for measuring exposure that provide different guidelines for estimating the magnitude, duration, and frequency of exposure. The EPA has a general guide that involves a screening and an advanced exposure assessment. The screening assessment uses readily available measured data and release and exposure estimates, whereas the more advanced assessment relies on direct sampling. Specific tools for estimating exposures such as E-FAST and the EPI Suite were described in the previous chapter.[30]

Exposure assessments for a "drop-in" chemical substitute might be narrowly defined to assume that the exposure of the chemical of concern and the alternative are similar; however, this assumption would ignore differences in exposures across the life cycle of each chemical. Exposure assessments could be useful here in determining differences among alternatives that do present different exposure scenarios. Risk assessments also can play a role where it is determined that all of the alternatives present significant hazards or there are no acceptable substitutes for a necessary chemical of high concern, and exposure controls may be necessary.

Material Management Assessment

Material management assessments evaluate whether the use of potential alternatives will impact natural resources and the generation of hazardous or nonhazardous waste and contribute to or hinder materials recycling and reuse. Assessing material management is based on material flow analysis, a technique for measuring the stocks and flows of materials in a system based on an input–output analysis or a simple mass balance. Material flow analysis starts with a fixed unit like a production facility, an economic sector, or a geographical region and identifies the material flows in terms of volumes and rates such that inputs, outputs, and residuals balance.[31]

In terms of chemical alternatives assessment, a material assessment is broadened to consider the various impacts that potential alternatives would have on the chemical production and disposal systems in terms of resource use, wastes and risks. Where the systems of production are highly integrated as in the production of petrochemicals, the assessment would consider the "embedded hazard characteristics" in the potentially impacted branches of chemical production. At the chemical level, this might involve analysis of yield efficiencies, chemical displacements, and net risk increases or decreases, whereas at the product level, this could involve any of several measures of waste and recycling per unit of product. Such an assessment might also consider the source of materials (renewable or nonrenewable) and the characteristics of the wastes in terms of their own degree of hazard.

Life Cycle Assessment

Conceptually, a life cycle assessment (LCA) involves four components: goal and scope definition, inventory analysis, impact assessment, and interpretation.[32] During the goal and scope definition, a functional unit is defined and bounded. The inventory analysis identifies the range of environmental impacts for each stage of the life cycle identified in terms of energy and resource consumption and waste generation, and the impact assessment then characterizes the functional unit in terms of its impacts or compares the impacts of different options. The actual calculation of impacts is assisted today by fairly elaborate LCA software packages, which are available from several private vendors. However, because of the range of potential impacts and the need to collect large amounts of data, a full LCA is often time consuming and expensive.

To reduce the resource burden of completing full LCAs, efforts have been made to create "streamlined LCAs" that rely on reduced levels of scale, depth and detail, or on qualitative "life cycle thinking" or "high-medium-low" rankings rather than quantitative measures. Whereas full LCAs may be too resource intensive for chemical alternative assessments, these more streamlined LCAs may provide sufficient differentiation among alternatives in terms of energy, resource, and waste impacts.[33]

Social Impact Assessment

Modeled on the conventionally accepted procedures of environmental impact assessment, social impact assessment has largely been developed and used to address the social, economic, demographic, cultural, and sometimes political impacts of international development projects. In 1993, an interorganizational committee of social scientists set out a framework composed of six broad principles and guidelines for conducting social impact assessments ranging from how to collect information to how to mitigate negative impacts.[34]

Social Life Cycle Assessment (S-LCA) has been proposed as a compliment to LCA to address the social impacts along the life cycle of a process or product. Guidelines for S-LCA are now under development to address the social conditions of workers, consumers, and local communities; however, these procedures are only recently developing, and there has been little field or empirical testing. Although it would be valuable to assess the social and economic consequences of shifting to safer chemistries, appropriate tools are just emerging, and to date they have not been adapted to the needs of safer chemical policies.[35]

12.7   Promoting Substitution

Chemical action planning and alternatives assessment are yet under development, and there has been no rush to include them in government policy. During negotiations over REACH in Europe, there was a strong push from the civil society advocacy community to require alternatives assessment (referred to as substitution planning). This was broadly resisted by industry, but some elements have been introduced into the implementation process for authorization. Applicants for authorization of substances of very high concern must provide an "analysis of alternatives" that considers their risks, the technical and economic feasibility of substitutes, and, in some cases, various socioeconomic factors. With less controversy, a formalized alternatives assessment procedure has been introduced into the proceedings of the Stockholm Convention on Persistent Organic Pollutants. Parties objecting to the inclusion of a chemical as a candidate for a global ban because there are no acceptable alternatives are encouraged to use a five-step assessment protocol to document their argument.[36]

Maine and the State of Washington can require alternatives assessment for priority chemicals in their new chemical policies. However, California may be leading the way. In 2008, the legislature passed a law on safer consumer products (AB 1789) that requires the Department of Toxic Substances Control to establish a process for identifying chemicals of concern and requiring that manufacturers of products containing those chemicals conduct an alternatives assessment. A parallel law passed at the same time required the state Office of Environmental Health Hazard Assessment to define the set of hazard trait criteria noted in chapter 10. Using these criteria, the Department prepared an initial list of some 2,300 candidate chemicals of concern, screened it down to 248, and began to identify priority products that contain one or more of those chemicals. Once a priority product is announced, the manufacturers of that product must prepare an alternatives assessment that includes a life cycle analysis to identify chemicals or nonchemical changes that could replace the targeted chemicals of concern. Based on this alternatives assessment, the Department then selects one or several regulatory responses that range from product labeling and end of life management requirements to a product sales prohibition. In 2013 the Department posted three priority chemicals in priority products: unreacted diisocyanates in spray polyurethane foam; Tris(1,3-dichloro-2-propyl) phosphate in children's foam padded mattresses; and methylene chloride in paint and varnish strippers. Debbie Raphael, who recently stepped down as director of the Department, is highly optimistic: "California is taking a leadership role in the movement to make products safer for consumers. I expect the lessons we learn from implementing this groundbreaking regulation will inform the future actions of governments and private companies in the adoption of safer alternatives."[37]

More recently, the National Academy of Sciences (NAS) has completed a selection guide for chemical alternatives that reviews the various alternative assessment frameworks and proposes a thirteen-step procedure that is an expanded version of the seven-step framework described above. The framework elevates exposure assessment, expands the range of physiochemical properties to be considered, and champions the use of data from high-throughput toxicity testing and computational analysis.[38]

Chemical action planning and alternatives assessment provide the ground-level mechanics for a higher level chemical conversion strategy. While leading corporations may invest in such substitution planning methods, it will take many companies working on many chemicals to achieve a scale large enough to shift the chemical production and consumption system. To date, alternatives assessments have largely been completed by individual companies or the EPA. There would be greater value if groups of firms or organizations within specific economic sectors began using these tools for the benefit of multiple companies or whole economic sectors. Where several firms in one sector or a collection of firms connected through a common supply chain work together to seek safer alternatives, they can more efficiently identify safer alternatives and send a strong message to chemical manufacturers and suppliers that these alternatives are needed. This is where a precautionary approach can be a driver for industry-wide innovation. As Joel Tickner from the University of Massachusetts Lowell explains, "The "look before you leap" orientation and the practice of seeking the safest option make alternatives assessment the operational embodiment of the Precautionary Principle."[39]

To ensure more successful substitutions, a planning process such as the toxics use reduction planning process and an alternatives assessment that cover hazards, life cycles, technical performance, exposures, costs, material impacts, and (possibly) social impacts provide the procedural mechanics. But this will not be enough if there are not continuously emerging a safer and, then, safer range of chemical and nonchemical alternatives. Here we turn to consider the innovative opportunities for green chemistry and engineering in broadening the chemical market and refocusing the chemical industry.

13

Developing Safer Alternatives

After Hurricane Katrina stuck New Orleans in 2005, thousands of the city's residents were left homeless. The Federal Emergency Management Agency brought in hundreds of trailers to provide temporary accommodations. Months after the trailers had been occupied, the Sierra Club released a study that showed exceptionally high concentrations of formaldehyde in the trailer's indoor air. The investigation that followed found that the source of the formaldehyde was off-gassing from the mastic used in the trailer's interior plywood panels.

This should not have been a surprise; formaldehyde-based glues are common in plywood panels. During the production of plywood panels, the wood layers are bonded together under high heat and pressure using urea-formaldehyde (for indoor use) and phenol-formaldehyde (for outdoor use) as an adhesive. Formaldehyde is used because it is tough, strong, and inexpensive. However, formaldehyde is classified by both the EPA and International Agency for Research on Cancer (IARC) as a probable human carcinogen.

Recognizing the rising public concern over formaldehyde in plywood, Columbia Forest Products enlisted Kaichang Li from the Department of Wood Science and Engineering at the University of Oregon to develop a formaldehyde-free adhesive for binding together the layers of its plywood panels. Professor Li had been studying the way that marine mussels stay tightly gripped to rocks in the midst of pounding ocean currents. He found that mussels secrete proteins known as byssal threads, which could also be developed by modifying the amino acids in soy proteins. Collaboration with Hercules, Incorporated provided a proprietary curing agent and led to the development of PureBond™, a formaldehyde-free adhesive derived from food-grade soy flour. By using this adhesive as a binder, Columbia Forest Products is able to market plywood that earns LEED green building credits and can be certified by the Forest Stewardship

Council. In 2007, the company received a Presidential Green Chemistry Award for its soy-based adhesive. In its award announcement, the EPA noted that the adhesive had been used to replace more than 47 million pounds of the firm's conventional formaldehyde-based resins. This useful chemical invention inspired by a natural process suggests the many opportunities for safer alternatives innovations and makes up the fourth building block of a safer chemical policy framework.[1]

13.1   Chemical Invention

The invention of new chemicals is one of the driving forces of chemistry. Papers are written, awards are presented, and careers are advanced when chemists invent new and useful chemicals. Chemical invention involves the creation of a new substance, compound, mixture, or chemical synthesis pathway. The term *chemical invention* differs from the term *chemical innovation*. Chemical innovation refers to the acceptance, adoption, or use of a chemical within a new application. The development of safer chemicals requires both activities.

There are several routes for chemical invention. Some new chemicals emerge as a response to a challenge to find a substance that fits a clear social need. The invention of the first thermoplastic, Celluloid, during the 1860s provides such an example. With a publicly defined need for a non-ivory billiard ball, a competition was announced for a substance that met well-defined cost and performance parameters. The winning nitrocellulose-based polymer provides an example of a "need-driven invention." Other inventions arise from a desire to do something with a chemical resource. Many chemical inventions emerge as chemists seek to develop commercially viable chemicals from the residuals of coal or oil distillation or other chemical production processes. The development of acrylonitrile from surplus propylene is an example of a "supply-driven invention." Supply-driven inventions have been one of the great hallmarks of the petrochemical revolution and have been one of the keys to the development of the highly integrated structure of their production systems. Safer chemical inventions do arise from supply-driven inventions, but most such inventions will be driven by the need to replace chemicals of high concern.

The development of alkaline copper quaternary (ACQ) provides a recent example of a need-driven invention. It was noted in chapter 12 that ACQ was developed to replace chromated copper arsenate (CCA) as a wood preservative in pressure-treated construction and landscape lumber.

The pressure-treated wood industry is a $4 billion industry producing more than 7 billion board feet of degradation-resistant wood each year. In 2001, more than 95 percent of pressure-treated wood used in the United States was treated with CCA. The use of pressure-treated wood widely dissipates arsenic and chromium across the environment. Arsenic and chromium present both human and aquatic hazards; therefore, there was a significant need to find a safer alternative to CCA. The development of ACQ by Chemical Specialties, Incorporated during the 1990s was one of the most significant chemical substitutions of the decade. Because more than 90 percent of the 44 million pounds of arsenic used in the United States each year goes into CCA treatment, the switch to ACQ promises to nearly eliminate the use of arsenic in the United States. In addition, the conversion to ACQ is expected to cut nearly 64 million pounds of chromium (VI) use per year.[2]

Government agencies, corporations, universities, and advocacy organizations could spur the development of these chemical inventions by periodically listing chemical applications where new, safer chemicals or synthetic routes are needed. Monetary awards could provide needed incentives. The federal government has several such "challenge" programs. The National Aeronautics and Space Administration (NASA) does this regularly. Recently, NASA has posted a challenge for the design of a returnable space robot and has completed another challenge for a highly energy-efficient, small, fast aircraft.[3]

Awards and public recognition can also play a role in promoting chemical invention. The EPA's Presidential Green Chemistry Challenge provides well-publicized annual awards. These awards are presented with much celebration at a reception at the National Academy of Sciences, followed by a national green chemistry and engineering conference sponsored by the American Chemicals Society's Green Chemistry Institute. The winners announced in 2013 included an academic award for nontoxic, biobased adhesives and composites, a small business award for the use of less toxic trivalent chromium replacement for hexavalent chromium in metal plating, a synthetic pathway award for a less toxic, low-waste synthesis process for making deoxyribonucleotide triphosphates used in genetic screening, a reaction award for a dispersive polymer that reduces the use of titanium dioxide in paints, and a safer chemical award for a biobased dielectric fluid for use as an insulator in transformers.[4] Over an 18-year period, the EPA has received nearly 1,500 award nominations and posted ninety-three awards. In a follow-up assessment, the agency has estimated that the awards alone reduced the use of 826 million pounds of hazardous chemicals, conserved 21 billion gallons of water, and reduced 7.8 billion pounds of carbon dioxide equivalents. Although these awards are given to applicants who have ready inventions, it would not be a big step for the EPA or a credible body such as the American Chemical Society (ACS) to announce chemical categories or problem substitutions where new chemical inventions are needed.[5]

13.2   Novel Chemicals

Even a quick review of current chemistry journals reveals hundreds of examples of chemical inventions. Biochemistry, biotechnology, and, more recently, nanotechnology have opened up new frontiers in chemistry and generated new chemical processes and compounds. Biochemistry that focuses on the makeup of proteins, carbohydrates, lipids, and nucleic acids lays the foundation for new biologic chemistries. Biotechnology has turned this knowledge to commercial use. Drugs for treating diabetes, arthritis, hepatitis, cancer, and infertility have all been developed through biotechnological processes. Commercial applications of biotechnology have also been important in producing biofuels, degradable plastics, and biobased oils and lubricants. Genetic engineering has developed novel cell lines with wide applications in pharmaceuticals and specialty chemicals. Biochemical processes based on genetic modifications have generated agricultural products for increasing crop yields, extending growth periods, and improving the taste, texture, and storage life of foods. The genetic modification of enzymes has produced a new class of catalysts such as subtilism, an enzyme used in detergents, which has been genetically modified to improve its performance with bleach.

Nanotechnology involves the design, production, and application of chemicals at the nanoscale. It is now possible to engineer novel nanoporous and nanocrystaline structures at the molecular level that greatly increase the capacity for catalytic reactions, structural performance, and energy efficiency. The applications are numerous across product categories. Nanomaterials such as nanotubes, nanowires, fullerines, and quantum dots have created better sensors, optics, electron transfer, microprocessors, magnets, coatings, and visual displays. Carbon nanotubes provide substantially improved strength characteristics for structural polymers used in tools and construction. The increased surface-to-volume ratios of nanomaterials present greater opportunities to capture sunlight and could soon develop thin film polymers based on fullerenes and titanium dioxide to replace silicon in solar cells.

The Project on Emerging Nanotechnologies lists some 1,600 consumer products or product lines manufactured with nanomaterials, with three to four new products entering commerce every week. These range from sunscreens and cosmetics to textile coatings, paints, varnishes, disinfectants, electronic products, and household appliances. The most common chemical elements used in these products include silver, carbon, silicon, titanium, zinc, and gold.[6]

However, both bioengineered and engineered nanosubstances present derived hazards—properties that impart the desirable reactivity but result in the harmful characteristics. The genetically modified cell lines of biotechnology provide both the beneficial performance and worrisome attributes of modified crops and pesticides and they have been actively resisted by the public in Europe. There appears to be less concern over genetic alteration of industrial enzymes and the use of recombinant cells in the production of drugs. However, genetically altered microorganisms can still create hazards in labs or in dissipative applications such as the treatment of wastes or the leaching of minerals. While genetically modified cell lines are designed to degrade outside industrial production processes, it is possible that some bioengineered substances may be released to the environment long enough to result in ecological effects. For all of their potential utility, nanoscaled particles appear to present a wide range of new hazards that cannot be easily generalized. Their greater specific surface area (surface area per unit of weight) leads to increased rates of absorption through the skin, lungs, and digestive tract. Early studies suggest that exposure to some nanomaterials can lead to adverse effects in the brain and lungs of test animals. Nanofibers appear to behave much like hazardous asbestos fibers. Although few health effects studies have been conducted on nanosilver and silver is a recognized aquatic toxin, nanosilver has been given a conditional registration as an antimicrobial pesticide by the EPA. Indeed, little study has been done on the environmental fate and transport of most of these nanomaterials.[7]

These new biochemistries and nanoscaled chemistries could be laying the foundations for safer chemicals. Advocates of biotechnology argue that genetically altered enzymes reduce the need for dangerous production catalysts, and genetically modified pest-resistant crops require less pesticide applications. In 2006, the European Parliament sponsored a project on the role of nanotechnologies in chemical substitution, noting a series of existing or proposed products that use nanomaterials to replace hazardous chemicals in areas such as coatings, flame retardants, plasticizers, catalysts, and solvents. About the same time, the U.S. EPA held a national conference on the "environmental benefits of nanotechnology." Since then, a dialogue about "green nano" has emerged that focuses on the many potential weight-reducing, fuel-enhancing, and pollution-reducing applications of nanotechnologies.[8]

However, these novel chemicals should not be deemed safer based on the benefits of their potential applications. The unintended or embedded hazard characteristics need to be addressed. Some potentially instructive efforts are being made. One strategy has been to adapt the Twelve Principles of Green Chemistry to nanosynthesis. Another strategy involves avoiding hazardous feedstocks (e.g., silver, cadmium, titanium dioxide, copper) and metal precursors and replacing acutely toxic reagents and hazardous solvents in nanotechnology production. Natural compounds, such as starches, sugars, proteins, and ascorbic acid, have been successfully introduced as safer alternatives for reducing agents and nanoparticle coatings.[9]

If biochemistries and nanotechnology are to develop products and processes that are both safe and functional, then the research programs that support their development need to link these two objectives. Both biotechnology and nanotechnology receive substantial federal encouragement and support. One biotechnology research program, the Therapeutic Discovery Project, distributed more than $1 billion per year in research grants. In addition, the government annually pumps some $2 billion in research support through the National Nanotechnology Initiative. This is big money, and it could be directed toward developing these novel chemicals as safer chemicals. To date, it has not.

13.3　Safer Chemicals

But what then do we mean by safer chemicals? A tentative definition was provided in chapter 5 by noting that for a chemical to be considered safer than a chemical of concern, it must be relatively safer than the chemical of concern in terms of a baseline of factors, including:

• its inherent toxicological profile;

• its physical characteristics in terms of acute hazards, flammability, corrosiveness, and explosive potential;

• its potential for persistence and accumulation in organisms and food chains; and

- its metabolic, ecologically transformative, and degradation characteristics.

In addition, it is useful to consider other (life cycle) factors that may directly affect the hazards of a chemical, including;

- its fate and transport in the environment;

- the characteristics of its synthetic production processes and the intermediary chemicals used in those processes; and

- its energy and resource consumption during production, use, and disposal.

There are three important things to note about this definition. First, there is always a baseline. A safer chemical is not an absolute concept—it is always defined relative to a chemical or condition that is more hazardous and for which it may be a preferred substitute. Second, there are always tradeoffs. While an alternatives assessment may reveal tradeoffs among cost, performance, and safety, even within the narrower concept of chemical safety, there are typically tradeoffs among various health end points and hazard traits. One chemical may pose no chronic health effects but may be highly flammable or explosive, whereas another chemical may present no carcinogenicity but be a well-recognized reproductive or hematological toxin. Third, there is always room for improvement. The search for a safer chemical is a continuous process where possibilities always exist that a safer chemical of the future may someday replace what is considered a safer chemical today.

In working to develop safer chemicals, it is useful to have both rules and principles as guides. Rules like regulations bound the chemical design space, restricting the development of certain chemicals or the use of those chemicals in chemical synthesis. Principles are more flexible, providing guidance inside the design space for preferred outcomes or avenues of development. Rules create hard boundaries, whereas principles provide for more flexible assessments that include comparisons among options and trade-offs among attributes. Rules are common in chemical regulations in both the United States and European countries; however, European counties are more likely to invoke principles in chemical policies (e.g., Substitution Principle, Precautionary Principle, Polluter Pays Principle). The prohibition strategy and the aversion strategy described in the previous chapter provide useful rules for chemical design, whereas the twelve green chemistry principles provide principles. Using rules and principles, it could be possible to develop a protocol for designing safer

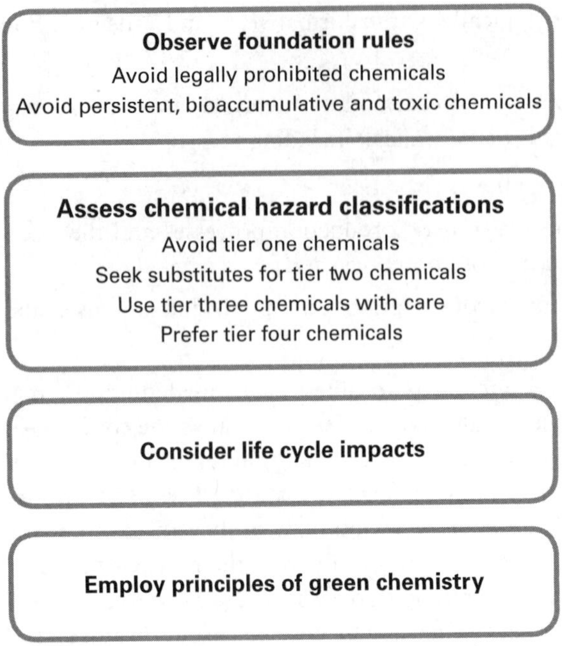

Observe foundation rules
Avoid legally prohibited chemicals
Avoid persistent, bioaccumulative and toxic chemicals

Assess chemical hazard classifications
Avoid tier one chemicals
Seek substitutes for tier two chemicals
Use tier three chemicals with care
Prefer tier four chemicals

Consider life cycle impacts

Employ principles of green chemistry

Figure 13.1

Steps to Developing Safer Chemicals

chemicals that would be similar to the priority setting protocol and the alternatives assessment protocol described earlier. Figure 13.1 suggests such a protocol with a series of steps or screens that could help guide chemists and others in designing safer chemicals.

It is useful to establish a foundation—fundamental rules that are basic to all safer chemicals. A prohibition strategy for avoiding legally banned chemicals is self-evident. The Montreal Protocol and Stockholm Convention provide global instruments for ensuring international prohibitions on the manufacture and use of chemicals that are internationally recognized as so hazardous that complete elimination is warranted. The European Union and many national governments have also used state authority to prohibit the manufacture and sale of chemical substances of very high concern. An aversion strategy is also useful here. There is a growing consensus that PBTs are worth avoiding even where they have strong market value. The EPA maintains a list of PBTs, as does the European Union. Indeed, the Scandinavian counties consider the two characteristics of persistence and bioaccumulation—two attributes for which there are well-recognized tests and measures—to be enough to place a chemical on an avoidance list even without addressing its relative toxicological characteristics, which are often more difficult to determine with broad consensus.

Screening out these two categories of chemicals leaves a wide array of chemicals to assess. Comparative chemical hazard assessment tools like the *Green Screen* could be used here; however, if the Universal Classification of Chemical Substances were populated with chemicals, it could provide a framework for setting appropriate precautionary principles. First-tier chemicals are not considered safe and should be avoided. Higher tier chemicals should be used with decreasing levels of caution in keeping with the progression of the tiers. Second-tier chemicals are safer and can be used but mostly where appropriate substitutes are unavailable. Third-tier chemicals are even safer and can be used but with suitable precautions. Fourth-tier chemicals are safest and should be used where possible. Where chemicals are not classified because of uncertainty, lack of consensus, or lack of sufficient information, chemical hazard assessments that provide well-defined guidelines for dealing with such data gaps could be used to benchmark chemicals as to their degree of hazard.

Once a chemical has been classified in terms of its hazardous properties, a review of its life cycle impacts should be conducted to recognize its embedded hazards. A chemical's life cycle energy and water consumption, greenhouse gas emission, environmental fate and transport, and waste generation should be considered from the feedstock source to its end-of-life management. Chemicals that require large energy inputs to produce or whose production generates proportionately large volumes of waste are less safe than lower energy input, lower waste output chemicals. Chemicals that are likely to cause hazardous conditions when they are recycled or disposed are less safe than those that do not. This need not be a full life cycle assessment—a life cycle inventory or simple life cycle thinking might be enough to note the presence or absence of areas of concern.

The remaining step is to consider the potential routes for synthesis and processing. Here the Twelve Principles provide a well-established guidance framework. Preventing wastes, achieving atom efficiency, avoiding auxiliary substances, minimizing derivitizations, using renewable source materials, and reducing chemical accidents—these are all principles for safer chemical synthesis.

13.4   Safer Chemistry Design Tools

With so many different factors to consider, a chemist setting out to synthesize safer chemicals could use some decision assistance. There are a range of measurement indicators and metrics for the quantifiable aspects of green chemistry. Reaction efficiency metrics that compare the actual yield against the ideal yield are the simplest. Atom economy and the "E-factor" can be used to measure yield efficiency. However, these are efficiency metrics; they do not account for the degree of hazard.[10]

The EPA has developed a set of tools that can be used to identify degrees of hazard. During the 1990s, the agency developed an interactive software program called the Green Chemistry Expert System (GCES) that can be used to select green chemicals and reactions. The GCES includes an interactive tool designed to quantify and categorize hazardous chemicals used or generated in a chemical production process as well as several information modules and searchable databases.[11] To help new chemical applicants preparing a Pre-manufacture Notification (PMN) under TSCA, the EPA developed the Synthetic Methodology Assessment for Reduction Techniques (SMART) to identify safer chemical selection and pollution prevention opportunities. To use the SMART module, a user enters the type of reaction plus, for each reactant, the chemical identification, the role of the chemical in the reaction, and the quantity to be used. Using an internal algorithm, SMART then calculates the amount of waste generated categorized into four hazard categories based on EPA criteria. The program provides a qualitative indication of the level of concern raised by the reaction based on the hazard classification of the reaction chemicals and amount of waste generated. Where the level of concern is high, SMART directs the user to various GCES databases such as Green Synthetic Reactions modules and Designing Safer Chemicals modules. The first presents alternative chemical processes for replacing a more hazardous chemical with a less hazardous one, whereas the second provides qualitative information regarding the toxicity of a compound within certain chemical classes, the mechanisms of its toxicity, and suggestions on structural modifications that could reduce the toxicity.

To wrap all this up, the EPA developed the Sustainable Futures Program to encourage companies to screen chemicals for potential hazards at the research and development stage. During the 1990s, the EPA packaged together several of its chemical assessment methods and made them available in order to encourage safer new chemical PMN applications. Eastman Kodak and PPG Industries experimented with the package and delivered positive results. Today, the program includes nine of the EPA's screening tools for developing a Sustainable Futures Assessment that can help companies predict and potentially avoid the hazards of a proposed chemical.[12]

New developments in computational toxicology offer an expanded health effects database on chemicals and a rapid screening capacity for assessing the hazards of new molecules. Tiered testing programs based on in vitro assays and in silico profiling combined with structure-activity assessments can provide the data needed for determining the hazard characteristics of chemical profiles. The EPA's ToxCast provides a wealth of publically available information on more than a 1,000 chemicals, while a new toxicological priority index called ToxPi can indicate visually how chemicals score on different tests, however chemists will need to be comfortable with toxicological terms and mechanisms to use these tools effectively.[13]

For example, a group of chemists and others have developed a tiered screening tool for use in chemical synthesis that identifies the likelihood that a new chemical could disrupt the endocrine system. The Tiered Protocol for Endocrine Disruption (TPED) consists of five tiers that include computational assessments, cellular-target assays, and mammalian testing. The first tier uses models of physical properties and chemical reactivity to predict endocrine disrupting potential. Tier 2 uses in vitro screens to test a chemical's capacity to modulate biological signaling pathways. Tier 3 incorporates cell- and tissue-based assays to determine whether those pathways could lead to cell division, differentiation, or death. Animal tests are used in the fourth and fifth tiers to examine the impact of a chemical for multiple end points and estimate potential human responses. The tiered framework allows chemists to enter the screening at the level that appears most appropriate and creates an integrated but flexible protocol that easily integrates with conventional chemical design processes.[14]

13.5   Nonchemical Alternatives

Safer chemicals are not the only solution to chemicals of concern. There are a wide range of safer nonchemical alternatives to hazardous chemicals. Products can be redesigned to eliminate the need for a hazardous chemical. Chemical processes such as degreasing can be avoided by substituting mechanical or physical processes such as air jets and hand wiping. Solid state and radiation-cured surface coatings can eliminate the need for solvent-based paints, and corrosion-protective paints can be avoided by using corrosion-resistant alloys. Household pests can be controlled with architectural barriers. Bothersome weeds can be managed by hand or mechanical weeding.

Similarly, production processes can be redesigned to eliminate dangerous chemicals by changing the speed, conditions, or operating parameters
of production operations. Years of experience in pollution prevention
have shown how entire production steps can be eliminated though better
process management. For example, tighter inventory management of
metal parts can eliminate the need for oiling components and thereby
eliminate the need for the cleaning solvents used to later remove the oil.
Sequencing batch paint mixing from lighter to darker colors can eliminate
the need for vat cleaning between product mixes.

In food and fiber production biological control agents such as naturally occurring fungi, bacteria and viruses can be used as substitutes for
hazardous chemical pesticides. *Bacillus thuringiensis*, neem, garlic, and
parasitic nematodes can be used in small farms as well as kaolin clay and
dormant oil sprays. There are also well-developed changes in agricultural
practices that can reduce the need for herbicides and insecticides. Organic
farming avoids the use of hazardous chemicals by substituting natural
chemicals and safer crop management practices. Integrated Pest Management (IPM) and Integrated Plant Nutrient Management (IPNM) employ
changes in farming practices that discourage pests by using careful field
monitoring and timed or tailored uses of pesticides and fertilizers. Sustainable agriculture that promotes low till agriculture, crop rotation, and
intercropping has been shown to reduce the use of pesticides and increase
agricultural yields.[15]

Substituting safer technologies suggests the application of "green engineering."[16] Green engineering has been proposed in the field of chemical
engineering to address improvements in unit operations and the design of
chemical reactors and separation devices. Process mass integration, process energy integration, and mass exchange networks can be used to convert wastes to feedstocks and optimize energy conservation by balancing
heating and cooling requirements. However, green engineering has been
defined more as a means of reducing the health and environmental effects
of chemical processes, rather than using engineering to develop products
and practices that reduce or eliminate the need for hazardous chemicals.
Simple, elegant engineering design could employ physical, mechanical, or
operational alternatives to products and practices that now require hazardous chemicals. For instance, mechanical fasteners and simple tabs and
inserts can be used instead of adhesives to hold things together, and well-
designed hand pumps can eliminate the need for chemical propellants.

Still, there are often trade-offs. Substituting mechanical for chemical
processes may increase energy requirements or generate ergonomic haz-

ards. Many chemically intensive products have been adopted to reduce physical work (e.g., industrial cleaning agents), decrease physical accidents (e.g., nonskid floor coatings), reduce fires (e.g., flame retardants), increase sanitation (e.g., chlorinated disinfectants), and conserve energy (e.g., isocyanate foam insulation). Before presenting nonchemical alternatives as replacements for such functions, it is important that they be evaluated carefully as part of an alternatives assessment.

13.6   Safer Alternatives Innovation

It is not enough to invent safer alternatives to chemicals of concern—innovation in the chemical economy requires that these alternatives be adopted at a scale necessary to achieve market conversion. This is a common challenge in all technological innovation. The redesign of a product or conversion of a process toward safer alternatives can be promoted by a range of drivers, including customer demand, market opportunities, dynamics within the supply chain, pressures from advocacy organizations, concern for worker safety, liability and litigation concerns, corporate public relations, and government regulations. However, there are also a range of barriers, including ignorance of alternatives, "up front" costs, inadequate payback, concern about product performance, market risks, management "conservatism," and lack of public, supply chain, or regulatory pressure. Different firms within the same sector will respond differently to these drivers and barriers. Some will adopt safer alternatives right off, whereas others will wait to consider the experience of the early adopters, and still others will lag and resist until well after the substitution has been widely accepted.

Everett Rogers and others who have studied the diffusion of innovation have documented these common patterns of behavior.[17] If the cumulative number of adopters is plotted over time, there appears an S-shaped curve with only a few adopters in the early period, then a rush of adopters until a significant number of the potential adopters have adopted the innovation and then a tailing off as the laggards acquiesce. The shift away from petroleum-based to soy-based inks in newspaper printing during the 1980s roughly followed this pattern.

The willingness to adopt innovations varies significantly across the chemical industry. As noted earlier, those producing basic chemicals are tied up in heavy capital investments and low profit margins. There is room for innovation in plant design in the big new refineries and processing plants being built in the Near East and Asia, but the older plants in

Europe and the United States offer limited opportunities. Some firms in the consumer product sector of the chemical industry that place a high value on innovation are adopting safer chemical inventions to meet customer demands, whereas others are tied to highly profitable brand name products with large market shares and are loath to embrace innovation. In the specialty chemical business, this orientation toward innovation is heavily influenced by the business model of the company. The customer-oriented specialty chemical companies are among the first to adopt new innovations, whereas the low-cost providers will be among the last. Those who embrace an "excellence in operations" model will change but only after much analysis and piloting.

However, Rogers goes on to note the importance of factors exogenous to the technology and management structure, such as diffusion networks and change agents that can act as catalysts, technical assistants, and facilitators. A classic study of physicians' adoption of tetracycline found that close connections among influential doctors resulted in a kind of "chain reaction" diffusion of the drug, and another study of farmers' acceptance of hybrid corn documents the critical role played by state agricultural extension agents in facilitating the adoption of the new corn.[18]

These social factors are often missing in advancing chemical innovation. The communication links among academic green chemists, chemical manufacturers, and influential product manufacturers are not well developed, and there are no green chemistry extension agents. What would be useful in advancing green chemistry invention and safer chemical innovation is the convening of professional networks of green chemists and more formally constructed "communities of practice" that could provide a means of interaction and dialogue among chemistry innovators and those likely to promote or adopt safer chemicals. Add to this a collection of "centers" (the current university centers and more) where green chemistry practice could expand on the Twelve Principles with more detailed corollaries, system changing experiments, and inventories of green chemistry successes and failures.

These linkages are emerging. The Green Chemistry and Commerce Council has launched an effort to build better communication among retailers, product manufacturers, and chemical suppliers. Green Center Canada, a government-initiated, interdisciplinary center that evaluates green chemistry proposals from academic researchers and links them up with funding and industrial partners, provides an attractive model. Washington State has just launched its own Northwest Green Chemistry Center with similar objectives. Two journals, *Green Chemistry* and *Green*

Chemistry Letters and Reviews, provide a resource for this dialogue, and there are increasing numbers of green chemistry blogs and "listservs" on the Internet. The American Chemical Society (ACS)'s Green Chemistry Institute sponsors the annual Green Chemistry and Engineering Conference, and plays a leading role in promoting such collaborations.

Just as funding from the National Nanotechnology Initiative drew researchers from around the nation to nanotechnology, a similar Green Chemistry and Engineering Initiative could provide a focus and funding for a much broader interest in green chemistry among mainstream scientists. Such a program might provide extramural funding for university research, conferences, and workshops for scientists and professionals and stipends for teachers and students. Funding might also support a few regional centers that could act as incubators and "hubs" for growing constructive networks and learning environments among chemists, health and environmental scientists, business professionals, teachers, and advocates. Substantial funding programs for research in chemistry and the material sciences already exist at the National Science Foundation and the Departments of Energy, Agriculture, Commerce, and Defense. The Department of Energy's National Research Labs might serve as such hubs. For several years, the Sandia National Laboratories hosted a useful internal research program on "environmentally conscious manufacturing" that could serve as a model.

13.7   The Business of Safer Alternatives

Innovation in the chemical market requires that safer alternatives be marketed, distributed, adopted, and valued. There needs to be an expanding market for safer chemicals and entrepreneurs eager to take green chemistry laboratory inventions, scale them up, and find the investors necessary to support new businesses or business units that can bring to market safer chemicals and products made with safer chemicals. However, the market offers a difficult supply and demand problem. Suppliers are reluctant to offer "greener products" if there is no evidence of demand, and consumers will not purchase such products if none is on the market. "A retailer can play an important part in educating consumers and specifying safer products from suppliers," Roger McFadden from Staples notes, "but safer products need to be there. We can't sell it, if no one makes it or puts it on the market."[19]

This is the same classic problem faced by any new product seeking to enter the market. The problem involves both market scale and product turnover. Conventional products with large market shares tend to persist in the market and reduce the market space for new entrants. Laundry detergents provide a good example. The market is dominated by a few old name products with significant brand loyalty. There are a host of new products touting superior environmental attributes, but they remain niche products because they cannot reach the scale necessary to win over customers with long-held purchasing behaviors. Scale here is often determined by the profitability of a chemical or product. Whereas green chemistry processes may lower the costs and improve the yield of chemical production, a safer, lower cost chemical may still not compete well against a higher cost but highly profitable chemical.

However, promising chemical inventions need not compete head to head with dominant technologies. Safer alternatives to chemicals of concern might best be built in niche and marginal markets, where they can develop, stumble, learn, grow, and achieve a level of experience and maturity that provides a solid base for larger ambitions. Indeed, there are many small- and medium-sized firms out there generating products intended for a specific safer alternatives market. Some are starting with renewable resources and biomass feedstocks. Others are working on chemicals with greener chemical synthesis processes, whereas still others are struggling to market products with less hazardous constituents. Table 13.1 suggests a sample of such firms.

These firms and many others that are just starting up are seeking to compete in markets not dominated by traditional products or to build new markets that go beyond the markets that traditional products satisfy. They need seasoned management and venture capital. They need the resources necessary to carry chemical inventions though piloting and prototyping to full-scale production. Some of these firms may be betting that good ideas and solid inventions will attract larger firms and investment companies interested in acquiring their intellectual property, process technologies, management capacity, and market share. The tendency for major chemical suppliers and brand product manufactures to forgo internal research and development means that if they are to acquire chemical inventions, then they need to look to licensing, mergers, or outright corporate purchases. Safer alternatives innovation then means growing chemical inventions attractive enough for both market growth and investor acquisition.

13.8   Getting to Scale

Promoting safer chemical invention and safer nonchemical alternatives and supporting the adoption of these innovations will be slow and far

Table 13.1
Businesses Making Safer Chemical Products

Company	Product	Characteristics
Segentis	Biobased plasticizers	Phthalate-free plasticizers compatible with polylactic acid and polyhydroxyalkanoate polymers
Metabolix	Biobased gamma-butyrolactone and butanediol	Chemical intermediates used in the production of resins, fibers, solvents, personal care products
SoyClean	Soy-based lubricants, greases, cleaners	Primarily derived from soy and citrus, vegetable and seed oils
Air Products	Tomadol™	Nonylphenol ethoxylates-free surfactants made from palm or coconut oil
Soy Technologies	Soyanol™	Ready-to-use formulations for cosmetics, personal care products, paints and coatings
Allylix	Renewable specialty chemicals	Terpenes and derivatives for crop protection, biocides, flavors, fragrances, and pharmaceuticals
SyntheZyme	Biobased ω-hydroxyfatty acid monomers	Polyhydroxyalkanoate polymers and biosurfactants
P2 Science	Unspecified green chemistry products	Company in formation

from scaled to have major impacts on the chemical economy without bold and ambitious steps.[20] These include:

Develop more interdisciplinary research and development initiatives. More university- and industry-sponsored research and development centers are needed. The DFE Formulators Program and the Green Chemistry and Commerce Council provide cross-sector collaborations, but what is now needed is well-funded sector-based or regional research centers that can support and catalyze longer term research and development programs.

Integrate green and safer chemistry into early product development. There is a growing field of sustainable product design with several new academic centers and training programs. A few corporations have led in integrating safer chemistry into product design, but the scale remains small. The Green Chemistry Pharmaceutical Roundtable has worked to consciously assist firms in integrating green chemistry into drug development. Such programs are needed in other economic sectors.

Develop metrics, tools, and methods for measuring safer alternatives development. "What gets measured gets managed" is an old management aphorism. The rules and principles of safer chemical development provide a basis for various measurement and comparison tools. It would be helpful to develop better metrics for measuring the energy, water, and resource demands of various chemicals over their life cycles. The *Green Screen* and other decision assisting tools need further development and adoption by governments, corporations, and educational institutions.

Aggregate market demand for safer chemical and nonchemical alternatives. The market for safer alternatives needs to be built by consciously coordinated purchasing initiatives. The vulnerability of government procurement programs that specify safer chemicals has been described earlier. However, large-scale institutional and retail purchasing has more independence. At the moment, it is too disorganized and ad hoc to send common and clear market signals.

Provide sufficient research funding and provide for safer alternatives venture development. Chemical manufacturing firms need to reinvest in robust research and development capacities. Chemical suppliers and product manufacturers need to provide funding for internal research and development. Government funding is needed for academic research, and venture capital is needed for entrepreneurs seeking to pilot and scale up production of safer alternatives.

Incorporate green chemistry into academic science and chemistry education. The development of green chemistry curriculum, training modules, and problem sets has been on a steady incline in the United States, however, it is still small. The Green Chemistry Commitment described in chapter 9 offers a significant step to document and promote the growth of green chemistry education.

13.9   An Innovative Safer Alternatives Market

A 2011 study completed by economists at the University of Massachusetts Amherst argues that the exporting of chemical production by U.S.

firms, the low level of R&D investment, and the weak regulatory environment in the United States work against the domestic chemical industry. Focusing this industry on safer and greener chemicals could create significant economic benefits. One way to document this is to look at the growth of firms and production lines that are self-consciously producing green chemical replacements for conventional chemicals, lower waste production technologies, and chemicals from renewable resources. Pike Research, a leading market research firm, projects that this market segment, which in 2011 had annual revenues around $2.8 billion, will grow to $98.5 billion by 2020—an impressive rate of growth. However, given that the global chemical industry should reach $5.3 trillion in annual revenues by 2020, this market segment is still small.[21]

Tracking investments in new green chemical plants and production lines only represents a part of the conversion of the chemical economy to safer chemicals. Some safer chemicals will be developed by larger companies for internal use and not marketed to external customers. Other safer chemicals will be developed and marketed as better performing or less costly products without promoting health or environmental benefits. From a marketing perspective, it is often more strategic to promote performance and cost and only note environmental attributes as subsidiary benefits. Other chemicals may simply be safer even though that is not why they are being developed.

Biobased feedstocks for chemicals could be a significant source of safer chemicals; however, the rate of development of biorefineries and the direction of their development is hard to predict. Biorefineries could compete with existing petroleum refineries if the price of oil and gas rise substantially; if the price stayed high and stable, biorefineries could also become models for the conversion of petroleum- to plant-based refineries. The U.S. Department of Energy has set a goal to replace 30 percent of current petroleum-based fuels with biofuels and 25 percent of industrial organic chemicals with biomass-based chemicals.[22]

However, the recent rapid growth of the U.S. natural gas industry may change all of this. The hydraulic fracturing ("fracking") of shale is opening up a steady source of natural gas that has stabilized the domestic natural gas market in the United States, leading to greater investment and industry growth. Fracking offers several benefits for the conventional chemical industry. First, it creates a growing, steady, and domestic feedstock source for organic chemicals, particularly for petrochemicals; second, the increasing flow of natural gas is lowering energy costs for the production of chemicals; third, there is a new market for chemicals used in the natural gas extraction processes (polymers, gelling agents, foaming agents, etc.).

The American Chemistry Council estimates that fracking will generate a 25 percent increase in the domestic ethane supply. This could lead to some $16.8 increase in capital investments by the chemical industry in petrochemical and derivatives capacity and a $38.2 billion increase in U.S. chemical production. Already, the availability of cheap natural gas is having effects on chemicals. In 2010, the Gulf Coast petrochemical and derivatives sectors moved up to be the second most competitive region in the world. U.S. polymer producers have experienced an increase of nearly 10 percent in plastics exports, and plans are already developing for new ethane "cracking" facilities in the Appalachian region.[23]

Innovation is both a technological and a social process. The introduction of safer chemicals and chemical processes requires decentralized but actively linked processes for communication, experimentation, and education. It needs support from social networks, social media, and the broader environmental advocacy movement. Innovation in popular music offers a model. Thousands of little bands, an influential set of music agents, and an active and critical audience linked up around the world set the context for a constantly evolving and innovative market where even small, marginal music inventions have an opportunity to be heard, evaluated, and commercialized.

So here are four core building blocks for a safer chemical policy framework. They parallel the policy components of the earlier chemical control policies, and they are here reconsidered as components of a new safer chemical policy framework. Such policies must identify and characterize the chemicals on the chemical market and set priorities for consideration. This requires that sufficient information is generated and made available on the inherent characteristics, production and use, and release and exposure of those chemicals. Based on this foundation, it is possible to consider how to reduce the production and use of chemicals of concern by replacing them with safer alternatives. Finally, efforts must be made to continually increase the supply of safer chemical and nonchemical alternatives to ensure the success of future chemical substitutions. Chapters ago, this book considered the chemical control policies of the last century. Now, it is time to consider the mechanics of policy in promoting safer chemicals and how these instruments and approaches can be woven together to create a comprehensive and integrated safer chemical policy framework.

14

Drafting Safer Chemical Policies

Agenda 21, the blueprint adopted at the 1992 United Nations Conference on Environment and Development, crafted a policy for toxic chemicals in chapter 19:

Adopt policies and regulatory and non-regulatory measures to identify and minimize exposures to toxic chemicals, by replacing them with less toxic substances and ultimately phasing out the chemicals that pose unreasonable and otherwise unmanageable risks to human health and the environment and those that are toxic, persistent and bioaccumulative and whose use cannot be adequately controlled.[1]

This policy set the stage for the development of the United Nations Environment Program's Strategy for the Sound Management of Chemicals. The 2002 World Summit on Sustainable Development reaffirmed this policy in committing that by 2020, "chemicals are produced and used in ways that lead to the minimization of significant adverse effects on human health and the environment."[2] The broad chemical policy transformation accomplished with the enactment of REACH in the European Union demonstrates the effectiveness of using government policy to promote safer chemicals. This ambitious regulation plus the new European product directives, the international chemicals conventions, SAICM, and recent state legislation in the United States provide illustrative models for a potential safer chemical policy framework in the United States.

None of these policies provides complete analogues for national policy in the United States, but most go beyond the current U.S. counterparts. How could chemical policies that are focused on converting the chemical market and transforming the chemical industry in the United States be structured? Such policies would need to be comprehensive, hazard-based, transformative, and effectively directed toward converting the current economy to a more sustainable economy. They would need to provide new goals that are broad and ambitious, better coordinated legal authorities to cover all chemicals, new instruments for managing chemicals, and new program infrastructures that are well integrated with private initiatives.[3]

A chemical conversion strategy must be broad and encompassing, and although the strategy needs to be inclusive and supportive of government, business, civil society, and scientific initiatives to be effective, there needs to be a coordinated, overarching safer chemical policy framework. The European Union's REACH regulation demonstrates the importance of a harmonized, common chemical policy framework that can link together and support the many parallel and compatible regional, sector, and institutional policies.

14.1   A Safer Chemical Policy Framework

So what does such a safer chemical policy framework look like? At this point, it is useful to assemble the various building blocks described in the preceding chapters. The framework presented in figure 14.1 locates the four building blocks as the center of the policy framework. However, as presented here, two additional elements are included: an introductory element on goals and plans and a concluding element on government responses. The history of the chemical control laws reveals how wandering and difficult to manage the implementation of these laws has been, in part, because they are largely without clear goals. It is also useful here to consider the instruments and procedures that would be necessary for a government to implement its part of this policy framework.

Although presented here as discrete units, the components of this framework could be overlapping and intermeshed and even implemented iteratively with some activities reviewed and redone on a periodic basis. Also left unclear here is the agency of these activities—who carries them out? Although it is easy to see this as a plan for federal government activity, this should not be the case. Many of these activities are currently conducted by private firms and NGOs, and some of these activities, which have conventionally been the responsibilities of federal agencies, can and should be shifted to these other parties. The next chapter will explore how various governments and other parties could share these tasks.

14.2   Establish Goals and Plans

A safer chemical policy framework needs national goals and a national plan. The EPA has created a national Clean Water Action Plan and a

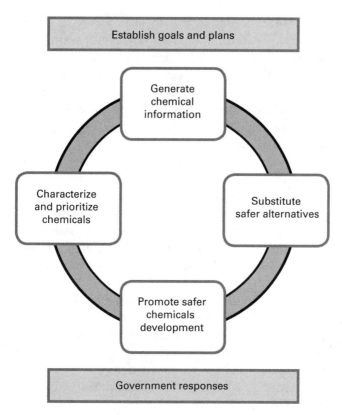

Figure 14.1

A Safer Chemical Policy Framework

Climate Change Action Plan. These national plans have involved a wide range of stakeholders and many federal agencies ranging from the Departments of Agriculture, Commerce, Defense, and Transportation to the U.S. Geological Society and the Postal Service. The national plan for water lays out more than 100 specific actions to be carried out by federal agencies to improve the nation's water quality. However, no comprehensive national action plan has been developed for chemicals.

Such a plan could provide a national roadmap for a comprehensive chemical conversion strategy. Ideally, the plan would provide a justification for public and private action, broad national goals, a set of strategies, objectives for specific agencies, businesses and other stakeholders, and benchmarks and metrics for measuring progress. The process of preparing such a plan would need to be broadly inclusive and, given the widely diverse perspectives that this would involve, carefully crafted.[4]

National goals would be essential. However, the federal government has seldom set national goals for environmental laws. A significant exception is the Clean Water Act, which offers a bold commitment "to restore and maintain the chemical, physical and biological integrity of the Nation's waters' and to achieve fishable and swimmable waters by 1985"—a commitment that has yet to be achieved. In 1998, the EPA established ten goals for clean air, water, food, and communities, but these were not carried forward into subsequent administrations. TSCA opens with a commitment to ensuring that adequate chemical effects data and adequate government authority exist so as to regulate substances that present an unreasonable risk. This provides objectives for the statute but not national goals for chemicals. Nowhere in its federal statutes or programs does the government state broad national goals for managing chemicals.[5]

In contrast, the European Union has made just such a commitment. In its 2001 Sustainable Development Strategy, the European Union adopted a goal for chemicals that has come to be called the "generational goal": "by 2020…chemicals are only produced and used in ways that do not pose significant threats to human health or the environment."[6] This same goal adopted under SAICM now forms the international standard for the sound management of chemicals worldwide.

Sweden has crafted even more definitive goals for chemicals in its national environmental policy. In the "Swedish Environmental Quality Objectives"[7] drafted by the government in 1997 and formally adopted into the National Environmental Code in 1999, Sweden committed to fifteen national environmental quality objectives. Among these is one focused on the achievement of a "non-toxic environment":

The environment must be free from man-made substances and metals that represent a threat to health or biological diversity. This means that:

• the levels of substances that occur naturally in the environment must be close to background levels,

• the levels of man-made substances in the environment must be close to zero.

To achieve this objective, the Swedish government issued guidelines on chemicals to assist companies in their product development and chemical strategies and to aide government agencies in implementing the environmental code:

1. New products introduced into the market are largely free from:

   • man-made organic substances that are persistent and liable to bioaccumulate and from substances that give rise to such substances,

   • man-made substances that are carcinogenic, mutagenic, and endocrine disruptive—including those that have adverse effects on the reproductive system, and

   • mercury, cadmium, lead, and their compounds.

2. Metals are used in such a way that they are not released into the environment to the degree that causes harm to the environment or human health.

3. Man-made organic substances that are persistent and bioaccumulative occur in production processes only if the producer can show that health and the environment will not be harmed.

These are hazard-based, precautionary goals. Were the United States to develop comprehensive chemicals policies, these policies might establish and promote practical and measurable goals such as these. Such goals would need to be broad and universal, addressing chemicals throughout the economy and across their life cycles. Given the established legal conventions for chemical control in the United States, a national chemical goal might best rely on the broadly accepted definitions for hazard and risk. For example, a national goal sympathetic to international goals might commit to establishing by a certain date an environment free of manufactured hazardous chemicals that present a threat to human health or the environment and establishing interim reduction goals for those substances that pose the highest and most concerning risks throughout their life cycles. Framing a national goal in these terms focuses action on chemicals based on the intrinsic properties of chemicals but establishes priorities based on risks and the likelihood of significant or disproportionate exposures.

14.3   Characterize, Classify, and Prioritize Chemicals

The first building block of a safer chemical policy framework involves characterizing, classifying, and prioritizing chemicals. Preparing a chemical profile for each chemical on the market would be the responsibility of each chemical manufacturer, importer, or supplier wishing to maintain a chemical on the market. A "Chemical Information Management Service" that could be either a government agency or a private body would be responsible for setting chemical testing standards and the standards for the minimum data set for delineating a sufficient chemical profile.

The characterization of chemicals in chemical profiles would form the basis for chemical classification. There needs to be a single and universal chemical classification scheme—what has been called here a Universal Classification of Chemical Substances. The Chemical Information Management Service would maintain this classification scheme. The classification scheme would be similar to the four-tiered model developed in chapter 10, and the criteria would be based on the GHS criteria with clear decision rules in order to screen chemicals into each category based on their intrinsic hazards. The classification process would begin with each chemical supplier locating their chemicals within the tier that they deem appropriate. The placement of a chemical into a specific category would be subject to public review and comment. A panel of experts might be available to assist where consensus cannot be reached, but, ultimately, the default for a chemical without consensus would be the category of chemicals of unknown concern.

Once a chemical has a complete chemical profile and has been located into the tiers of the Universal Classification of Chemical Substances, it would be given a chemical registration number—a "UCCS number." A distinct chemical registration provides a clear identification of each chemical and a guarantee in the market that a chemical has a sufficient chemical profile. Registration has been fundamental to the effectiveness of FIFRA's pesticide regulations. Extending registration to all industrial and agricultural chemicals and all chemicals used in products would create a harmonized information base for all chemical suppliers selling chemicals in the United States. This would also dismiss the artificial distinction between new and existing chemicals by requiring a common minimum data set for both new and existing chemicals.

The registration of chemicals provides the baseline inventory for ensuring that all chemicals are fully characterized and classified. All pesticides in the United States are already registered and have many of the basic elements of a chemical profile. Because the World Health Organization harmonized its international pesticide classification system with the GHS, pesticide classification should not be difficult. The characterization of industrial chemicals in the United States would appear more burdensome were it not that REACH already requires registration and a dossier for all chemicals produced and used in the European Union. The task would only be burdensome on smaller firms and those that only produce for the domestic markets. The first step for higher production industrial chemicals, notification to the TSCA inventory, is routinely completed every four years. Such notification could be converted to a registration once a chemical profile has been completed and a chemical has been classified.

A government agency or a collaboration of firms in a given sector or region would set about to identify one or several high-priority chemicals that require action. In some cases, individual chemicals would be prioritized; in other cases, groups of chemicals (e.g., phthalates) might serve as a priority. A well-developed prioritization protocol would be useful in identifying priorities, and the chemical prioritization scheme presented in chapter 10 suggests a model. Such prioritizing would require analysis of chemical hazards, human and environmental exposures, and the systems in which those chemicals are manufactured and used.

14.4   Develop and Make Accessible Chemical Information

Good and accessible chemical information provides the second building block of a safer chemical policy framework. Generating such information has been greatly advanced by the large number of chemical databases and chemical assessment tools and methods that have been developed by various government and nongovernment organizations in the United States and elsewhere. Many databases on chemical hazards have been created by authoritative bodies, and there is a limited collection of data on chemical production and emissions. None of this is complete, and there are many data gaps to contend with.

For substances such as heavy metals, halogens and aromatic hydrocarbons, sufficient information for characterizing and classifying chemicals is currently available. However, for many other chemicals with insufficient data, generating such data should be a continuous and ongoing process. A formalized process could begin with rapid screening procedures, using indicative flags to identify points where more detailed assessment is warranted, and proceed with assessment to a point where sufficient information exists to characterize and classify a chemical and only proceed to more thorough studies where previous studies are contradictory, the controversies are vociferous, or the stakes are high. Legitimate confidential business information could be protected in chemical profiles with clear criteria for defining such information and a means to protect against its disclosure. To keep chemical profiles current and respectful of the most recent science, some period could be specified, whereby chemical profiles need to be updated.

There are institutional and Internet vehicles for promoting access to this information including the proposed Chemical Information

Management Service noted above. To be effective this information must reach chemical suppliers, brand-identified product manufacturers, civil society organizations, retailers, institutional procurement institutions and consumers. Transparency and the free flow of trustworthy chemical information in well-structured feedback loops is the key for both driving and facilitating the transition to safer chemicals.

14.5   Substitute Safer Chemicals

Identifying and selecting safer alternatives makes up the third building block. Here the focus is on converting the chemical market by encouraging the substitution of priority chemicals of concern with safer alternatives. Once a chemical or chemical group has been designated as a high priority, the next step would be the preparation of a "chemical action plan." These chemical action plans would take a broad systems perspective on the priority chemicals, identifying the functional use of the chemical and the role the chemical plays within chemical production systems or use patterns that are likely to change if safer alternatives are adopted. These plans would also require alternatives assessments and life cycle considerations. Like the California Safer Consumer Product program, the plans would identify a range of feasible alternatives or indicate the absence of such alternatives. Where feasible alternatives are identified, the chemical action plan would describe those selected for implementation, set goals and timelines for substitution, and identify the needed resources for implementation. Where no feasible alternatives are identified, the plan would provide documentation such that the analysis could be reviewed. The review of chemical action plans could be conducted by government agencies, independent third-parties hired by firms, or the general public if the plans were to be publicly posted on the Internet.

The process of preparing chemical action plans could vary depending on the nature of the chemicals or groups of chemicals prioritized. Where one chemical manufacturer or importer is the dominant source of a chemical, that firm might prepare a chemical action plan alone. In some cases, firms commonly using a chemical might work with their supply chains and involve vendors, suppliers, and customers in the generation of chemical action plans. In other cases, collections of firms in specific economic sectors might join together in collaborative "Sector Chemical Workgroups" to develop chemical action plans. Such collective planning could be organized around any economic sector where highly hazardous chemicals are heavily used (see table 14.1).

Table 14.1
Economic Sectors with Opportunities for Safer Chemicals

Sector	Common Chemicals of Concern
Health care	Mercury, latex, PVC, phthalates, disinfectants, pharmaceuticals
Cosmetics	Parabens, phthalates, diethanolamine, formaldehyde, glycol ethers, mercury, lead
Household cleaning products	Phthalates, tricolsan, nonylphenol ethoxylates, phosphates, synthetic musks, pesticides
Electronics	Lead, cadmium, mercury, hexavalent chromium, antimony, beryllium, halogenated flame retardants, PVC, perfluorinated compounds, "conflict minerals"
Outdoor apparel	Flame retardants, biocides, nonylphenol ethoxylates, perfluorinated compounds
Building construction	Formaldehyde, isocyanates, toluene, xylene, lead, cadmium, PVC, halogenated flame retardants, phthalates,
Agriculture	Pesticides, heavy metals, artificial hormones
Automobile	Lead, cadmium, mercury, toluene, xylene
Retail	Hazardous chemicals in products
Education	Chemicals in art supplies and science labs

Focusing on economic sectors or common chemical user groups can reveal the complex systems by which chemicals flow among producers and users. The pesticide control laws primarily focused on the agricultural sector and FDA's drug control authorities focused on the pharmaceutical sector provide economic sector models that identify the sector determinants of chemical use, the efficacy of products, and the necessary restrictions on use. Working within a sector makes sense because the drivers within the sector create an economic and a technical logic on what and how chemicals are used across many firms. A broad look across a sector can identify the barriers to innovation and the unexplored chemical and nonchemical alternatives that may be appropriate across many firms within the sector. A useful lever for system change appears in economic sectors where alternatives may be more readily diffused when early adopters within a sector provide models of change-oriented behavior for later adopters. Working to build collaborative Sector Chemical Workgroups could set chemical policies in the context of multiple objectives.

For instance, working within a sector to convert the chemicals of a sector can parallel efforts to convert that sector's carbon emissions, water consumption, or working conditions and reduce the sense that these are competing objectives.

14.6   Develop Safer Alternatives

The fourth building block involves the development of safer alternatives. Converting to safer chemicals requires the development of many currently unavailable alternatives. The broadly emerging efforts of chemists and engineers to develop green chemistry and engineering solutions are critical. The issue here is twofold: how best to encourage the development of inherently safer chemicals and technologies, and how to help these inventions reach the market in a manner that leads to widespread adoption. This means not only supporting chemical research but also business entrepreneurs and private venture and socially directed investors. Because innovation is a social process that relies on the encouragement and support of many colleagues, facilitators, and investors, it is important to better develop the chemical production and consumption system's linkages and communication pathways.

To facilitate the development of those system linkages, the government or a collaboration of universities and investors might create a "Green Chemistry and Engineering Initiative" that could support research and early product development. This would be broader than a research grants program; it might involve regional or sector-specific hubs where laboratories and incubators could be set up, including groups of corporate partners and an assortment of early investors with an interest in green business development.

14.7   Government Responses

Chemical action plans provide direction for both industry and government responses. The response from government agencies would depend on the substance of the chemical action plan. Where the plan is well documented and reveals suitable alternatives and plans for their adoption, the government may simply acknowledge and encourage the plan. Where the plan identifies feasible alternatives but also identifies limitations and barriers to successful adoption, the government might offer technical assistance, economic inducements, or research programs to encourage implementation. Where there is no plan or plans present no alternatives, governments could default to conventional chemical control policies.

Regulation

The most compelling government responses involve government regulatory instruments. Clear and specifically directed regulations can restrict the use of the most concerning chemicals, raise the costs of dangerous chemicals, and create a common floor of limits on emissions, wastes, and exposures. The current chemicals control laws provide authority for a range of regulatory responses to hazardous chemicals, including outright prohibitions, conditions on use, conditions on release, disposal restrictions, suspensions or cancellation of registrations, exposure protections for workers or users, labels and warnings, and simple notifications.

Current laws allow for direct restrictions of chemicals of highest concern (the prohibition strategy). The limited use of the powers under TSCA to ban chemical use was described in chapter 3; however, the cancellation and suspension of hazardous pesticides under FIFRA has been more common, and restrictions on the use of the significant ozone-depleting chemicals under the Clean Air Act amendments were completed during the 1980s. For priority chemicals in the lowest tier of the Universal Classification of Chemical Substances or in the category of unknown concern, government agencies could similarly use a direct restriction to reduce or suspend production or use. Such actions could be called for where producers refuse to supply information needed to complete chemical profiles (what the Europeans call "No Data; No Market") or where a chemical is so clearly dangerous as to warrant significant restrictions.

More common would be the use of regulations to respond to chemical action plans. In some cases, the plan might call for changes in regulations that could facilitate the substitution of safer alternatives. Where a plan calls for changes in exposure standards or renegotiated permits, regulations could be adjusted. Where regulations need modifications to allow for safer alternatives or pilot periods to try out safer alternatives, regulations could be temporarily suspended. In other cases, regulations could be used to define the types of substitutes that could be used as safer alternatives.

One good model can be found in the amendments to the Clean Air Act of 1990. The Strategic Ozone Protection Program establishes the requirements for phasing out the use of ozone-depleting substances. Section 612 creates a "Safer Alternatives Policy" by declaring:

To the maximum extent practicable, class I and class II substances shall be replaced by chemicals, product substitutes, or alternative manufacturing processes that reduce overall risks to human health and the environment.[8]

The section continues by providing guidance to the EPA:

Within 2 years after enactment…the Administrator shall promulgate rules under this section providing that it shall be unlawful to replace any class I or class II substances with any substitute substance which the Administrator determines may present adverse effects to human health or the environment, where the Administrator has identified an alternative to such replacement that:

1. reduces the overall hazard to human health and the environment; and
2. is currently or potentially available.

Regulations always change markets, and when regulations are properly designed, markets can and do respond with innovations. Government regulations can drive firms to innovate by developing or converting to new chemicals or technologies. Experience with the prohibitions on polychlorinated biphenyls and ozone-depleting chemicals demonstrated how regulations have driven innovation where the regulations are clear and forceful and reasonable timelines are provided. Regulations can also encourage market innovation more indirectly by requiring public reporting, substitution planning, and product labeling. The Massachusetts Toxics Use Reduction Program and California's Proposition 65 provide valuable examples and lessons. Regulations must be well crafted because they may have unintended effects. For instance, government prohibitions can drive chemicals out of the market but may displace them onto unregulated (foreign or illegal) markets. Poorly designed and enforced regulations can discourage compliance, lead to costly government policing, and stall innovation.[9]

Even the threat of regulations can have telling effects in a market system. Sweden and Denmark have been successful in using the prospects of chemical regulation by publishing "observation lists" of hazardous chemicals that suggest future action and encourage substitution. Indeed, the U.S. Federal Reserve Bank and other nations' central banks often hint at policy changes long before they enact them to test how effectively mere suggestions might move markets where investors are highly sensitive to predicting the future.[10]

Requiring firms to prepare chemical profiles, conduct alternatives assessments, or complete chemical action plans can have similar effects. Without mandating the substitution of a chemical of high concern, requiring the managers of a firm to consciously examine the risks and costs of the continued use of the chemical can provide meaningful incentives to finding substitutes. This has been a key to the success of the Massachusetts Toxics Use Reduction Program.

Economic Instruments

Economic policies to deter the use of hazardous chemicals have been less common than regulatory policies. In part, this is because fiscal, tax, and subsidy policies are typically outside the jurisdiction of environmental agencies and the legislative committees that fashion environmental laws. However, economic instruments could be quite valuable in shifting chemical markets.

As a starter, the existing economic policies that support and protect hazardous chemical production could be redesigned. Government tax policies and direct subsidies that advantage fossil fuel extraction and processing could be adjusted to better accommodate alternative feedstocks for chemical synthesis. In addition, the current agricultural subsidies that underwrite the low price of corn-based ethanol could be revised to encourage a wider range of biomass sources of chemical feedstocks.

Likewise, the economic drivers that lead to cheap and disposable products such as municipally funded waste management could be replaced with policies that promote remanufactured or recycled products. Because the significant costs of product waste management fall on local governments, these costs do not appear in the retail price of products. Internalizing those costs by product take back programs, such as the extended producer responsibility programs implemented in Europe on product packaging and automobiles, can redirect the management costs of end-of-life products back onto product manufacturers.[11] By assuming responsibility for the waste management costs of products containing hazardous chemicals, product manufacturers realize an incentive for reconsidering those chemicals. Several states have implemented such product take back programs on used tires, batteries, and electronic products. The successful programs that collect and return some 94 percent of lead acid batteries to smelters internalize the costs and improve the management of the lead that would otherwise be released as wastes.[12]

Government responses to chemical action plans could include economic instruments. Price-based economic instruments such as pollution taxes and chemical user fees offer potential tools for driving up the price of chemicals identified as priorities. Several states use waste generation or waste "tipping" fees to recover some of the costs of waste treatment and discourage the generation of wastes. Insurance requirements directly imposed or considered necessary to do business raise chemical prices where chemical risks and the likelihood of damage are high. There are also two-tiered instruments such as deposit/refund rebate programs, tax/subsidy programs, and user fee/service assistance programs that could combine negative and positive incentives for raising hazardous chemical prices while lowering safer chemicals costs. Tariffs or import duties on hazardous chemicals or products containing undesirable chemicals have also been employed; however, such fees need to be carefully constructed so as not to present a discriminatory restriction on trade under international trading rules.

The most direct instrument for internalizing costs and driving chemicals of concern out of the market is a use fee or tax levied at the point of chemical registration or sale. The 1980 amendments to FIRA allowed the EPA to charge an annual registration maintenance fee for pesticides and many producers decided then to cancel registrations rather than to meet the registration requirements. Such a fee (a graded tax, actually) was placed on chlorofluorocarbons and other ozone-depleting gases under the 1990 Clean Air Act Amendments as a means of ensuring that the United States met its obligations under the Montreal Protocol. This tool is the least difficult to administer, and its directness ensures its effectiveness.

Indeed, a graduated chemical use fee could be linked to the Universal Classification of Chemical Substances, such that chemicals in the lowest tier would have the highest fees, and the fees would diminish progressively in each higher tier until there is no fee on preferred chemicals.

Special taxes on pesticides have been employed in Denmark, Sweden, Belgium, and Norway since the mid-1980s. In Denmark, a universal fee was levied based on the volume of the pesticide purchased. The levy is credited with generating a 50 percent reduction in pesticide use in ten years. Thereafter, the fee structure was altered to create a differentiated fee structure based on classes of substances differentiated by potential health and environmental effects. Revenues from the fees, estimated to be some $11 billion over the life of the program, have been used to support farmer education, organic agriculture, and research in safer pest management.[13]

Finally, there are lessons from the emission trading programs established under the Clean Air Act to reduce acid rain by reducing the amount of sulfur dioxide released from electric power-generating utilities. The program is like a game of musical chairs. The government issues a certain number of emission permits nationally, such that every utility gets an allotment. Over the years, the government then lowers the number of permits. In the meantime, utilities may buy and sell permits, such that those making the most progress on sulfur reduction can earn revenue from selling their no longer needed permits to those making the least progress. The program has been successful in reducing aggregate levels of sodium dioxide, but a range of criticisms focus on how the concept benefits polluting companies and offers little protection to communities that abut facilities that are slow to upgrade. However, adapting such a cap and trade program to specific high-priority chemicals might have market-wide effects that hasten the substitution of safer alternatives.[14]

Technical Assistance
While regulatory and economic instruments can drive change, technical training and specialized assistance can help to ensure that the resulting changes are more likely to be preferred outcomes. The best outcomes come where regulations are paired with nonregulatory technical assistance and capacity-building services. Such assistance can come from governments, universities, nongovernment organizations, or business associations and include:

- site-specific technical assistance,
- education and training programs, and
- demonstration programs.

The national experience with state pollution prevention programs demonstrates how reductions in the generation of pollution and waste can be achieved where competent technical assistance programs work successfully with firms at the facility level. To assist firms with toxics use reduction, the Massachusetts Toxics Use Reduction Program Office of Technical Assistance has provided workshops, demonstrations, and trade fairs and conducted hundreds of onsite technical consultations. The Toxics Use Reduction Institute has set up a library, a web-based information service, and a highly successful toxics use reduction planner training program. The Institute's Surface Solutions Laboratory provides testing services to small firms in Massachusetts that are under various regulatory and market pressures to seek safer surface cleaning technologies. Over a fifteen-year period, this lab assistance has been credited with a decrease of more than 100,000 pounds of chlorinated solvents by Massachusetts firms.[15] Such technical assistance programs are moderately expensive because they are labor intensive, so it is useful, as in Massachusetts, to fund them through fees.

Technical workshops, professional conferences, technology fairs, guidance manuals, workbooks, videos, Webinars, and certificate training programs have been used in many states to promote waste reduction, energy efficiency, and hazardous chemical substitutions. Demonstration programs that permit firms to see successful conversions augment technical assistance programs by providing real-time, practical examples of transitions that work. Not only do such demonstrations provide public recognition, they also serve to promote technology diffusion among firms.

Federal agencies also provide technical services. The Federal Food Quality Protection Act established a technical services program at the EPA for promoting safer pesticides and nonchemical pest management practices in the agriculture sector. The agency's Design for Environment program has sponsored sector-based work groups in the electronics, furniture manufacturing, and household cleaning products sectors to develop protocols for alternatives assessment and tools for comparative chemical hazard assessment.

Government Investments

Where chemical action plans are unable to identify suitable alternatives to priority chemicals, efforts could be made to encourage the development of appropriate substitutes. Thus, a government response to a chemical action plan might be to trigger further research on the properties of hazardous chemicals or research for the development of safer alternatives. This could involve financial support for targeted health effects research, investments in green chemistry and engineering, and policy interventions to open and protect early market entry for safer alternatives. This can be accomplished with several instruments, including:

- government investments in research,
- government support in the market, and
- support for science education.

Government support for environmental health research can promote safer alternatives in three ways: first, by expanding the toxicological and pharmacological understanding of how chemicals affect the environment and public health; second, by developing the standard tests and methods needed for determining the hazardous characteristics of chemicals; and third, by supporting graduate training programs in toxicology, pharmacology, epidemiology, and the environmental sciences. The research on chemical effects and the development of standard tests and protocols is fairly well organized through the federal grants programs at the National Institute for Environmental Health and Safety and the various National Institutes of Health, but research methods for environmental and ecological effects still need to be developed.

Direct investments in targeted green chemistry could spur the development of safer alternatives. Much more could be accomplished through the regional or sector-based green chemistry and engineering hubs, and the funding programs suggested in chapter 13. There is a tiny research program for green chemistry at the EPA, but additional programs could be developed at the National Research Labs and through the National Science Foundation to increase the amount of research and number of researchers.

Government can also play an important role in fostering and aiding new chemicals and chemical processes in the market. Preferred government and institutional procurement policies, product certification and labeling programs, and public education and awareness initiatives can further advance the transition to safer chemistry. For example, the EPA manages an environmental technologies program that supports research on the development of environmental technologies (e.g., pollution control, environmental assessment, waste reduction, and waste management technologies) and also assists in the verification, certification, and market entry of those technologies.[16]

Education and professional training provide an informed public and the technically trained scientists and technicians needed to advance a transition to safer chemicals. Research support and educational stipends for college science students could encourage interdisciplinary research projects well informed by health, environmental, and sustainability perspectives. Public resources are needed to promote green chemistry and engineering curriculum development and education programs and to support innovations in chemistry teaching at the primary and secondary educational levels.

Surveillance, Monitoring, and Evaluation

Unlike the environmental protection laws of the 1970s, the chemical control laws offered little requirement for monitoring and program evaluation. Significant resources went into developing regulations, establishing operational programs, and installing the infrastructure to ensure compliance. However, the statutes provided few procedures for measuring progress on chemical management programs or for periodic program evaluation. The EPA collects data on certain program indicators, such as number of pesticide registrations and number of new chemical premanufacture notices received each year, but this is not a substitute for program evaluations that demonstrate program effectiveness. Indeed, without national goals for chemical management, measuring progress on chemical management has been elusive.

If there were national goals for chemicals, then there could be a set of indicators and metrics and a process for collecting and reporting appropriate data. The capacity for tracking chemical production and chemical releases is already established nationally through the EPA's Chemical Data Reporting (CDR) and Toxic Release Inventory (TRI) programs. Because of recent revisions, the CDR requirements now include some data on chemical exposure and chemical processing and use as well as basic production data. Although the thresholds are too high and reporting intervals are too long, the basic framework is in place to track chemical manufacture and importation. The TRI is more than twenty-five years old and a model of program success. It could be expanded to include priority chemicals where lower thresholds might be useful in a manner that could be used to monitor program performance on chemical releases.

Performance indicators based on environmental or human monitoring could also be used. The field of environmental monitoring has developed over the past decades and today there are on-going efforts to monitor primary air pollutants and contaminants in water resources. It would be possible to track changes in the presence of priority chemicals in air, water, and soil samples to determine the effects of chemical reduction and substitution programs. Freshwater and air sampling protocols and technologies are common in practice, and new equipment allows for both passive sampling and remote surveillance that is cost effective for continual environmental monitoring. Biomonitoring of natural flora and fauna could also be used to track changes in the presence of priority chemicals.

Human biomonitoring provides another means of tracking the effects of chemical policies. The National Biomonitoring Program has established the procedures for its broad, national survey, and although it currently covers fewer than 300 chemicals, it could be expanded to include priority chemicals where reductions and substitutions are being addressed. The California Environmental Contaminant Biomonitoring Program provides for the collection and chemical analysis of human biological specimens from a broad sample of California residents that demonstrates how states could engage in such surveillance.

Although both environmental monitoring and biomonitoring provide instruments for measuring changes in chemical presence, they only measure those substances not rapidly metabolized in organisms or transformed or decomposed in the environment. They are better at measuring long-term trends than addressing short-term goals; however, they may be particularly valuable in identifying anomalies and unexpected spikes and recurrences of chemicals of concern. Although interpreting such data and communicating to the public can be complex, this does not limit its usefulness in tracking program performance.[17]

14.8   Moving the Policies

An ambitious range of new initiatives is needed to create safer chemical policies for the United States. The components described here could provide a starting framework. National goals and a plan for safer chemicals would provide a clear set of policy and programmatic directions. The registration of chemicals and a common protocol for chemical profiles would create a consistent floor of information for all chemicals. A tiered framework for chemical classification would lay the basis for chemical prioritization and drive the development of more chemical information. Chemical action plans, increased chemical information transparency, and programs developed to promote safer chemical innovations could provide new means for encouraging the substitution of chemicals of concern and government regulations, economic tools, technical assistance, and government investments could be more strategically coordinated to support the transition to a safer chemical market and a more sustainable chemical industry.

Some of these activities could be conducted by federal agencies; however, many of them are better conducted by state government, private industry, professional bodies, and nongovernment organizations. Table 14.2 suggests how these various functions might be distributed across possible organizational structures.

Some states are performing these functions today, and more could be encouraged. Some firms have well-developed chemical action plans. Collaborations among firms within some sectors are already working to substitute safer alternatives for priority chemicals of concern. Nongovernment consumer and environmental advocacy organizations are

Table 14.2
Options for Chemical Policy Roles and Responsibilities

Function	Federal Agency	State Agency	International Agency	Private Corporations	Industry Collaborations	Nongovernmental Organizations
Setting national goals and plans	○					
Registering chemicals	○		○			
Creating chemical profiles			○	○	○	○
Establishing a chemical classification system	○		○		○	○
Classifying chemicals	○	○	○	○	○	○
Drafting chemical action plans	○	○		○	○	
Conducting alternatives assessments	○	○		○	○	○
Requiring compliance	○	○				
Providing technical assistance	○	○		○	○	○
Supporting research	○	○		○	○	○
Surveillance and monitoring	○	○		○	○	○

convening business groups and developing alternatives assessment tools and methods. Universities, standard-setting bodies, new business associations, and private consultants could all play important roles in these policies, as well as private firms, industry collaborations, and nongovernment organizations. Encouraging such leadership and coordinating these activities will require some rethinking among government agencies, some new authorities, and, in some cases, some government restructuring. The next chapter considers these possibilities.

V

Chemicals without Harm

Achieving a world in which chemicals are no longer produced or used in ways that harm human health or the environment, and where POPs and chemicals of equivalent concern no longer pollute our local or global environments, and no longer contaminate our communities, our food, our bodies, or the bodies of our children and future generations.

—International POPs Elimination Network, *Toxics Free Future* (2011)

15

Reconstructing Government Capacity

When the European Union negotiated the REACH Regulation, it not only created new roles and responsibilities for industry and the governments of the Member States, it also created a new chemical agency. Located in Helsinki, the European Chemicals Agency (ECHA) has primary responsibility for the implementation of REACH. While the Director General for Environment, located in Brussels, retains authority for major policy issues, ECHA receives and evaluates the individual chemical registrations under REACH and proposes substances of very high concern for authorization. However, since its creation, ECHA's role has been steadily growing. The agency, today, addresses a broad range of chemical issues, including the Classification, Labeling, and Packaging (CLP) Regulation, the Biocidal Products Directive, and the Prior Informed Consent Regulation.

To be effective, a chemical conversion strategy will require government support and guidance. Given the long period of stalled federal initiative on chemical policy in the United States, this will require redesigning and rebuilding federal capacities. However, there is increasing sophistication in addressing chemical policies among state governments, and there are emerging many new ways that government agencies—both federal and state—can work with industries, economic sectors, and civil society organizations. The previous chapters have noted the host of ongoing activities that can be built on. Some involve adapting existing regulatory and economic policy instruments, whereas others involve building new federal relations with industry, nongovernmental organizations, and international institutions.

15.1 Government Activism

Effective safer chemical policies require government support, guidance, and regulation. This concept is not broadly accepted. Today, the role of government, particularly the federal government, is hotly debated in the United States. There are strong voices for a free market economy unfettered by government involvement. These arguments come from well-developed ideologies that have long defended Americans from unwanted government intervention and censorship. This tenet is an important part of the country's political history.

However, Americans have never embraced a fully free market. From the earliest settlements, the church and the town constable have set effective brakes on economic activity. Early government restrictions on alcohol sales, prostitution, and gambling were followed by state regulations on land use, property transfers, and commercial transactions. Indeed, private firms have often sought government authority to protect markets, level playing fields, and discipline those involved in fraud, misrepresentation, and intellectual property theft. The first state and federal laws requiring regulations on pesticides and drugs were driven by legitimate businesses seeking protection from fraudulent manufacturers.

Just as government regulations set driving speeds on highways and designate flight corridors for air traffic, regulations are needed to create a fair platform for chemicals in commerce. The environmental emission and effluent standards, the occupational health and safety standards, and the pesticide tolerance standards have all served to frame a lower threshold barrier for hazardous chemical exposures. This governmental role is important because it helps to internalize the social costs of private actions and create a common floor below which private actions cannot be permitted. Children's toys, for example, should not be painted with lead, and hair shampoos should not contain carcinogens. Admittedly, these standards are often too low, not always well enforced, and, for many chemicals, absent. More effort on standard setting could produce a broader array of standards. However, with the exception of pesticide tolerances, the progress in reviewing, expanding, or tightening these standards over the past twenty-five years has been slow and torturous. OSHA, for instance, has taken an average of seven years to write a single chemical occupational health standard. In 2006, the agency promulgated a standard for hexavalent chromium that took more than twelve years to draft.[1]

Even if these standards could be broadened and tightened, in themselves and alone, they would not lead to safer chemicals. Regulations that can prevent the worst behaviors are often too broad and clumsy to produce the best behaviors. However, a more solution-driven chemical conversion strategy could be built on the existing regulatory foundation and expand its reach to include a wider assemblage of instruments and approaches. By carefully selecting a range of instruments, the restrictive measures of regulations could be married with positive incentives for substitution and innovation, and technical and financial assistance could be added to ease the shift to safer chemicals.

Here, several different approaches could be employed. The progressive chemical policies of some product manufacturers and chemical suppliers noted in chapter 8 suggest the possibility of expanding the self-regulation capacities of high-performance firms as a means of moving economic sectors toward chemical transitions. Nike, S.C. Johnson, and HP lead their industries with sophisticated programs for evaluating chemicals, restricting the most dangerous, and substituting safer alternatives. However, a firm-by-firm approach to improved chemical management will never reach the scale needed to transform the chemical market.

Alternatively, government agencies could initiate voluntary "partnership" programs that encourage "beyond compliance" environmental protection performance by offering awards, public recognition, or the easing of regulatory burdens. Some such programs have been employed over the past twenty-five years. In 1991, the EPA launched its "33-50 Program" to encourage large firms to voluntarily reduce pollution from seventeen chemicals of high concern first by 33 percent and then by 50 percent. The EPA invited some 8,000 firms to participate, and eventually 1,300 did. Releases of the seventeen chemicals among the firms did significantly decrease between 1991 and 1998, when the program was closed, but the several studies that have tried to evaluate the program provide mixed conclusions on how much of the reductions were due to the program.[2]

During the 1990s, the EPA embarked on several voluntary partnership programs, including Project XL, the National Environmental Leadership Program, WasteWise, Energy Star, and StarTrack, to promote improved environmental performance among willing firms. In 2000, the EPA wrapped up several of these programs to create the National Environmental Performance Track Program, which offered public recognition, speedy permit services, and some leniency on regulatory responsibilities in return for measurable commitments to improve environmental performance in at least some of twenty environmental categories. Because the program only admitted firms that already had implemented environmental management systems, these firms tended to be leaders within their economic sectors. Although the program enrolled some 550 members, it was criticized for its low performance and regulatory leniency and was closed in 2009.[3]

The PFOA Stewardship Program and the Pesticide Environmental Stewardship Program noted earlier are examples of voluntary programs specifically focused on hazardous chemicals. Another EPA program, the National Partnership for Environmental Priorities, encouraged voluntary reduction in the use of hazardous chemicals such as lead and mercury. By design, all of these programs rely on the voluntary commitments of the most responsible firms. Left out are those firms that are either too small or too unwilling to participate, and it is not unreasonable to assume that among these firms are some of the most in need of better chemical management.

So, if private self-regulation and voluntary partnership programs are not sufficiently robust or scaled to effectively move the chemical market toward safer chemicals, are there other instruments worth considering? A third, more promising approach appears in programs focused on moving economic sectors.

15.2   Engaging Economic Sectors

The EPA has initiated several sector-based approaches to environmental protection. In 1993, the EPA created the Sustainable Industry Program to explore the value of working with specific industrial sectors, and this led to the creation of the Sector Facility Indexing Project, which developed a series of common national indicators and metrics that could be used to compare environmental performance among various industrial sectors.[4]

Between 1994 and 1998, the EPA sponsored the Common Sense Initiative (CSI), which convened multistakeholder working groups in six economic sectors (automobile manufacturing, printing, computers and electronics, metal finishing, iron and steel production, and petroleum refining) to develop "cleaner, cheaper, smarter" consensus-supported approaches to "regulatory reinvention." The CSI was an ambitious project involving substantial commitments of agency staff and resources. However, its effects on regulations were modest—of some thirty work-group recommendations (involving forty-five different projects), only five resulted in direct revisions to agency regulations. On this basis, reviewers have been critical of the project.[5]

Yet there is more to learn from the CSI. By its objectives, the project was largely focused on improving the EPA's regulatory approach, and that it did poorly. Instead, the CSI could have been focused on improving the economic sector's overall environmental performance. Many of the participants and the ideas they promoted focused more on improving environmental management in the sector than improving the regulations. For instance, the largest share of recommendations made by the workgroups was aimed at changing behaviors in the sector: nearly 50 percent of the projects were focused on research and information collection, 24 percent were focused on education, and only 20 percent recommended policy changes. When interviewed during two evaluations, participants, even when critical of the progress made, praised the level of stakeholder dialogue, learning, and constructive engagement. Indeed, collaboration in the metal finishing industry was so strong that the workgroup developed a *Metal Finishing Strategic Goals Program* and a *National Metal Finishing Environmental Research and Development Plan*, and recruited 250 firms to work with the EPA to achieve the goals.[6]

Even more promising lessons can be drawn from the EPA's Design for Environment (DfE) program. The DfE program was established in 1992 as a support for the agency's pollution prevention programs. It provides information on safer chemical use in products and packaging and end-of-life product disposal. The program has worked successfully to bring together broadly based stakeholder dialogues on difficult chemical and product design problems in electronics, hazardous chemicals in cleaning formulations, and flame retardants in furniture and plastics. The stakeholder dialogues are crafted to identify and evaluate a broad array of safer alternatives that could be adopted by firms in each of the affected economic sectors. Clive Davies, the director of the program, claims, "The DfE program works because it is open and inclusive and because we work to find solutions relevant to firms' needs. We build effective partnerships in specific sectors to find cooperative solutions to chemicals problems in that sector. It's voluntary, but participation has been strong and we can often document success in terms of the shift to safer chemicals."[7]

These economic sector programs could be more directly informed by the negotiated agreement programs of Japan and the Netherlands.[8] Under these programs, government authorities have set goals and priorities and convened negotiating dialogues among groups of firms in either a region or an economic sector. Once the dialogue reaches agreement, a contract or "covenant" is developed and signed, and firms are held to meeting their specific objectives. Regardless of whether such legal agreements are needed, these negotiated dialogues offer lessons for sector-based programs in the United States. Once priority chemicals have been identified, "Sector Chemical Workgroups" could be formed to identify and assess safer alternatives and draft collective agreements into chemical action plans. In some cases, this would require strong government encouragement; however, in those sectors where a chemical management program has already begun, the government's initiatives could be graded and tailored to meet the specific needs of the leading firms. Table 15.1 lists several sectors that already have ongoing collaborations working on safer chemicals.

Collaborative planning processes to develop sector-based chemical action plans could be developed where leading firms within a sector or chemical user group have already established chemical management programs. In sectors such as the automobile sector, the health care sector, the electronics sector, the cleaning products sector, or the clothing and apparel sector, where efforts to create multifirm chemical databases and guidelines have already advanced, the programs already in place could be used as models that could be expanded across the sector. In these sectors, there may be little need for government engagement. In other sectors, government agents might need to convene leading firms and important stakeholders into Sector Chemical Workgroups and identify the priority chemicals and possible avenues for cooperative agreements but then withdraw to allow the participants to develop chemical action plans and draft the agreements. In still other sectors, government initiatives might need both to initiate and maintain sector collaborations and default to conventional government regulations where cooperation is unlikely or unproductive.

The formation of Sector Chemical Workgroups in the chemical industry could provide the most effective of these collaborations. The Green Chemistry Institute's Pharmaceutical Roundtable already offers an initial workgroup for drug manufacturers. Similar workgroups could be developed for the bulk chemicals and some of the specialty and consumer product chemicals. The 1990's chemical industry technology roadmaps and the National Academy of Science's *Sustainability in the Chemical Industry* report could provide an initial starting point for this work.

In each of these scenarios, it would be important to identify leading firms and business associations as leverage points for building the cooperative agreements and finding effective stakeholders that might work with the industries. In such sector-based approaches, there is always the possibility of "free riders"—firms that benefit, without sharing the costs— but this is potentially off-set by the capacity of the most engaged firms gaining a "first mover" advantage. The benefit of a sector-tailored approach also means that successes of sector workgroups could be rewarded with increasing amounts of flexibility and decreasing government involvement leaving government oversight to be focused on sectors with less successful progress.

Table 15.1

Economic Sectors with Ongoing Safer Chemical Initiatives

Sector	Promoters	Programs
Health care	Health Care without Harm, Practice Greenhealth	CleanMed, Green Guide to Health Care
Cosmetics and personal care products	Environmental Working Group, GoodGuide, Campaign for Safe Cosmetics	Skin Deep, Safe Cosmetics Compact, Safe Cosmetics Business Network
Cleaning products	EPA DfE, GreenBlue Institute	CleanGredients, Safer Detergents Stewardship Initiative
Electronics	Green Electronics Council, Institute for Printed Circuits, Lead Free Soldering Partnership, International Electronics Manufacturing Initiative	EPEAT, JIG, Environmental Product Declarations
Clothing and apparel	American Apparel and Footwear Association, Sustainable Apparel Coalition, Outdoor Industry Association	Eco-labels, Eco-Index, Higg Index, Joint Roadmap for Zero Discharge, Environmental Product Declarations
Building construction	USGBC, Healthy Building Network, Swedish Construction Federation	LEED, BASTA, Pharos, Environmental Product Declarations, Eco-labels
Agriculture	National Sustainable Agriculture Coalition, Northeast Organic Farming Association	Organic Farming, Integrated Pest Management, Integrated Nutrient Management
Automobile	Automobile assembly companies, Ecology Center	IMDS, GADSL, Consumer Action Guides
Pharmaceuticals	ACS Green Chemistry Institute	Green Chemistry Pharmaceutical Roundtable

15.3   Engaging the States

Some states offer fertile ground for statutory initiatives and new administrative programs, and some of these states could be effective partners for federal agencies in implementing comprehensive chemical policies. States such as Maine, Massachusetts, New Jersey, Minnesota, New York, Washington, Vermount, Oregon, and California—those states with more developed environmental and public health constituencies—have long been leaders in environmental policy. These states have established some of the strongest public health protection and pollution prevention programs in the country, and during the past decade, these and several other states have passed laws banning or prohibiting the use of chemicals such as mercury, lead, cadmium, bisphenol A, and brominated flame retardants in consumer products.[9]

The California legislature has been particularly active in restricting the use of dangerous chemicals, with laws focused on regulating perchloroethylene and formaldehyde. In 2003, California passed a law banning the use of two flame retardants—penta- and octa-brominated diphenyl ethers—in commercial products, and thereafter another eleven states passed similar legislation. In 2005, California went on to pass legislation requiring the labeling of cosmetic products that contain ingredients known to cause cancer or birth defects. More recently, the California Department of Community Affairs redrafted its flame retardant standard for upholstered furniture that had essentially required the use of brominated compounds. In 2007, the state of Washington passed a law restricting the use of the flame retardant, deca-brominated diphenyl ether, followed the next year by a toy safety standard law that sharply restricts the use of phthalates in children's products.[10]

These are all chemical-by-chemical chemical control laws. Recognizing the limits of this approach, some states have passed more comprehensive chemical policy laws during the past several years (see table 15.2). The Washington State law was described in chapter 10. The Minnesota and Maine laws have required state authorities to prioritize groups of chemicals that might endanger children and infants. Each of these laws authorizes state agencies to develop lists of chemicals of concern that are likely to expose children. From these lists, the state agencies can then identify a small number of chemicals each year as priority chemicals. In Maine and Washington, manufacturers of products containing those chemicals must either substitute the chemicals or present alternative assessments that then guide government responses. Washington has developed a priority

Table 15.2
Comprehensive State Chemicals Policies

State Law	Comprehensive	Number of Chemicals	Number of Priority Chemicals	Requires Chemical Use Reporting	Requires Alternatives Assessment	Provides for Government Response
Massachusetts, *Toxics Use Reduction Act of 1989*	Only listed chemicals	1300	TRI PBTs plus 4 highly hazardous substances	For all listed chemical meeting a threshold	As a part of toxics use reduction plans	State and university technical assistance
Oregon, *Senate Bill 737* (2007)	Persistent pollutants	2000	118	-	-	State makes recommendation for action
Washington, *Children's Safe Products Act of 2008*	All chemicals of concern to children	476	66	For products containing priority chemicals	State does alternatives assessments for priority chemicals	-
Maine, *Kids Safe Products Act of 2008*	All chemicals of concern in children's products	1700	49	For products containing priority chemicals	State does alternatives assessment for priority chemicals	State may promulgate a prohibition on product marketing
Minnesota, *Toxic Free Kids Act of 2009*	All chemicals of concern to children	1700	9	-	-	State makes recommendations for use reduction
California, *Safe Consumer Products Regulation (2008)*	All chemicals of concern in products	3000	May do 2 to 4 per year	For priority chemicals in priority products	For priority chemicals in priority products	State responses range from product labeling to phase outs

list of sixty-six chemicals that are of high concern for children, while Maine has identified forty-nine as priority chemicals of concern and selected two for regulatory attention: bisphenol A and nonylphenol ethoxylates.[11]

These state laws are similar in function to the safer chemical policy framework proposed here and demonstrate the emerging capacity at the state level to manage chemicals in a comprehensive manner. In the states where there currently exists infrastructure for prioritizing chemicals and requiring firms to conduct alternatives assessments, the capacity for encouraging chemical action plans already exists. In addition, some states have begun to work cooperatively to form common chemical databases and inventory chemical use. The Interstate Mercury Education and Reduction Clearinghouse (IMERC) described earlier provides a central clearinghouse for fifteen states to collect and manage information on the mercury-containing products regulated by those states. In addition, the clearinghouse provides information and guidance materials on requirements for mercury-added product notification and labeling, state mercury bans and restrictions, and mercury collection programs. More recently, twelve states have established the Interstate Chemicals Clearinghouse (IC2), a cooperative chemical information sharing network for state environmental agency staff charged with implementing new chemical policy laws. The IC2 is working to set up a common Internet access portal for state information on chemical hazards and a common format for collecting information on chemicals in products. The IC2 *Guidance for Alternatives Assessment* was described in chapter 12.[12]

Although most states (with the exception of California) are too small to broadly affect national chemical markets, they are well positioned to guide and assist in drafting chemical action plans and assist with chemical substitutions at the facility level. State agencies and legislatures are close to local industries, and they are often better prepared than their federal counterparts to tailor chemical programs to the specific problems and needs of those sectors prominent in their jurisdictions. Over the past two decades, the state pollution prevention programs have been effective in assisting local enterprises in phasing out chlorofluorocarbons and other ozone-depleting substances, replacing chlorinated solvents used in metal parts cleaning, and reducing the use of high-VOC paints and lacquers in furniture coatings.

State activism also offers a means to experiment with new policy ideas before they are adopted nationally. In doing so, the multitude of state laws can drive industry to seek federal policy in order to harmonize the

diverse and sometimes contradictory requirements of different state statutes that can create a complicated national market. Such was the case
with the multiple state worker right-to-know laws of the 1980s, which
led industry associations to reconsider their opposition to and instead
support a national OSHA hazard communication standard. As Mike Belliveau, the founder of the Environmental Health Strategy Center in Maine,
says, "The states have been the true leaders in environmental health policy. Between 2003 and 2011, 19 states adopted 93 chemical safety policies
and the majority of these passed with bipartisan support."[13]

15.4   Engaging International Initiatives

The U.S. government need not set chemical policy in isolation. The chemical registration process in REACH is driving the generation of a basic set
of chemical information for all chemicals in international commerce. The
GHS and the CLP are standardizing the procedures for the characterization and labeling of hazardous chemicals worldwide. Even with major
gaps, the international chemical treaties and SAICM have established
global processes for chemical management and a goal for the sound management of chemicals. U.S. initiatives do not need to duplicate these
efforts. Instead, the chemical strategy in this country should, where possible, adopt international standards and procedures and harmonize program elements with international policies.

A good starting point would be the ratification of the Basel, Rotterdam, and Stockholm Conventions. The progress being made under these
conventions is slow, but these agreements are setting standards for global
chemical manufacture and trade. The federal government has little role in
the development of these agreements because the U.S. Congress has not
ratified them. Indeed, because there is no need to internationally account
for trade in chemicals or hazardous chemical wastes, the EPA has made
little effort to quantify or track chemical and chemical waste exports,
even where such tracking is required under FIFRA and the federal
Resource Conservation and Recovery Act.[14]

U.S. chemical profiles could be harmonized with the data required for
the chemical safety dossiers under the Europe Union's REACH Regulation and similar Asian chemical policies. Chemical registrations implemented in the United States could harmonize with REACH registrations.
Those substances of very high concern identified under REACH for
authorization could also be identified as Tier One Chemicals of Very High
Concern under the Universal Classification of Chemical Substances.

There already exist effective international chemical information data-bases and chemical information sharing programs among nations through cooperative agreements and the OECD's E-Chem Portal. Ongoing efforts to harmonize chemical testing protocols and consider protective policies on nano-scaled chemicals under the OECD can continue to set standards for similar policies here.

Where the United States could take a lead among nations is in the development of chemical action plans and government responses to them. The broad systems orientation suggested here for chemical action plans, economic sector workgroups, and the development of alternatives assessments and comparative chemical hazard assessments are innovative features that could move international chemical policy forward.

The long isolation of the United States from participation in international environmental conventions works against U.S. interests and limits the service that U.S. policymaking could provide in shaping international strategies such as SAICM. Senate ratification of the Basel, Rotterdam, and Stockholm treaties and the changes that would thereby be needed in U.S. law would help to harmonize U.S. law with international law. Although such ratification would be difficult in the current Congress, this should not limit administrative efforts to coordinate programs with intergovernmental treaties and agreements. The recently announced agreement to harmonize trade and regulatory policy between the United States and the European Union provides opportunities for improving both U.S. and European policies, although this will require a concerted effort by those who are seeking to harmonize to the highest standards.[15]

15.5   Engaging Civil Society Organizations

The United States has a large and highly active assortment of nongovernmental organizations (NGOs) focused on environmental and public health protections. Many of these organizations address chemical hazards as environmental media pollutants, consumer product risks, or legacy waste site contaminants. The Center for Health and Environmental Justice has spearheaded a campaign to reduce the use of PVC, particularly in schools. West Harlem Action for Environmental Justice has campaigned in New York City for clean air, healthy communities, and toxic-free products for low-income families. The legislative campaigns around lead, mercury, aldrin, dieldrin, PVC, brominated flame retardants, and bisphenol A have been largely driven by NGO advocacy organizations.

However, many of these organizations have gone beyond single chemical opposition to more comprehensive policies. The spirited and well-crafted campaigns of the environmental advocacy organizations drove the enactment of California's Proposition 65 and Massachusetts' Toxics Use Reduction Act. As noted earlier, Health Care without Harm has worked with the health care sector to reduce the use of chemicals of concern in hospitals and medical services. Campaigns by the Environmental Health Strategy Center in Maine and Washington Toxics Coalition in the State of Washington are responsible for those states' children's safe chemicals laws, and the advocacy of the Ecology Center in Michigan led to a Governor's Green Chemistry Executive Order. And, it is the international environmental advocacy campaigns that have popularized globally the Swedish concept of a "toxics-free environment."

Most of the U.S.-based campaigns have focused on state and local governments. Influencing national debates over chemical policy requires a significant amount of resources, and with the exception of a few large Washington-based environmental advocacy organizations, this means the federal political arena is dominated by national business associations, well-funded lobbying firms, and substantially endowed Washington policy analysis centers. However, state legislatures and state and local agencies are more accessible to community-based organizations and the less-well-funded public health and environmental advocacy organizations. State-based advocacy organizations loosely organized through a network called Safer States have been particularly effective in lobbying for state chemical policies that have ranged from marketing prohibitions on products containing bisphenol A residues to restrictions on lead and phthalates in children's products. More than one hundred environmental and environmental justice advocacy organizations participate in a national network called Coming Clean that has workgroups on topics ranging from hazardous dump site remediation to emerging chemical technologies.

Although these organizations have focused on state government authority, they have also been successful with economic sector chemical campaigns that by-pass government mandates. The apparel and footwear industry was pressed to collaborate on reducing chemical wastes by Greenpeace. The Campaign for Safe Cosmetics has negotiated directly with the cosmetics industry. Health Care without Harm's campaign to press health care group purchasing organizations to avoid mercury and PVC in product purchases was described earlier. By advocating more directly with industry, these NGOs have discovered the benefits of direct engagement. This has led to a deeper understanding of the leverage that can be achieved by working with key firms or progressive leaders in an economic sector and pressing and assisting them in seeking safer alternatives. Tools such as Clean Production Action's *Green Screen* and the Healthy Building Network's *Pharos* were developed to be used by industry to assist in adopting safer chemicals. Environmental Defense Fund set up an office in Bentonville, Arkansas, to work directly with Walmart's corporate staff to implement environmental objectives into Walmart policy, including a shift toward safer chemical ingredients in products.

Such close working relations provide firms with early insight into the ideas and emerging concerns of civil society organizations and provides the advocacy organizations with an intimate relationship in helping firms make desired changes. Such relationships require trust and a careful respect for roles and responsibilities because there is always the risk that objectives and actions could be misunderstood or manipulated.

The nation's long history of open government provides opportunity for these civil society organizations to be involved in the drafting and implementation of government chemical policies. Such involvement offers many benefits, such as increased public awareness, additional information and ideas, pubic legitimacy, dedicated constituencies for new laws and regulations, and popular advocacy to counterbalance resistance to new policies. Engaging civil society organizations requires special attention to resources because many community-based and environmental justice organizations operate on quite limited budgets. This need not create a barrier to participation. During the 1990s, the EPA organized several community-based programs to advance environmental justice and local environmental protection, and some of these programs provided financial aid to citizen groups to assist them in participating in these efforts.[16]

The years ahead will see growing sophistication and influence from civil society organizations. The need for these organizations to educate activists, mobilize public sentiment, create innovative ideas and tools, and provide political constituency will grow as their demand for a "toxic-free environment" grows. Government authorities need to build stronger relations with these organizations and benefit from their ideas and energy.

15.6   Restructuring Federal Agencies

Safer chemical policies require a coordinated approach among a range of agencies and programs and a harmonized hub for chemical information generation, analysis, and promotion. Governments are composed of inter-related systems that function best when they are well coordinated and consistently focused on concrete goals with real timetables. With so many different federal and state agencies engaged in chemical policy, this will require commonly shared goals and new integrative structures. For this, it is reasonable to consider the need to restructure existing government regulatory agencies and create a new central agency that could promote cooperation and reduce inefficiencies.

The 1970 FIFRA amendment that transferred responsibility for pesticide safety to the EPA was a first good step in federal program integration. Not only did this free the Department of Agriculture from both regulating and promoting food and fiber production, it merged the competence of two scientific and regulatory staffs into a more integrated service. But this was only the first step. A similar transfer of responsibility could merge responsibility for product safety, workplace safety, and environmental safety into one unified division. The safety standards for chemicals contained in occupational health and safety and consumer product regulations are often different from and less protective than the standards in environmental regulations even where they involve the same chemicals. The budgets are even more wildly unequal. The current EPA is a substantial agency with an annual budget over $8 billion, whereas the two agencies charged with occupational health and consumer safety are miniscule (with a combined budget under $700 million). A bigger agency with more cross-functional integration could reduce the current discrepancies, create one harmonized set of standards, and coordinate inspection and enforcement activities into a more efficient and less inconsistent enterprise.

J. Clarence Davies, a past administrator at the EPA, suggests such a merger of safety agencies which would create a new "Department of Environmental and Consumer Protection." This would be a new cabinet-level department that includes the current EPA and other agencies, such as the Occupational Safety and Health Administration, the U.S. Geological Survey, the National Oceanic and Atmospheric Administration, the National Institute of Occupational Safety and Health, and the Consumer Product Safety Commission. This would be an agency of 43,000 full-time equivalent employees with a budget of some $18 billion, making it about half the size of the Treasury Department and a quarter the size of the Department of Homeland Security. Such a new agency could more centrally integrate those government services responsible for environmental, consumer, and occupational health science and more effectively coordinate the standard setting and regulatory functions that are now so ill matched. In Davies' words, "it would be a science agency with a strong regulatory component rather than a regulatory agency with a science component."[17]

Although a big, new department-level, science-based regulatory agency might offer a more robust and effective safety agency, it might also become an even bigger bureaucracy that resists rather than encourages innovation and flexibility. In addition to such a serious overhaul of the regulatory agencies, it would be useful to create a smaller, lighter agency to monitor and oversee the transition of the chemical market. This might be a new National Chemicals Agency. The Swedish Chemicals Agency (KemI) in Sweden and the European Chemicals Agency in Helsinki offer interesting models of government divisions that oversee chemical information and management programs (see box 15.1).

Such a new federal agency could be a non-regulatory division that serves as a focus for collecting and managing chemical information, providing public access to chemical information, maintaining information on safer alternatives to hazardous chemicals, receiving and cataloguing alternatives assessments, and periodically reviewing and evaluating the effectiveness of chemical policies and programs. Such an agency could be the

Box 15.1
Models of Governmental Chemical Agencies

Swedish Chemicals Agency (KemI)

KemI is a supervisory authority under the Ministry of the Environment that works in Sweden and the European Union to promote legislation and rules that contribute to achieving "a nontoxic environment." KemI keeps a products registry, maintains chemical hazard databases, assesses the risk of chemicals, provides support to local authorities, and maintains statistical information on chemicals.

European Union: Chemicals Agency

The new European Chemicals Agency's mission is to ensure consistency in chemical management across the European Union and provide technical and scientific advice, guidance, and information on chemicals. ECHA is responsible for managing the chemical registrations and authorizations under REACH and notifications under the CLP (Classification, Labeling, and Packaging Regulation) and the Biocide Directive.

Chemical Information Management Service noted earlier that collects and evaluates the chemical profiles and develops and manages the Universal Classification of Chemical Substances. In addition, the chemicals agency could serve to coordinate funding for the Green Chemistry and Engineering Initiative to promote new chemistry and engineering research and education.

A new chemicals agency would not replace the current regulatory agencies responsible for chemical management and regulation and health and environmental protection. These responsibilities would lie with divisions for environmental protection, consumer product safety, and occupational health and safety within the new Department of Environmental and Consumer Protection. Rather, a central chemicals agency would offer a parallel entity that serves as a focal point and resource for chemical information, characterization and classification, and safer chemical research and promotion.

15.7   Statutory Reform

Reconstructing government capacity could be advanced with changes in federal statutes. Amendments would be needed to FIFRA, FFDCA, the Consumer Product Safety Act, and TSCA, as well as changes to OSHA and some of the media environmental protection laws. These laws need to better balance command and control regulatory features with incentives for chemical conversion. Although work needs to continue on the slow procedures for setting safety standards and exposure controls, more attention needs to be given to generating chemical information, characterizing chemicals, and developing safer alternatives.

The chemical control statutes need to be more clearly directed with national goals, better integrated and coordinated, and less burdened by procedural complexities. Even with industry carrying more of the burden for generating chemical information and converting to safer alternatives, the government could use more authorities and capacity. More authority to register, test, and regulate cosmetics under the FFDCA would bring products containing chemical ingredients that are directly applied to the body in line with products that are intended to be ingested. A better means of tracking and evaluating consumer products in the economy is needed, and amendments to the CPSA could eliminate the law's current procedural burdens, provide national standards for chemical ingredient disclosure for both formulated and assembled products, and shift the responsibility for identifying and declaring chemical hazards in products from the government agencies to product suppliers.

The Food Quality Protection Act (FQPA) has done much to update pesticide tolerance levels and encourage the use of safer pesticides and plant protection programs. However, the risk-based approach to permitting pesticide use needs to be augmented with a hazard-based approach that requires chemical action plans for pesticides deemed to be priority chemicals. The recent amendments to the Consumer Product Safety Act provided more regulatory authority on lead and phthalates in children's products but offered little reform to the operational features of a very crippled statute. A good reform of the statute might close out the appointed board, which has encouraged rather than insulated the agency from political pressures and merge its functions into the new consolidated Department of Environmental and Consumer Protection. Additional authorities could provide for government responses to chemical action plans that involve product labeling, restrictions for priority chemicals, and product recalls.

TSCA needs reform. The EPA could use more authority and less procedural hurdles to require the registration of chemicals, the development and submittal of chemical profiles (including a minimum data set), the classification of chemicals into the Universal Classification of Chemical Substances, the drafting of chemical action plans and the setting of agency responses not bound by the "least burdensome" procedural test, or the requirement to defer responses to other agencies. In parallel with the pesticide law, the legal standard for action needs to be changed from "unreasonable risk" to "a reasonable certainty of no harm." The agency needs more authority and more conducive procedures for collecting chemical use and exposure data and better balanced procedures for maximizing public access to chemical information while protecting confidential business (CBI) information. A common registration for all chemicals could dispense with the current distinction between new and existing chemicals and substitute the current insufficient PMNs with more robust chemical profiles.[18]

There may also be a need for an entirely new statute that focuses more centrally on the conversion of the chemical economy. Such a statute might create the new chemical agency and set up procedures for government agencies to work directly with leading firms and economic sectors in developing chemical action plans. The Departments of Energy and Commerce have such authorities and use them effectively to develop sector-based energy conservation and national business development plans and programs.

Over the past several years, there has been an active coalition of environmental, labor, environmental justice, and public health organizations mobilized into a national campaign to reform TSCA. The coalition, named "Safer Chemicals, Healthy Families," has worked with leaders in Congress to draft a bill that was approved by the lead Senate committee in 2012.[19] Under this bill, firms would provide a minimum set of health and environmental effects data for all existing chemical substances and mixtures and for all new chemicals along with the premanufacture notice. Firms wishing to claim CBI protection would be required to substantiate their claims and achieve agency approval. The EPA would have authority to establish a list of priority chemicals and, for each priority chemical, establish a health-based safety standard that ensures a reasonable certainty of no harm. For a priority chemical to stay on the market, chemical manufacturers or suppliers would be required to determine and declare that their chemical meets the safety standard. Tailored safety standards would be set for persistent and bioaccumulative chemicals, and expedited risk management action would be required for those chemicals deemed to be of highest concern. Special programs would be established to address the risks to vulnerable populations, such as children and people living in highly contaminated "hot spots," and funding would be available to support a network of green chemistry research centers.

In 2013, a second bill, largely sponsored by chemical industry lobbyists, was introduced that goes nowhere near as far. The "Chemical Safety Improvement Act" strips away many of the reporting features of the earlier bill, makes the EPA responsible for determining whether a chemical meets the safety standard, removes the expedited action for priority chemicals, deletes the "hot spots" provision and the green chemistry funding, and gives the EPA authority to preempt state chemical management activity if the agency lists a chemical as a "high priority" or "low priority." There is little of reform here, and the state preemption provision places a disturbing chill on state regulatory actions.

It is too early at this writing to predict whether TSCA will see reform. The preexistence of TSCA and the hostile resistance of the chemical industry trade associations, combined with a historically conservative Congress, has restrained what even the most ardent reformers have been able to accomplish. Similar campaigns to amend or reform other federal statutes or draft completely new laws would most likely meet the same hurdles. Although such campaigns are needed and will surely continue, it is important to consider changes that do not require major changes in federal statutes.

15.8   Providing Adequate Government Resources

Resources for chemical policy at the EPA have always been marginal. Shortly after passage of TSCA, the budget for chemicals grew by the year, but after 1980, as the EPA budget was scaled back, funding for TSCA was substantially reduced. From 1981 to 1986, the budget for chemical management at the EPA was cut by 27 percent. Throughout the 1990s, the resource level for chemicals remained constant even as new responsibilities for lead, the voluntary programs (the 33/50 Program, etc.), pollution prevention, and the TRI expansion were added. In 1999, the agency budget for chemicals was $30 million and supported a staff of 270. By 2013, the budget for chemical safety had risen to $54 million, but the staffing levels had changed little. Compared with the significant resources that the EPA dedicates annually to air quality and climate change ($1.1 billion) and protecting water quality ($3.8 billion), the resources allocated for chemicals are miniscule. There is some irony in this, as converting to safer chemicals could substantially reduce the release of hazardous chemicals and diminish the compliance responsibilities of those firms that currently maintain facility air emission and water discharge permits.[20]

A safer chemical policy requires resources for chemical information management, working with firms and sectors, and maintaining regulatory compliance. Resources are also needed for advancing research on chemical effects and safer alternatives. When the federal government has committed to advancing science, it has been willing to commit large amounts of funding. The National Science Foundation provided $3 billion to advance the Human Genome Project, and the National Nanotechnology Initiative has channeled some $12 billion in research funding over the past decade through ten federal agencies. A safer chemical policy framework should include a significant chemical research program. A bill to support green chemistry research, the Green Chemistry Research and Development Act, has passed the House of Representatives; however, the bill has never made it through the Senate. This bill would provide $165 million over three years to support green chemistry research. Although this would be a beginning, a more appropriately scaled federal Green Chemistry and Engineering Initiative should involve several times this allocation.

Generating such funds need not strap the federal budget. The registration fees suggested in chapter 14 could be used to generate the revenue for such expenditures. Registration fees are already assessed on pesticides and drugs. A simple annual registration fee of $5,000 on the 7,600 chemicals with production or importation above 25,000 pounds (based on the EPA's 2012 Chemical Data Reporting reporting) spread across 1,500 chemical suppliers would generate $38 to $60 million per year. If a higher annual registration fee of $20,000 were set for the 1,169 high-production-volume chemicals (reported under the EPA High Production Volume Challenge) spread across some 250 chemical suppliers, an additional $23 million would be generated annually. The most effective fee might be based on the degree of hazard of a chemical (represented by the tiers in the Universal Classification of Chemical Substances) and the production volume. Assuming, conservatively, that some 2,000 chemicals are classified as Tier One: Chemicals of Very High Concern (just the CMRs) and 10 percent are manufactured or imported in volumes over a million pounds per year, a fee of 10 cents per pound would generate at a minimum $20 million annually. Another $15 million might come from lower per pound fees on chemicals in Tiers Two and Three.

A fee on chemical production could have a negative impact on domestic chemical production. A fee on chemical importation would avoid this but have an impact on domestic product production, even if it could sustain a challenge under the World Trade Organization rules. A more effective fee, and one more appropriate to a consumer economy, would be a sales tax or fee placed on products that contain chemicals of high concern. Although such a fee would be less likely to burden U.S. chemical production, it would require a sophisticated means of accurately reporting the chemical constituency of products.

However, an even more complex scheme was used during the 1990s for funding the "Superfund" hazardous waste cleanup program under the Comprehensive Environmental Response, Compensation, and Liability Act of 1980 (CERCLA). Under that law, a 9.7 cent annual tax per barrel of crude oil, a variable annual per barrel tax on a defined set of hazardous chemicals, an annual corporate "environmental tax," and an appropriation from general federal revenues were used to generate a fund of some $1.6 billion by 1996.[21]

15.9   Reestablishing Government Authority

Government authority is important in affecting a safer chemical policy framework. The kind of government functions suggested here could be deemed "interventionist" by those who champion a "hands-off, laissez faire" approach to market performance. However, as noted earlier, without adequate and accessible chemical information and the means to make chemical sales reflect the full social and environmental costs of chemicals, the market alone cannot promote safer chemicals. The ideas presented here suggest the kind of "industrial policy" that was debated during the 1980s and avoided by the many administrations that followed. However, new interest in market intervention has gingerly emerged since the economic recession of 2008-2009. That recession gave rise to some fairly dramatic government interventions to bail out financial institutions, stabilize banks, all but nationalize the auto industry, and create an enormous economic stimulus program. Pressure to reduce unemployment, stimulate growth, "rebalance" the economy away from financial services, and respond to the ever increasing economic competitiveness of China have stemmed some of the conventional opposition to government programs. The reconstruction of government chemical policy envisioned here may sound more compelling than it would have several years ago.

Whether Congress will enact broad new chemical policies or substantially upgrade current ones remains uncertain at this time. However, much can be accomplished within current structures. The federal government has significant authority now, although often procedurally or politically restrained. The EPA could develop a national plan for chemicals and set national goals without new statutory authority. New statutory authority is not needed for government agencies to work more closely and cooperatively with industry sectors, nongovernmental organizations, the states, and international organizations. The conventional science-based research funding agencies and the National Research Laboratories could increase funding for green chemistry without new legislation. Tools such as life cycle assessment and alternatives assessment are being developed without government involvement. Product labeling programs, retail and institutional procurement initiatives, and industry standards and codes of conduct are all advancing without government involvement.

Bolder initiatives such as the creation of a national chemical agency or a new Department of Environmental and Consumer Protection would require congressional approval. However, many other more immediate changes, such as enhancing data collection, reforming CBI procedures, creating a tiered chemical classification framework, prioritizing chemicals, and designing chemical action plans, could go forward without new legislation. Such efforts would need to be carefully crafted to withstand industry challenges, congressional oversight, and court reviews, but they could be launched as pilot programs or timed experiments or set up where there are cooperative industry user groups or willing economic sectors.

The environmental and consumer advocacy campaigns, the ongoing initiatives inside leading chemical and product manufacturing industries, and the several initiatives in universities and firms to redirect chemistry toward safer chemicals need to be united with these opportunities for a more activated government. This coordination may emerge gradually, but it could be more effectively developed with a conscious effort to bring together leaders from each of the strategic fronts to create an integrated means for ongoing collaboration.

In time, the chemical market will be a driver of government initiatives. As REACH expands with comprehensive registration for well over a hundred thousand chemicals and authorization becomes mandatory for a host of chemicals of high concern and as the European Union directives force the redesign of products, the chemical market will change. As legislation harmonized with European laws continues to appear across Asia and other developing countries and leading retailers and product manufacturers continue to "deselect" some of the chemicals of highest concern, further shifts will occur in the chemical market. Even the slow pace of the international treaties—the Basel, Stockholm, and Rotterdam conventions—will continue to put pressure on and shape the broader chemical market. Such changes will most likely reduce resistances to chemical policy initiatives in the United States, and, eventually, what is not possible today will look more attractive in the future if for no other reason than promoting national competitiveness.

Existing government authority will continue to provide a basic platform for chemical control policies in the United States, and where possible, new safer chemical programs and authorities will be achieved. The struggle for safer chemicals by environmental and public health advocates, state legislators, government program directors, and leading business and business collaborations will continue. Reforming government authority may not be the most compelling driver for a chemical conversion strategy, but it will continue, and even gradual changes will lay the groundwork for better days.

16

Solving the Chemicals Problem

The scale of changes envisioned here is significant. Converting the chemical market, restructuring the chemical industry, redesigning the practice of chemistry, and reforming government laws, policies, and agencies—none of these can be accomplished without enormous effort.

The current chemical control policies are insufficient and too often ineffective. New approaches to chemical management are needed in the United States because a large body of science has documented the harmful potential of too many chemicals on the current chemical market; because biomonitoring studies now reveal how many of these chemicals are present in human bodies; and because the science of chemistry can produce safer alternatives. Risk assessments might show that the small doses of synthetic chemicals in humans are "acceptable." That misses the point. This is not a judgment about science; it is simply pragmatic—if technically sufficient, cost competitive, and safer alternatives to hazardous chemicals can be developed and made available, they should be adopted. Indeed, there is an economic imperative here as well. New chemical policies are needed in the United States to keep U.S. industry and the U.S. economy competitive among those nations that are driving the international chemical market.

In many cases, this will require rewriting and amending existing chemical control policies, but in other cases, new chemical management approaches could be built on the foundations of existing chemical control laws by redesigning and expanding programs and coordinating government initiatives with industry, civil society, state, and international initiatives. Federal legislative reform may be possible within the current national political climate, although efforts such as the ongoing campaign to reform TSCA suggest just how difficult and potentially counterproductive this may be.

However, much more than government policy reform is envisioned here. Only by recognizing the changes now emerging in the market, in industry, and in science and working to coordinate, harmonize, scale-up, and encourage those changes can a force large enough to effect really big system-changing transformations be assembled. Government authorities need to be seen as one of the drivers and facilitators of change, but only one among several.

16.1   Charting a Way Forward

This book has attempted to suggest a way forward that recognizes current initiatives and suggests ways to extend and better focus them. This will require both a new policy framework—what is here called a safer chemical policy framework—and a more integrated and aggressively pursued strategy—a chemical conversion strategy. Rather than a single government policy, the policy framework envisioned here is composed of many policies among many private and public institutions.

The first and most fundamental recommendation is *to reframe the chemicals problem* from a focus on a small—or even a large—number of hazardous chemicals to a broader focus on the system of chemical production and consumption—the chemical industry and the chemical market. This broader focus opens up important opportunities for shifting chemical policy and redirecting science and government agencies.

The second recommendation is *to recognize the multiple forces that are now driving the development and adoption of safer chemicals and to integrate and coordinate these strategic drivers into a more effective vehicle for change.* Restructuring government capacity needs to be coordinated with the forces that are driving the chemical market, changing the chemical and product manufacturing industries, and redirecting chemistry.

The third recommendation is *to address this broader definition of the chemicals problem with solutions that are more comprehensive, hazard-based, and system changing.* Fully addressing the chemical production and consumption system means addressing all chemicals from those of most concern to those of least concern and employing new instruments and tools to guide the conversion of the chemical production and consumption system from hazardous chemicals to safer alternatives.

So those are the big recommendations. The text has also noted a range of more specific tasks that need to be accomplished to move the chemical conversion strategy forward. These include:

1. All chemicals on the commercial market need to be characterized with a minimum base set of chemical information and classified as to their degree of hazard.

2. A harmonized chemical information reporting system needs to be established to increase the flow of information in production supply chains and to provide product consumers, end users, and end-of-life managers with the information that they need on chemicals in products.

3. Retailers and institutional and government procurement organizations need to expand their focus and reframe their business models to ensure customers that the products that they offer do not contain chemicals of concern.

4. Leading product manufacturing firms need to formalize their chemical policy programs, increase their chemical information transparency, actively avoid chemicals of concern, and identify and adopt safer chemicals in production processes and products.

5. Firms need to collaborate across economic sectors to develop plans, databases, and tools for identifying safer alternatives to chemicals of concern and working to adopt those alternatives.

6. Chemical manufacturers and suppliers need to identify opportunities for restructuring production processes and products to reduce or eliminate the use or production of chemicals of concern.

7. Chemical action planning methods and tools such as alternatives assessment, comparative chemical hazard assessment, life cycle assessment, and risk assessment need to be more fully developed and used to promote the adoption of safer chemicals.

8. Environmental and consumer advocacy organizations need to expand their market campaigns and more strategically target chemicals, companies, and economic sectors where pressures would exert system-changing outcomes.

9. States and local governments need to develop policies and programs to promote reduction in the use of hazardous chemicals in production processes and products. This might involve new laws and regulations, tailored fees, expanded technical assistance services, targeted procurement policies, and direct investments in safer product development.

10. Federal government capacity and collaboration needs to be strengthened and better focused on the development and implementation of safer chemical policies and programs. This should include amending and reforming the chemical control laws (TSCA and CPSA, in particular).

11. Chemists need to be encouraged and supported in designing safer chemicals that avoid the use or release of hazardous chemicals in chemical synthesis and processing.

12. Broader awareness and understanding of the health and environmental effects of chemicals needs to be integrated into the teaching and curriculum of the sciences, engineering, and business management.

Along the way, the text has identified new or reformed organizations, programs, and tools that could serve to facilitate this new policy framework:

• New Chemical Information Management Services are needed to manage the profiling and characterization of chemicals, promote the generation and diffusion of chemical information, support the development of safer chemicals and nonchemical alternatives, and promote the adoption of these alternatives. This could be a new National Chemicals Agency, such as the European Union's European Chemicals Agency, or parts of this service, or all of it, could be implemented through new or existing private bodies, such as professional or scientific organizations, standard-making bodies, nongovernment institutions, or private corporations.

• A Universal Classification of Chemical Substances is proposed to create a map of all chemicals, ranking them in terms of their hazards and preference. Rather than relying on one entity to characterize and classify chemicals, this classification scheme would be populated over time by chemical producers and users voluntarily locating chemicals into the classification tiers. The classification framework and the process of populating it could be managed by the Chemical Information Management Service.

• For those chemicals or chemical groups identified as priorities, individual corporations, groups of firms, or economic sectors should prepare "chemical action plans" that identify and evaluate safer alternatives and set timetables and objectives for adopting them. Alternatives assessments and comparative chemical hazard assessments would be critical to these plans.

• Although individual companies are now taking the lead on moving toward safer chemicals, more progress could be achieved where leading firms in economic sectors work together to create databases, tools, and plans. Sector Chemical Workgroups are proposed here to coordinate these sector-based collaborations and oversee the generation of chemical action plans.

• Where safer alternatives are not available, research should be encouraged to support the development of safer chemicals and technologies. A National Green Chemistry and Engineering Initiative is proposed to fund and support research into safer chemicals and nonchemical alternatives.

Is this enough to change a system as massive as the chemical economy? A comprehensive approach to all chemicals classified into a broad map of chemical preferences, increased chemical information that is accessible and flows more easily within supply chains and commercial product markets, new tools for supporting the adoption of safer chemicals, new linkages among progressive corporate managers, advocates and chemists, a vigorous program for developing safer and greener chemistry, and a strengthened government capacity to manage chemical policy are all means for affecting system change. These initiatives have strategic value, but they can only achieve system-changing scale where they are focused at sensitive points in the chemical production and consumption system. The previous sections have identified several places in this system where thoughtful attention or effective pressures could be exerted that would lead to potentially system changing outcomes. These include:

• Chemical and product component supply chains. Improving supply chain information flow and public transparency offers the foundation for more precautious decision making.

• The point of contact between chemical suppliers and product manufacturers. Product manufacturers offer the most significant drivers for changing chemicals in products, and brand-identified manufacturers are the customers that chemical suppliers are most likely to respond to.

• The point of contact between product manufacturers and retailers and institutional procurement organizations. Large-volume retailers and procurement organizations determine the range of products offered to end-use consumers and can effectively demand product manufacturers produce products free of chemicals of concern.

• *Investors* Large-scale investors can shape public corporation policies and could promote the production and use of safer chemicals.

• *Economic sectors* When firms work together in economic sectors on chemical management, they have greater capacity for collective innovation and productive market development.

• *Consumer networks* Consumers connected through social media can inform one another and coordinate collective responses large enough to pressure markets.

• *Government regulations* Regulations that are clear, forceful, and proceeded with enough time to adapt have historically been effective in changing chemical markets.

• *Research grant making* Scientific development is significantly influenced by government grants and investments, and these programs could be designed to promote safer chemical development.

• *Science education* Training students shapes the future and provides the intellectual and skill resources for long-term system change.

Identifying these and other points of leverage and being strategic in shaping interventions can make the many initiatives identified in the strategic fronts most effective. Those working inside and outside of business and industry and those working inside and outside of government all have opportunities for moving strategic changes forward. A chemical conversion strategy requires a campaign that links together these many initiatives. The task ahead is not to start such a campaign—there are many activities already engaged—but to organize and coordinate the activities that already exist, further develop their effectiveness, and fill in gaps where necessary.

16.2   Mobilizing for Change

A change as broad as what is imagined here cannot be realized through private negotiations or driven by a small group of elite individuals. It takes large and influential social movements to achieve major policy change in the United States. The struggle for passage of the Wilderness Act during the 1960s and the environmental protection laws of the 1970s were the direct result of broadly based national movements. More recently, the 2010 enactment of the Patient Protection and Affordable Care Act was propelled by an extensive grassroots campaign. Health Care for America Now, the campaign's central coalition, mobilized more than 1,000 organizations in all fifty states representing thirty million people and spent more than $53 million over a twenty-month period.[1] The absence of such a broadly based grassroots constituency is seen as one of the key factors that led to the failure to enact national climate change policy in 2010.[2]

Federal conditions are not likely to become more conducive to major policy initiatives any time soon. However, lessons for more promising broad-scale change can be drawn from the increasingly successful campaigns to codify gay marriage and legalize marijuana use. Both of these initiatives have arisen from broadly based grassroots movements that have grown in the face of strong opposition. Both campaigns have avoided the U.S. Congress and focused instead on private corporate policy, municipal ordinances, court decisions, and state legislation. Although there is no immediate possibility for new federal legislation in either of these two issue areas, the emerging patchwork of state, local, and private policies is fundamentally shifting national thinking on both gay rights and recreational drug use.

This reality suggests the value of a large and diverse movement that recognizes and includes all the various initiatives pressing for safer chemicals. Over the past forty years, a broadly based grassroots environmental health movement has emerged that focuses on the health and environmental harm caused by toxic and hazardous chemicals in the environment. Although this movement has roots deep in the early twentieth-century public health movement, its modern formation was galvanized by the popular protests during the late 1970s that arose around the hazardous waste landfilled at Love Canal, New York. Originally mobilized around hazardous wastes, the movement has grown to address chemical hazards in workplaces, schools, farms, neighborhoods, drinking water, food supplies, and, most recently, commercial products. It has had various names, such as the "toxics movement" and the "anti-toxics movement," and over the years it has grown to support and align with the broadly based environmental justice movement.[3]

Today, the environmental health movement is diverse and inclusive, with participants ranging from low-income members of communities of color, labor activists, and those disabled by chemical exposure to scientists, professionals, and economically advantaged consumers. It is a decentralized movement with hundreds of locally based and health-related organizations that focus on campaigns for waste site cleanup, pollution control, chemical accident prevention, pesticide reduction, regulatory enforcement, chemical control laws, and legal bans on chemicals of concern. This movement has supported the chemical-focused market campaigns, the new state chemical policy laws, and the campaign to reform TSCA described earlier.

This environmental health movement could carry forward the chemical conversion strategy envisioned here. It has the constituency, resources, and experience that would be needed to achieve the scale of changes envisioned. However, this would require accepting the reframing of the chemicals problem to focus on the systems of production and consumption that generate an economy based on hazardous chemicals. Taking this

systems perspective would offer a new perspective on the multiple linkages among the organizations and initiatives working to replace hazardous chemicals and invite an examination of the systems of chemical
production and consumption to identify the most promising levers for
change. This perspective would encourage recognition of new allies
among those inside government and businesses that would be needed to
more fundamentally reform the chemical market and the chemical industry. Scale matters. It will take a large and powerful movement to effect the
ambitious objectives of a chemical conversion strategy. However, time
maters as well. It takes time for the multiple activities of hundreds of
groups to be linked together and for solidarity to emerge among those
finding a common purpose. For now, the environmental health movement
would need to focus on increasing awareness, integrating the multiple
initiatives, and developing the policies.

Increasing Awareness

The key to success in reducing the production and use of hazardous
chemicals lies in viewing the chemicals problem in a systems perspective.
This requires recognizing that shifting the market toward safer chemicals
requires transforming the chemical industry—both its technical infrastructure and corporate mission. Whole branches of chemical production
may need to be redesigned to change the end-use chemicals that are distributed for manufacturing products. The retailers, product manufacturers, and chemical distributors in the supply chain would need to
communicate to chemical manufacturers that the increasing consumer
demand for safer products is important enough that the chemical industry
sees value in adopting green chemistry principles and adds health and
environmental protection to its chemical design parameters.

Seeing the problem in this more expanded vista does not mean that
well-focused initiatives on specific chemicals or at specific firms and economic sectors are not useful. Rather, it means that such specific initiatives
are encouraged and strengthened, such that the cumulative effects of these
many initiatives reaches the scale needed to bring about broader system-
transforming changes. This involves integrating the strategy and developing the policies.

Integrating the Strategy

Converting the systems of chemical production and consumption requires
identifying the points of leverage within those systems, designing initiatives that press those levers, and integrating the initiatives such that

changes driven by one initiative compliment and support changes driven by other initiatives. Although this does not require a tight coordinating structure, it does require cross-system awareness, institutional structures for collaboration, and mutual support among initiatives.

The pace of this development need not be rushed. There is value in the flowering of many initiatives. A rush to institutionalize the diverse energies could encumber rather than facilitate the search for appropriate organization. It would be more productive at this time to invest in networks and clusters of activities and conscious efforts to link together people and activities across the different strategic fronts.[4]

Developing the Policies

Safer chemical policies make up the central framework for implementing the chemical conversion strategy. These policies cannot provide a single plan. They offer a collection of activities and authorities; some are already in place, and others are in need of design and installation. Chemical registration, a universal chemical classification framework, identification of priority chemicals, chemical regulation, economic incentives, and monitoring and evaluation are all conventional government functions. However, if not federal agencies, the states or some of them (e.g., the Interstate Chemicals Clearinghouse) could perform these functions. Already some of these functions (e.g., chemical registration and classification) have been completed by the European Union. Corporations, corporate associations (the Sustainable Apparel Coalition, the U.S. Green Building Council), nongovernmental organizations, and advocacy coalitions could be performing these functions as well.

Only time and history can determine whether TSCA, FIFRA, FFDCA, and the CPSA can be overhauled. Even then, the constraints under which the federal government operates in terms of budget, resources, and political will shall still persist. The political context that settled into Washington during the 1980s endures today. That is why the safer chemical policy framework proposed here is multi-organizational and multifaceted. It is why chemical information management, priority setting, and the methods and tools of chemical substitution are presented as government and/or nongovernment functions and why a sector-based approach that works with those firms that are ready to move is proposed. It is also why this text has searched through the various strategic fronts for drivers and initiatives that are market-based, industry-based, science-based, and based on partnerships with states, international institutions, and civil society organizations. Converting to safer chemicals will take many organizations and initiatives from these strategic fronts working together. But one way or another, a chemical conversion strategy needs to be advanced. That is why a safer chemical movement is crucial.[5]

This movement should reach beyond environmental health to find allies in other similarly directed movements, the labor, environmental justice, and climate change movements, in particular. The labor movement has a long history in advocating for workplaces that are safe for workers and offers a wealth of experience in negotiating chemical safety standards. Labor leaders could provide valuable input to the Sector Chemical Workgroups and support workers in negotiating for chemically safer working conditions. The environmental justice movement based in the nation's communities of color has long struggled with the threat of hazardous chemicals in the form of dangerous pollutants and mismanaged wastes. Expanding the alliance between the two movements would strengthen the capacity to protect the most vulnerable members of society and provide the moral strength of a truly diverse constituency in the drive for safer chemicals. The use of fossil fuels is central in generating both the chemicals problem and the threat of massive climate disturbance. Coal, oil, and natural gas provide the feedstock for both hazardous chemicals and carbon emissions. The highly integrated technological and corporate structure of the chemical and energy industries creates a massive obstruction to change, but also suggests that at a systems level, converting to safer chemicals and reducing carbon emissions should be one strategy. There is a ripe opportunity here to strategically align the environmental health and climate change movements and build a force large enough to bring about a truly sound approach to managing the planet's carbon resources.

Government action needs to be supported with private action. Where governments may be inhibited by dominant corporate interests, the multiplicity of civil society organizations, leading product manufacturers, retailers, institutional buyers, chemists, and product designers acting in the "open space" of the public commons can adopt chemical policies and carry out programs to promote safer chemicals. Internet transparency and the "crowd sourcing" potential of social media offer new means to assess complex information and make decisions based on broad public input and debate. There will continue to be a need for advocacy organizations raising public and professional awareness and identifying and putting pressure on companies and chemicals of concern. But there will also be a need for other nongovernment organizations made up of technical professionals working with industry to identify safer alternatives, prepare chemical action plans, and convene sector-based programs to implement chemical conversion programs. This requires a presence that is more intimately involved with the chemical-producing and chemical-using industries, more technically savvy and more openly engaged in direct dialogue with the technical managers of industry.

There is a new environmentalism emerging—an approach less focused on protecting endangered air sheds, water bodies, landscapes, and species and more focused on creating an economy that does not jeopardize the planet.[6] This new environmentalism is promoted by several far-sighted civil society organizations but shared with those in industry and government who see that redesigning the economy is the clearest path to achieving a sustainable society. This environmentalism is aligned with those who propose a "new economy" and seek to build sustainable production and consumption systems that are restorative, regenerative, and resilient.[7] The tools for this include economic policies that create a more locally based economy, where products are appropriate and bear the full cost of their production, transportation is minimized, energy and resources are conserved and enhanced, and wastes are eliminated. Within this perspective the need for an ever-flowing torrent of cheap and disposable consumer products can be addressed and a vibrant economy considered that does not require draining the earth of its remaining high-quality material and energy reserves. The environmental health movement focused on safer chemicals needs to find and align with these sustainable economy formations.

There will be resistance—there is today. Any change at the scale considered here will engender resistance. Change at this scale will benefit some directly, open up opportunities for many others, and potentially damage some that cannot change or stubbornly resist change. The status quo, even a malfunctioning status quo, creates its own rigidity. Gigantic systems do not change easily—many internal forces are designed for system homeostasis. There will be resistances from the leaders of businesses—both large and small—that see increased responsibilities and costs for collecting chemical information, identifying chemicals of concern, and considering alternatives. There will be resistances from firms in sectors that have little experience in positive collaborations with potential competitors, governments, and other stakeholders. There will be resistances from government agencies, the constituencies they serve or protect, and the legislative committees that review them. There will be resistance even among allies. Some who agree with the overarching goals of achieving a safer system of chemical production and consumption will resist because they do not like the processes, they think the changes are too fast or too slow, they don't like the potential partners, they disagree with the strategy, they take personal offense, they feel competitive or left out, or they foresee that they will suffer personal losses.

However, there is another kind of resistance—a resistance built deeply into the chemical industry and the business community more broadly. This is an ideological resistance that does not value the assertive role of government or believe that change should be directed by conscious forces other than the market. Those who share this political philosophy are currently well connected, powerful, and heavily financed. The chemical industry trade associations, many of the traditional business associations, and both the banking and financial investment sectors are likely to put up this kind of resistance. Capitalist institutions may be short sighted and obsessed with wealth generation, but they can present powerful resistance to changes that challenge their hegemony and affront their ideology.

16.3   The Costs of Inaction

Although the scale of the changes considered here is daunting, delaying these changes or proceeding too slowly presents significant costs. In 2000 (and again in 2013), the European Environment Agency released a report called *Late Lessons from Early Warnings*.[8] In a well-detailed series of case studies, this compendium documents the costs of waiting too long for science to establish beyond a reasonable doubt the dangers of a chemical. Too often not heeding early warnings has resulted in unnecessary harms and significant costs. Far too little attention has been paid to these costs because many of them are hidden and difficult to calculate. The long, costly process of cleaning up the legacy of the past mismanagement of hazardous chemical wastes has only begun. Current environmental releases of those wastes and emissions only continue these costs, as do the personal costs of illnesses and injuries caused by exposures to those releases. But there are also broad social and economic costs.

Recent studies suggest that the social costs of continuing the manufacture and use of hazardous chemicals are much larger than conventionally expected. Studies on occupational health and safety in California estimate that chemical exposure is a significant determinant of mortality. Of the deaths directly linked to occupational health and safety, 100 percent of pneumoconiosis deaths, 80 to 90 percent of cancer deaths, 40 to 50 percent of deaths associated with neurological disorders, and 40 to 50 percent of deaths associated with renal disease are attributable to chemical exposures.[9] Studies of children's health further reveal the health burden of hazardous chemical exposure. One study suggests that chemical exposure plays a significant role in 10 to 35 percent of asthma cases, 5 to 20 percent of neurological disorders, 2 to 10 percent of some cancers, and all lead poisoning.[10]

A 2011 study prepared by the World Health Organization has estimated that the total number of lives lost globally each year due to chemical exposure is 4.9 million (8.3 percent of total global deaths), and the costs in terms of years of disabled life due to chemical exposure is 86 million life years. Of special note was the finding that children bear 54 percent of this burden. The indicators of chemical exposure include direct exposure to chemicals at work or in products plus exposure to indoor air pollution and second-hand smoke. However, because these estimates include only a small number of chemicals (mercury, dioxin, organic solvents, PCBs, and chronic exposure to pesticides were not included) for which there are data, these numbers underestimate the health burden of chemical exposures. The United Nations Environment Program has tried to disaggregate these numbers to determine the costs just from occupational and outdoor air exposures to chemicals. Here the number of annual deaths from chemical exposures is estimated at 2.3 million, and the number of years of disabled life due to chemical exposure is 23.7 million. This is still greater than similar statistics for diarrheal disease, HIV/AIDS, or road traffic accidents.[11]

These social costs translate into economic costs. Consider the effects of just two chemicals on the projected economic lifetime productivity of U.S. children. A 2005 study found that somewhere between 317,00 and 637,00 children are born in the United States with blood levels of mercury and lead over levels associated with the loss of IQ. Projecting out the cost of loss of intelligence in terms of diminished economic productivity over the expected lifetime of those children, the study found the costs to society to be $8.7 billion annually.[12]

Several efforts to consider the economic costs of hazardous chemical exposure were carried out in Europe as a prelude to passage of REACH. In 2005, an analysis was done to determine the projected economic benefits of reducing human exposure to hazardous chemicals by the proposed REACH regulation. Several economic methods were used to determine the benefits, including direct avoided costs, willingness to pay assessments, and extrapolation from studies of specific substances. The direct costs approach that relied on just five avoided measurable costs, including sewage plant construction, water purification, and sewage sludge treatment, was deemed the most conservative. Using this method, the study found that the potential benefit of REACH on human health was estimated to be at a minimum €150–€500 ($190–$630 USD) million in 2017, with a potential long-term benefit over the succeeding 25 years of €2.8–€9.0 ($3.5–$11.5 USD) billion. Alternatively, the (less reliable) extrapolation form existing substances approach yielded potential benefits arising from saved health costs estimated at €200–€2,500 ($250–$3,200 USD) million in 2017, which aggregated over twenty-five years corresponded to €4.0–€50.0 ($5–$63 USD) billion.[13]

The social and economic costs of hazardous chemical exposure presented by such studies are underestimates. Because many of the costs of hazardous chemicals are externalized onto exposed consumers and workers, ecological systems, and state and local governments, the true costs of hazardous chemicals are hidden and not integrated into the price of chemical-based products or national government accounts. This might be acceptable if these costs were relatively minor or if there was little that could be done to lower them. However, as this text has demonstrated, much can be done to change the chemicals in use today, and the costs of not managing those chemicals are high—potentially, very high.

16.4   Making It Happen

Somewhere in the future lays a less hazardous chemical market and a more sustainable chemical industry. Production workplaces here in the United States and those in far locations will be cleaner and safer. Customers will buy and use products that are safer and less likely to pollute and threaten ecosystems when they are disposed. A robust chemical economy will deliver high-quality products that meet social needs and enhance and protect human health and the environment. That economy will never be perfectly safe, but it will be valuing safety and continually improving.

The task of converting the current systems of production and consumption to a safer economy needs to be framed as a national, system-wide strategy, the way that eradicating polio, building the interstate highway system, and putting a man on the moon was framed. The chemical conversion strategy envisioned here requires many campaigns and many initiatives focused on a common goal—a safer system of chemical production and consumption.

An enormous challenge and opportunity exists here for the U.S. chemical industry. Not only does it face intense global competition but it also faces a continuing credibility problem with the public. However, the industry holds the critical keys for achieving a safer chemical economy.

Green chemistry, renewable feedstocks, biorefineries, biobased processing, chemical leasing, green nanotechnology, supply chain integration, microscale production, and modular batch processing—there are many promising directions for technological innovation and new business models. Instead of waiting for market, government, and advocacy pressures to force shifts toward safer chemistries, the chemical industry could assume a proactive posture that recognizes the growing global desire for safer chemicals and commits to investing in the research and infrastructure that are needed to convert production processes and substitute hazardous chemicals with safer alternatives. The risk-based approach that has long guarded the industry against demands for fundamental change has been a great disservice to the industry. Within the industry today, there are many people who see the needs and are pushing for substantive changes: they just need to be recognized and encouraged. The search for highly functional, cost-effective, and inherently safer chemicals could frame the challenge for creating a truly sustainable chemical industry in the twenty-first century.

How fast can chemical conversion occur? Not very. Building a bigger environmental health movement and bringing its capacity to scale will take time, certainly a decade. There is much that can be done now at the product level. The chemicals in pharmaceuticals, cosmetics, personal care products, domestic household products, garments, textiles, building materials, toys, packaging, and many plastic products and components can probably be converted to safer alternatives soon. There are already many initiatives. The chemicals in solvents, reagents, coatings, inks, dyes, mastics and hundreds of specialty chemicals could take a while longer, but there are many possibilities. The basic platform chemicals will take longest, as they are fundamental building blocks. For some decades the best that can be hoped for is sound chemical management, production efficiencies that lead toward zero-waste, zero-emission processes, effective and well-maintained exposure controls, and a gradually shrinking market.

This strategy takes leaders. Chemists such as John Warner, business professionals such as Helen Holder and Roger McFadden, advocates such as Charlotte Brody, and entrepreneurs such as Dara O'Rouke provide the energy, ideas, and vision for this strategy. But hundreds of others could be named. They are today working in chemical corporations, product manufacturing firms, retailer stores, advocacy organizations, university laboratories, and government agencies throughout the country. History books herald significant individuals as change makers, but today (and probably throughout history) social transformations—big, fundamental transformations—are driven by the dreams and dedication of thousands of people. That is why mobilizing the environmental health movement is so important.

Mustering the environmental health movement to transform the chemical economy in the United States needs to unite with and support the emergence of the global safer chemical movement. The international value of the chemical and product policies that have occurred in Europe has been enormous, and it is important that these initiatives continue to provide new directions. The civil society activism on pollution, occupational safety, and chemical hazards that is emerging in Asia and the new Asian chemical policies will play a significant role in the future. Much of the international activism on chemicals is now focused on achieving the United Nation's goal for the sound management of chemicals by 2020. For now this single focus provides a broad international umbrella for many chemical initiatives around the world. However, 2020 will soon be here. The goal may or may not be reached, but it is the mobilization that is most promising. The global movement for safer chemicals will need to build beyond 2020, and SAICM for that matter, because developing a sustainable chemical economy will take decades.

Sustainability requires a global economy that respects the limits of the planet's resources—its materials, energy sources, ecological services, and assimilative capacities—provides adequate support for all of the earth's people, and takes into account the needs of future generations. Technologies can greatly expand the efficiencies with which these resources are used and broaden the range of people who benefit. Synthetic chemicals as a technology have demonstrated their great capacity for providing such benefits. Where synthetic chemicals have eased human suffering, replaced dangerous and menial labor, and magnified the functional performance of the earth's limited resources, these chemicals have and should continue to play a critical role in a sustainable future.

However, the promise of the synthetic chemical revolution has not been fulfilled. After some century and a half, we have dramatically reworked the earth's chemistries to create products and services that extend and improve human life. But along the way, we have generated too many health- and environment-threatening chemicals. We have much to show as we have advanced our material affluence, but we have settled too early in our search for acceptable chemistries. We want chemicals that are useful and effective—but also chemicals without harm. The challenge of converting to chemistries that meet the needs of the present without jeopardizing the capacities of future generations—that is, sustainable chemistries—still remains ahead of us.

Notes

Chapter 1. The Problem with Chemicals

1. Charlotte Brody, Vice President for Health Initiatives, BlueGreen Alliance, personal interview, January 16, 2014.

2. Roger McFadden, Senior Scientist, Staples, personal interview, October 6, 2010.

3. Anthony Lewis (quoting C. P. Snow) in "Dear Scoop Jackson," *New York Times,* March 15, 1971, p. 37.

4. The classic history of this period can be found in Ludwig F. Haber, *The Chemical Industry in the Nineteenth Century* (New York: Oxford University Press, 1958).

5. Several recent popular reviews document the complex hazards that many of these widely used chemicals pose. See Nena Baker, *The Body Toxic: How the Hazardous Chemistry of Everyday Things Threatens Our Health and Well-being* (New York: North Point Press, 2008); and Elizabeth Grossman, *Chasing Molecules: Poisonous Products, Human Health and the Promise of Green Chemistry* (Washington, DC: Island Press, 2009). Nancy Langston covers the scientific and regulatory struggle over diethylstilbestrol in *Toxic Bodies: Hormone Disruptors and the Legacy of DES* (New Haven, CT: Yale University Press, 2010); and Sarah Vogel describes similar struggles over bisphenol A in *Is It Safe: BPA and the Struggle to Define the Safety of Chemicals* (Berkeley: University of California Press, 2013).

6. Jeanette Mulvey, "Toxic Toys Create Silver Lining for Green Toy Companies," *Business News Daily*, December 19, 2010, see http://www.businessnewsdaily. com/512-toxic-toys-create-silver-lining-for-green-companies.html, accessed January 15, 2014. For a good source of information on hazardous chemicals in common domestic products, see U.S. Department of Health and Human Services, National Library of Medicine, *Household Products Database*, see http:// householdproducts.nlm.nih.gov/, accessed February 12, 2011.

7. See U.S. Environmental Protection Agency, Office of Research and Development, *Total Exposure Methodology Study*, Volumes I–IV, Washington, DC, 1985, see http://exposurescience.org/pub/reports/TEAM_Study_book_1987.

pdf, accessed February 11, 2011. A 2005 study sampled seventy homes in seven states and found the presence of all forty-four hazardous chemicals tested for. See Pat Costner, Beverley Thorpe, and Alexandra McPhearson, *Sick of Dust: Chemicals in Common Products—A Needless Health Risk in Our Homes*, Clean Production Action, unpublished report, March, 2005, see http://www.cleanproduction.org/static/ee_images/uploads/resources/Dust_Report.pdf, accessed October 1, 2014; and W. Butte and B. Heinzow, "Pollutants in House Dust as Indicators of Indoor Contamination," *Review of Environmental Contamination and Toxicology*, 175, 2002, pp. 1–46. See also Environmental Working Group, *Skin Deep* website, see http://www.ewg.org/skindeep/, all accessed October 1, 2014.

8. U.S. Centers for Disease Control, National Center for Environmental Health, *Fourth National Report on Human Exposure to Environmental Chemicals*, Atlanta, GA, July 2009, see http://www.cdc.gov/exposurereport/pdf/FourthReport. pdf, accessed January 4, 2014.

9. Alvin C. Bronstein et al., "2001, Annual Report of the American Association of Poison Control Center's National Poison Data System, 29th Annual Report," *Clinical Toxicology*, 50, 2012, pp. 911–1164, Table 16A, p. 933, see https://aapcc.s3.amazonaws.com/pdfs/annual_reports/2011_NPDS_Annual_Report.pdf, accessed March 22, 2012.

10. For example, see Richard Clapp, Genevieve Howe, and Molly Jacobs, *Environmental and Occupational Causes of Cancer: A Review of Recent Scientific Literature*, Lowell Center for Sustainable Production, University of Massachusetts Lowell, Lowell, MA, September 2005, see http://www.sustainableproduction.org/downloads/Causes%20of%20Cancer.pdf, accessed January 4, 2014; Lynn Goldman and S. Koduru, "Chemicals in the Environment and Developmental Toxicity in Children: A Public Health and Policy Perspective," *Environmental Health Perspectives*, 108 (supp. 3), 2000, pp. 443–448; Ted Schettler, Gina Solomon, Maria Valenti, and Annette Huddle, *Generations at Risk: Reproductive Health and the Environment* (Cambridge, MA: MIT Press, 2000); and World Health Organization, *Global Burden of Disease, 2004 Update* (Geneva, 2008), see www.who.int/healthinfo/global_burden_disease/GBD_report_2004update_full.pdf, accessed March 7, 2013.

11. See the U.S. President's Cancer Panel, *2008–2009 Annual Report*, Washington, DC, 2010, see http://deainfo.nci.nih.gov/advisory/pcp/annualReports/pcp08-09rpt/PCP_Report_08-09_508.pdf; and National Research Council, Commission on Life Sciences, *Scientific Frontiers in Developmental Toxicology and Risk Assessment*, Washington, D.C, 2000, p. 1, see http://www.nap.edu/catalog.php?record_id=9871, both accessed January 4, 2014.

12. See Robert D. Bullard, Paul Mohai, Robin Saha, and Beverly Wright, *Toxic Waste and Race at Twenty: 1987–2007*, United Church of Christ, March 2007, see http://www.ucc.org/justice/pdfs/toxic20.pdf, accessed February 25, 2013; and Sly J. Leith and D. O. Carpenter, "Special Vulnerability of Children to Environmental Exposure," *Reviews on Environmental Health*, 27(4), 2012, pp. 151–157.

13. D. H. Landers, S. J. Simonich, D. A. Jaffe, and L. H. Geiser, *The Fate, Transport and Ecological Impacts of Airborne Contaminants in Western National Parks*, U.S. Environmental Protection Agency, Office of Research and Development, NHEERL, Western Ecology Division, Corvallis, OR, EPA/600/R-07/138, see http://www.cfr.washington.edu/research.cesu/reports/J9088020046-WACAP-Report-Volume-I-Main.pdf; ; and Wing Goodale and Chris DeSorbo, *Chemical Screening of Osprey Eggs in Casco Bay, Maine: 2009 Field Season*, Report 2010–09, Biodiversity Research Institute, Gorham, Maine, 2010, see http://www.cascobayestuary.org/wp-content/uploads/2014/08/2010_goodale_osprey_2009_season_report.pdf, both accessed October 1, 2014.

14. See Mark Schapiro, *Exposed: The Toxic Chemistry of Everyday Products and What's at Stake for American Power* (White River Junction, VT: Chelsea Green, 2007).

15. See Theo Colburn, Diane Dumanoski, and John Peterson Meyers, *Our Stolen Future: Are We Threatening Our Fertility, Intelligence and Survival: A Scientific Detective Story* (New York: Dutton, 1996); Joe Thornton, *Pandora's Poison: Chlorine, Health and the New Environmental Strategy* (Cambridge, MA: MIT Press, 2001); Linda S. Birnbaum and Daniele F. Staskal, "Brominated Flame Retardants: Cause for Concern?", *Environmental Health Perspectives*, 112(1), 2004, pp. 9–17; P. Grandjean and P. J. Landigren, "Developmental Neurotoxicity of Industrial Chemicals—A Silent Pandemic," *Lancet*, 368:9353, December 2006, pp. 2167–2178; and Carl F. Cranor, *Legally Poisoned: How the Law Puts Us at Risk from Toxicants* (Cambridge, MA: Harvard University Press, 2011).

16. The term "chemicals" is used broadly throughout the book to mean elements, chemical substances, and chemical compounds. In general, the term refers to synthetic chemicals made through industrial processes. "Hazardous chemicals" are those chemicals whose inherent physical and chemical structures may cause harm to human health and the environment. The term "chemicals of concern" is a social construct developed by an organization, government, or institution to mean those chemicals that raise safety, health, or environmental concerns. Because of the unique regulatory provisions regarding pharmaceuticals, they are not considered here, other than to note those provisions.

Chapter 2.   Regulating Hazardous Chemicals

1. The American history of chemical pollution appears in several good texts: Martin Melosi, *Pollution and Reform in American Cities* (Austin, TX: University of Texas Press, 1980); C. E. Colton and P. N. Skinner, *The Road to Love Canal: Managing Industrial Waste Before EPA* (Austin, TX: University of Texas Press, 1996); and Bejamin Ross and Steven Amter, *The Polluters: The Making of Our Chemically Altered Environment* (New York: Oxford University Press, 2010).

2. See Ashish Arora and Nathan Rosenberg, "Chemicals: A U.S. Success Story," in Ashish Arora, Ralph Landeau, and Nathan Rosenberg, eds., *Chemicals and Long Term Economic Growth; Insights from the Chemical Industry* (New York: John

Wiley, 1998) pp. 71–102; and Kenneth Geiser, *Materials Matter, Toward a Sustainable Materials Policy* (Cambridge, MA: MIT Press, 2001) p. 45.

3. Rachel Carson, *Silent Spring* (New York: Ballantine, 1962) p. 25. For a brief history of these pesticides, see James Whorton, *Before Silent Spring: Pesticides and Public Health in Pre-DDT America* (Princeton, NJ: Princeton University Press, 1974).

4. Fred Aftaion, *A History of the International Chemical Industry: From the "Early Days" to 2000* (Philadelphia, PA: Chemical Heritage Press, 2001) p. 82. See also Alfred J. Chandler, Jr., *Shaping the Industrial Century: The Remarkable Story of the Evolution of the Modern Chemical and Pharmaceutical Industries* (Cambridge, MA: Harvard University Press, 2005). On the use of lead and cadmium, see Gerald Marowitz and David Rosner, *Deceit and Denial: The Deadly Politics of Industrial Pollution* (Berkeley: University of California Press, 2002) pp. 179–181.

5. Peter Spitz quotes Sanford Research Institute figures; see Peter H. Spitz, *Petrochemicals: The Rise of an Industry* (New York: Wiley, 1988) p. 233.

6. For example, see Susannah Handley, *Nylon: The Story of a Fashion Revolution* (Baltimore: Johns Hopkins University Press, 1999); and Susan Frienkel, *Plastics: A Toxic Love Story* (New York: Houghton-Mifflin Harcourt, 2011). For specifics, see Geiser, 2001, p. 45.

7. See Isadore Kallett and F. I. Schink, *100,000,000 Guinea Pigs* (New York: Grosset and Dunlop, 1933); Ruth deForest Lamb, *American Chamber of Horrors: The Truth about Food and Drugs* (New York: Farrar and Reinhart, 1936); and Marowitz and Rosner, 2002, pp. 179–181.

8. For the regulatory history of diethylstilbestrol, see Langston, 2010. For a readable history of the popular movement aroused by concerns over hazardous chemicals, see Kate Davies, *The Rise of the U.S. Environmental Health Movement* (New York: Rowman and Littlefield, 2013).

9. For a good review, see Sam Gussman, Konrad von Moltke, Francis Irwin, and Cynthia Whitehead, *Public Policy for Chemicals: National and International Issues* (Washington, DC: The Conservation Foundation, 1980).

10. Federal Food Drug and Cosmetic Act, 52 U.S. Stat. 1040.

11. For this history, see I. D. Barkan, "Industry Invites Regulation: The Passage of the Pure Food and Drug Act of 1906," *American Journal of Public Health*, 75:1, 1985, pp. 18–26; and David F. Cavers, "The Food Drug and Cosmetics Act of 1938: Its Legislative History and its Substantive Provisions," *Law and Contemporary Problems*, 10, 1939, pp. 2–42.

12. The history of this amendment is described in Vogel, 2013, pp. 34–63.

13. See Federal Insecticide, Fungicide and Rodenticide Act, 61 U.S. Stat. 163, P.L. 80–104; and Federal Environmental Pesticide Control Act, 86 U.S. Stat. 973, P.L. 92–516.

14. National Research Council, *Pesticides in the Diets of Infants and Children* (Washington, DC: National Academy Press, 1993); and Food Quality Protection Act of 1996, P.L. 104–170.

15. Toxic Substances Control Act, P.L. 94–469.

16. U.S. Congress, *Toxic Substances Control Act: Report by the House of Representatives Committee on Interstate and Foreign Commerce*, Report No. 94–1341, July 4, 1976, 94th Congress, 2nd Session, p. 1.

17. Quote taken from Robert F. Service, "A New Wave of Chemical Regulations Just Ahead?", *Science*, 235, August 7, 2009, pp. 692–693.

18. U.S. Office of the President, Council on Environmental Quality, *Toxic Substances* (Washington, DC, 1971) pp. 1–2.

19. These include the Flammable Fabrics Act, 67 U.S. Stat. 111; Refrigerator Safety Act, 70 U.S. Stat. 953; Federal Hazardous Substances Act, 74 U.S. Stat. 372; Poison Prevention Packaging Act, 84 U.S. Stat. 1670; Federal Hazardous Substances Act, 74 U.S. Stat. 372; and the Consumer Product Safety Act, 86 U.S. Stat. 1207.

20. See M. R. Fise, "Consumer Product Safety Regulation," in K. J. Meier, E. T. Garman, and L. R. Keiser, eds., *Regulation and Consumer Protection: Politics, Bureaucracy and Economics* (Houston: DAME Publications, 1998) pp. 259–279.

21. *Industrial Union Department, AFL-CIO v American Petroleum Institute*, 448 U.S. 607, 1980.

22. National Research Council, *Risk Assessment in the Federal Government: Managing the Process* (Washington, DC: National Academy Press, 1983).

23. California Office of Environmental Health Hazard Assessment, *Proposition 65 in Plain Language*, see, www.oehha.ca.gov/prop65/background/p65plain.html, accessed June 12, 2009.

24. Emergency Planning and Community Right to Know Act of 1986, 100 U.S. Stat. 1733.

25. National Pollution Prevention Roundtable, *Facility Pollution Prevention Planning Requirements: An Overview of State Program Evaluations* (Washington, DC, 1997).

26. For a review, see Kenneth Zarker and Robert Kerr, "Pollution Prevention through Performance-Based Initiatives and Regulation in the United States," *Journal of Cleaner Production*, 16, 2008, pp. 673–685.

27. Massachusetts Toxics Use Reduction Act, see http://www.mass.gov/eea/agencies/massdep/toxics/tur/, accessed September 16, 2014.

28. Massachusetts Toxics Use Reduction Institute, *Toxics Use Reduction Act Program Assessment*, University of Massachusetts Lowell, June 9, 2009; and Massachusetts Toxics Use Reduction Institute, *Trends in the Use and Release of Carcinogens in Massachusetts*, University of Massachusetts Lowell, June 2013.

29. Safer Chemicals, Healthy Families, *Healthy States: Protecting Families from Toxic Chemicals while Congress Lags Behind*, November 2010, see http://saferchemicals.org/wp-content/uploads/pdf/HealthyStates.pdf?465b45, accessed September 16, 2014. Information on state chemicals regulation statutes can be found on a periodically updated registry maintained by the Interstate Chemicals

Clearinghouse. See U.S. State-Level Chemical Policy Database, www.theic2.org/chemical-policy, accessed September 30, 2014.

Chapter 3.   Reassessing Chemical Control Policies

1. See Joel Tickner and Yve Torre, *Presumption of Safety: Limits of Federal Policies on Toxic Substances in Consumer Products*, Lowell Center for Sustainable Production, University of Massachusetts Lowell, February 2008, p. 5, see www.chemicalspolicy.org/downloads/UMassLowellConsumerProductBrief.pdf, accessed February 3, 2011.

2. Ronald Brickman, Shiela Jasanoff, and Thomas Ilgen provide one of the earliest studies of chemical policy implementation in *Controlling Chemicals, The Politics of Regulation in Europe and the United States* (Ithaca, NY: Cornell University Press, 1985). Carl Cranor extends this assessment by focusing on the relationship between science and law in *Regulating Toxic Substances: A Philosophy of Science and the Law* (New York: Oxford University Press, 1993); and Robert Gottlieb and his colleagues considers programs for reducing toxic chemical use in *Reducing Toxics: A New Approach to Policy and Industrial Decision Making* (Washington, DC: Island Press, 1995).

3. See Barkan, 1985, pp. 18–26; and P. Hilts, *Protecting America's Health: The FDA, Business and One Hundred Years of Regulation* (New York: Knopf, 2003).

4. Juhana Karha and Eric J. Topal, "The Sad Story of Vioxx, and What We Should Learn from It," *Cleveland Clinic Journal of Medicine*, 74(12), 2005, pp. 933–939.

5. Institute of Medicine, Board of Population Health and Public Health Practice, *The Future of Drug Safety: Promoting and Protecting the Health of the Public* (Washington, DC National Academy Press, 2007), see http://www.iom.edu/~/media/Files/Report%20Files/2006/The-Future-of-Drug-Safety/futureofdrugsafety_reportbrief.pdf, accessed February 11, 2013.

6. Nutrition Labeling and Education Act of 1990, 104 Stat. 2356.

7. U.S. Food and Drug Administration, *FDA's Authority Over Cosmetics*, see http://www.fda.gov/Cosmetics/GuidanceRegulation/LawsRegulations/ucm074162.htm , accessed September 16, 2014.

8. Fair Packaging and Labeling Act, 80 Stat. 1296.

9. See U.S. Government Accounting Office, *Chemical Risk Assessment: Selected Federal Agencies' Procedures, Assumptions and Policies*, GAO-01–810 (Washington, DC, 2001); and D. C. Christiani and K. T. Kelsey, eds., *Chemical Risk Assessment and Occupational Health* (Westport, CT: Auburn House, 1994).

10. Mary O'Brien reviews the various limits and misuses of risk assessment in *Making Better Environmental Decisions: An Alternative to Risk Assessment* (Cambridge, MA: MIT Press, 2000).

11. For two contrasting views on cost-benefit assessment, see Cass R. Sunstein, *The Cost–Benefit State: The Future of Regulatory Protection* (Chicago: American

Bar Association, 2002); and Frank Ackerman and Lisa Hienzerling, *Priceless: On Knowing the Price of Everything and the Value of Nothing* (New York: New Press, 2004).

12. U.S. Environmental Protection Agency, Office of Chemical Safety and Pollution Prevention, *Pesticide Industry Sales and Usage, 2006 and 2007 Market Estimates* (Washington, DC, 2011), see www.epa.gov/opp00001/pestsales/07pestsales/market_estimates2007.pdf, accessed March 7, 2013.

13. U.S. Environmental Protection Agency, Office of Prevention, Pesticides and Toxic Substances, *Implementing the Food Quality Protection Act: Progress Report* (Washington, DC, August 1999), see http://www.epa.gov/pesticides/regulating/laws/fqpa/fqpareport.pdf, accessed March 7, 2013.

14. U.S. Environmental Protection Agency, *Evaluation of U.S. EPA's Pesticide Product Registration Process: Opportunities for Efficiency and Innovation* (Washington, DC, March 2007), see http://www.epa.gov/evaluate/pdf/pesticides/eval-epa-pesticide-product-reregistration-process.pdf, accessed September 16, 2014. The data are from, "Pesticide Reregistration Performance, Measures and Goals," *Federal Register*, 74:91, May 13, 2009, pp. 22541–22547, see www.epa.gov/fedrgstr/EPA-PEST/2009/May/Day-13/p10758.pdf, accessed March 7, 2013. See also Jennifer Sass and Mae Wu, *Superficial Safeguards: Most Pesticides are Approved by Flawed EPA Process*, Natural Resources Defense Council, Washington, DC, March 2013.

15. John Wargo, *Our Children's Toxic Legacy: How Science and Law Fail to Protect Us from Pesticides*, Second Edition. (New Haven, CT: Yale University Press, 1998).

16. See John S. Applegate, "Synthesizing TSCA and REACH: Practical Principles for Chemical Regulation Reform," *Ecology Law Quarterly*, 35, 2008, pp. 721–769. See also Jessica N. Schifano, Ken Geiser, and Joel A. Tickner, "The Importance of Implementation in Rethinking Chemicals Management Policies: The Toxics Substances Control Act," *Environmental Law Reporter*, 41, 2011, pp. 10527–10543.

17. U.S. Environmental Protection Agency, Office of Pollution Prevention and Toxics, *Overview: Office of Pollution Prevention and Toxics Programs* (Washington, DC, December 24, 2003), see http://www.chemicalspolicy.org/downloads/TSCA10112-24-03.pdf, accessed March 7, 2013.

18. See Richard Dennison, "Ten Essential Elements in TSCA Reform," *Environmental Law Reporter*, 39, 2009, pp. 10020–10028.

19. U.S. Environmental Protection Agency, Office of Pollution Prevention and Toxics, *Overview: Office of Pollution Prevention and Toxics Programs* (Washington, DC, January 2007) p. 4, see http://www.epa.gov/oppt/pubs/oppt101c2.pdf, accessed March 7, 2013.

20. U.S. Environmental Protection Agency, Office of Pollution Prevention and Toxics, *2008 Inventory Update Rule Data Summary* (Washington, DC, 2009), see http://www.epa.gov/cdr/pubs/2006_data_summary.pdf, accessed March 7, 2013.

21. David Andrews and Richard Wiles, *Off the Books: Industry's Secret Chemicals* (Washington, DC: Environmental Working Group, December 2009), see http://www.ewg.org/sites/default/files/report/secret-chemicals.pdf, accessed February 3, 2011.

22. Environmental Defense Fund, *Toxic Ignorance: The Continuing Absence of Basic Health Testing for TOP-Selling Chemicals in the United States* (Washington, DC, 1997); and U.S. Environmental Protection Agency, Office of Pollution Prevention and Toxics, *HPV Chemical Hazard Data Availability Study* (Washington, DC, 1998), see http://www.epa.gov/chemrtk/pubs/general/hazchem.htm, accessed February 3, 2011. For current HPV Program status, see U.S. Environmental Protection Agency, Office of Pollution Prevention and Toxics, *High Production (HPV) Challenge*, see www.epa.gov/chemrtk/pubs/general/basicinfo.htm, accessed February 3, 2011; and Richard Dennison, *High Hopes, Low Marks, A Final Report Card on the High Production Volume Chemical Challenge* (Washington, DC: Environmental Defense Fund, July, 2007).

23. National Academy of Sciences, *Science and Decisions: Advancing Risk Assessment* (Washington, DC, 2008) p. 3.

24. EPA's chemical restrictions under TSCA Section 6 include restricting polychlorinated biphenyls (which was specifically mandated by the law), restricting ozone-depleting chlorofluorocarbons in aerosols, regulating disposal of dioxin-contaminated wastes, controlling asbestos used in schools, and banning hexavalent chromium in the circulating waters of commercial cooling towers.

25. U.S. Fifth Circuit Court of Appeals, *Corrosion Proof Fittings v. EPA* (947 F.2d 1201), 1991.

26. U.S. Government Accounting Office, *EPA's Chemical Testing Program has made Little Progress* (T-RCED-90–88) (Washington, DC, June 20, 1990); U.S. Government Accounting Office, *Toxic Substances, EPA's Chemical Testing Program Has Not Resolved Safety Concerns* (T-RCED-91–136) (Washington, DC, June 19, 1991); and U.S. Government Accounting Office, *Chemical Regulation: Options Exist to Improve EPA's Ability to Assess Health Risks and Manage its Chemical Review Program* (GAO-05–458) (Washington, DC, June 2005).

27. Mark A. Greenwood, "TSCA Reform: Building a Program That Can Work," *Environmental Law Review*, 39, 2009, pp. 10034–10041; and Lynn Goldman, "Preventing Pollution? US Toxic Chemicals and Pesticides Policies and Sustainable Development," *Environmental Law Reporter*, 32, 2002, pp. 11018–11041.

28. Teresa M. Schwartz, "The Consumer Product Safety Act: A Flawed Product of the Consumer Decade," in *George Washington Law Review*, 51(1), 1982, pp. 43–44. The quote is at p. 44. For a more expanded argument, see Tickner and Torre (2008) and E. Marla Felcher, *The Consumer Product Safety Commission and Nanotechnology* (Washington, DC: Woodrow Wilson International Center for Scholars, August 2008) p. 9, see http://www.nanotechproject.org/publications/archive/pen14/, accessed March 22, 2013.

29. OMB Watch, *Product Safety Regulator Hobbled by Decades of Negligence* (Washington, DC, February 5, 2008), see http://dev.ombwatch.org/node/3599, accessed March 22, 2013; and E. Marla Felcher, *The Consumer Product Safety Commission and Nanotechnology*, 2008, p. 9.

30. For the quotation, see Thomas H. Moore, "Statement Submitted to Senate Committee on Commerce, Science and Transportation's Subcommittee on Consumer Affairs, Insurance, and Automotive Safety," U.S. Congress, Washington, DC, March 21, 2007, see www.cpsc.gov//PageFiles/121174/moore2007.pdf, accessed March 22, 2013; and Public Citizen, *Hazardous Waits: CPSC Lets Crucial Time Pass Before Warning Public about Dangerous Products*, Washington, DC, January, 2008, see www.citizen.org/documents/HazardousWaits.pdf, accessed March 7, 2013.

31. U.S. Government Accounting Office, *Consumer Product Safety Commission: Better Data Needed to Help Identify and Analyze Hazards* (Washington, DC, 1997).

32. Robert Adler, "From Model Agency to Basket Case—Can the Product Safety Commission Be Redeemed," in *Administrative Law Review*, 41(1), 1989, pp. 61–129.

33. Martha Mendoza, "How Government Decided Lunch Box Lead Levels," *Washington Post*, February 18, 2007.

34. Mary Douglas and Aaron Wildavsky offer an early consideration of what today is called the "cultural theory of risk," in *Risk and Culture: An Essay on the Selection of Technological and Environmental Dangers* (Berkeley, CA: University of California Press, 1983).

35. Greenwood, 2009, p. 10036; and U.S. Environmental Protection Agency, *FY 2013: Budget in Brief* (Washington, DC, 2012), see http://yosemite.epa.gov/sab%5CSABPRODUCT.NSF/2B686066C751F34A852579A4007023C2/$File/FY2013_BIB.pdf, accessed January 14, 2014.

36. The regulatory philosophy of the federal government is spelled out clearly under Executive Order 12866 of September 30, 1993: "Federal agencies should promulgate only such regulations as are required by law, are necessary to interpret the law, or are made necessary by compelling public need, such as material failures of the private markets to protect or improve the health and safety of the public, the environment, or the well being of the American people. In deciding whether and how to regulate, agencies should assess all costs and benefits of available regulatory alternatives, including the alternative of not regulating." See U.S. Office of the President Executive Order 12866, *Federal Register*, 58:190, October 4, 1993, see http://www.archives.gov/federal-register/executive-orders/pdf/12866.pdf, accessed December 14, 2012.

37. For examples of how business lobbyists discredit the science on which government regulations can be based, see David Michaels, *Doubt Is Their Product: How Industry's Assault on Science Threatens Your Health* (New York: Oxford University Press, 2008).

Chapter 4.   Considering New Initiatives

1. For a fuller review, see Joel Tickner and Ken Geiser, *New Directions in European Chemicals Policies: Drivers, Scope and Status*, Lowell Center for Sustainable Production, University of Massachusetts Lowell, September 2003.

2. See Mark Schapiro, 2007, p. 30. For insightful reviews of differences between European and U.S. chemical policies, see Ragmar E. Lofstedt and David Vogel, "The Changing Character of Regulation: A Comparison of Europe and the United States," *Risk Analysis*, 21(3), 2001, pp. 399–416; and Richard A. Dennison, *Not So Innocent: A Comparative Analysis of Canadian, European Union and United States Policies on Industrial Chemicals*, unpublished report, Environmental Defense Fund, Washington, DC, April 2007.

3. United Nations, *Rio Declaration on Environment and Development*, Rio de Janeiro, Brazil, June 1992, Principle 15, see www.unep.org/Documents.Multilingual/Default.asp?documentid=78&articleid=1163, accessed February 2, 2011.

4. *Wingspread Statement on the Precautionary Principle*, January 25, 1998, see www.gdrc.org/u-gov/precaution-3.html, accessed February 2, 2011. For a broad review of the uses of the Precautionary Principle, see Carolyn Raffensperger and Joel Tickner, eds., *Protecting Public Health and the Environment: Implementing the Precautionary Principle* (Washington, DC: Island Press, 1999).

5. Joel Tickner and Ken Geiser, "The Precautionary Principle Stimulus for Solutions- and Alternatives-based Environmental Policy," *Environmental Impact Assessment Review*, 24, 2004, pp. 801–824.

6. Joel Tickner, Associate Professor, University of Massachusetts Lowell, personal interview, January 31, 2011.

7. European Commission, *European White Paper on a Strategy for Future Chemicals Policy*, Brussels, COM 88 Final, February 2001, see http://www.isopa.org/isopa/uploads/Documents/documents/White%20Paper.pdf, accessed September 30, 2014.

8. For a review of the U.S. role, see U.S. House of Representatives, Committee on Government Reform, *The Chemical Industry, the Bush Administration and the European Effort to Regulate Chemicals: A Special Interest Case Study*, Washington, DC, April 1, 2004, see http://oversight-archive.waxman.house.gov/documents/20040817125807-75305.pdf, accessed February 2, 2011.

9. See European Commission, *Registration, Evaluation, Authorization of Chemicals (REACH)*, Regulation (EC) No. 1907/2006, December 18, 2006, see http://ec.europa.eu/enterprise/sectors/chemicals/documents/reach/review-annexes/index_en.htm, accessed February 11, 2012.

10. European Chemicals Agency, *REACH Statistics*, Helsinki, see http://echa.europa.eu/information-on-chemicals/registration-statistics, accessed March 22, 2013.

11. REACH, Article 5.

12. REACH, Article 1(3).

13. For a similar argument, see Steffan Foss Hansen, Lars Carlsen, and Joel Tickner, "Chemicals Regulation and Precaution: Does REACH Really Incorporate the Precautionary Principle," *Environmental Science and Policy*, 10, 2007, pp. 395–404.

14. The European Commission announced the restriction of six substances under REACH in February 2011. These include bis (2-ethylexyl) phthalate, benzyl butyl

phthalate, dibutyl phthalate, hexabromocyclododecane, 5-tert-butyl-2,4,6-trinito-m-xylene (musk xylene), and 4,4'-diaminodiphenylmethane. See Cheryl Hogue, "Pushing for a Phthalate," *Chemical and Engineering News*, 89(9), February 28, 2011, p. 10. For a comparison of TSCA and REACH, see Applegate, 2008, pp. 721–750.

15. European Parliament and Council of the European Union, *Council Directive 202/96/EC on Waste Electrical and Electronic Equipment*, see http://eur-lex. europa.eu/LexUriServ/LexUriServ.do?uri=OJ:L:2003:037:0024:0038:en:PDF, accessed February 11, 2012; and European Parliament and Council of the European Union, *Council Directive 2002/95/EC on the Restriction of Certain Hazardous Substances in Electrical and Electronic Equipment*, see http://eur-lex.europa.eu/legal-content/EN/TXT/PDF/?uri=CELEX:32002L0095&from=EN, accessed September 30, 2014.

16. European Parliament and Council of the European Union, *Council Regulation (EC) 1223/2009 on Cosmetics Products Regulation*, December 22, 2009, see http://eur-lex.europa.eu/LexUriServ/LexUriServ.do?uri=OJ:L:2009:342:0059:02 09:en:PDF, accessed February 11, 2012.

17. A. J. Karabelas, K. V. Plakas, E. S. Solomou, V. Drossou, and D. A. Sarigiannis, "Impact of European Legislation on Marketed Pesticides—A View from the Standpoint of Health Impact Assessment Studies," *Environment International*, 35, 2009, pp. 1096–1107.

18. European Parliament and Council of the European Union, *Toy Safety Directive* (2009/48/EC), 2009, see http://eur-lex.europa.eu/LexUriServ/LexUriServ.do?uri=OJ:L:2009:170:0001:0037:en:PDF, accessed February 25, 2013.

19. See Henrik Selin and Stacy VanDeveer, "Raising Global Standards: Hazardous Substances and E-Waste Management in the European Union," *Environment*, 48:10, December 2006, pp. 7–18.

20. Much of this information is developed from an article prepared by Dae Young Park, *REACHING Asia Continued*, Department of Public International Law, unpublished, Gent University, Gent, Belgium, September 16, 2009, see http://papers. ssrn.com/sol3/papers.cfm?abstract_id=1474504, accessed February 9, 2012.

21. Of the chemicals on the IECSC, 3, 270 are listed as confidential; see CHEM-Linked, "China New Chemical Substance Notification—China REACH," see https://chemlinked.com/chempedia/china-reach#sthash.z1ateSCI.dpbs, accessed September 30, 2014.

22. Chemical Inspection and Regulation Service, *Chemicals Regulated by Japan Chemicals Substance Control Law*, see www.cirs-reach.com/Japan_CSCL/ Chemicals%20Regulated%20by%20Japan%20Chemical%20Substances%20 Control%20Law.pdf, accessed December 28, 2013.

23. Chemical Inspection and Regulation Service, *Overview of Chemical Management Policies in China*, updated 2013, see www.cirs-reach.com/China_ Chemical_Regulation/Overview_Chemical_Legislation_Regulations_in_China. html, accessed December 28, 2013. Progress on chemical notification has been complicated and for a time was halted because field inspections revealed many errors in submitted chemical information; see presentation by Nie Jinlei, 3rd

International Chemical Regulation REACH Workshop, Hangzhou, 2011, see http://www.reach24h.com/workshop2011/en/partners/press-release/109-the-status-quo-of-china-reach-implementation.html, accessed December 12, 2013.

24. For an excellent review of these initiatives, see Phillip Wexler, Jan van der Kolk, Asish Mohapatra, and Ravi Agarwal, *Chemicals, Environment, Health: A Global Management Perspective* (New York: CRC Press, 2011).

25. The original twelve substances included aldrin, dieldrin, endrin, heptaclor, toxaphene, hexochlorobenzene, chlordane, mirex, polychlorinated biphenyls, DDT, dioxins, and furans. The new chemicals include chorodecone, hexabromo-biphenyl, hexabromobiphenyl ether/ heptabromobiphenyl ether, pentachloroben-zene, tetrbromodiphenyl ether/pentabromodiphenyl ether, alpha hexachlorocyclo-hezane, beta hexachlorocyclohezane, lindane, and perfluorooctane sulfonic acid.

26. United Nations Environment Program, Rotterdam Convention, *Annex III Chemicals*, see www.pic.int/TheConvention/Chemicals/AnnexIIIChemicals/tabid/ 1132/language/en-US/Default.aspx, accessed March 22, 2013.

27. United Nations, Food and Agriculture Organization, *International Code of Conduct on the Distribution and Use of Pesticides*, see ftp://ftp.fao.org/docrep/ fao/009/a0220e/a0220e00.pdf, accessed March 22, 2013.

28. "Mercury-Emissions Treaty Is Adopted after Years of Negotiations," *New York Times*, January 19, 2013.

29. United Nations Economic Commission for Europe, *Globally Harmonized System of the Classification and Labeling of Chemicals (GHS)*, Third Revised Edition, see www.unece.org/trans/danger/publi/ghs/ghs_rev03/03files_e.html, accessed March 7, 2013.

30. European Commission, The *Classification, Labeling and Packaging of Chemical Substances and Mixtures*, see http://ec.europa.eu/enterprise/sectors/chemicals/ classification/index_en.htm, accessed September 30, 2014.

31. United Nations Environment Program, World Summit on Sustainable Development, *Plan of Implementation of the World Summit on Sustainable Development*, Johannesburg, 2002, Section 23, p. 13, see www.un.org/esa/sustdev/ documents/WSSD_POI_PD/English/WSSD_PlanImpl.pdf, accessed October 1, 2014.

32. See *Dubai Declaration on International Chemicals Management*, in United Nations, *Strategic Approach to International Chemicals Management (SAICM)*, *Dubai Declaration on International Chemicals Management, Overarching Policy Statement, Global Plan of Action*, Dubai 2006, p. 14, see http:// sustainabledevelopment.un.org/content/documents/SAICM_publication_ENG. pdf, accessed January 12, 2012.

33. The quote is from Ted Smith, Director of the International Campaign for Responsible Technology, personal interview, January 28, 2011. See Suzanne Deffree, "China RoHS: Ready or Not, It's Here," *Electronics News*, March 12, 2007, p. 2; and Noah Sachs, "Planning the Funeral at the Birth: Extended Producer Responsibility in the European Union and the United States," *Harvard Environmental Law Review*, 30:51, 2006, pp. 51–98.

34. Noah M. Sachs identifies four possible effects. See "Jumping the Pond: Transnational Law and the Future of Chemical Regulation," *Vanderbilt Law Review*, 62:6, November 2009, pp. 1817–1869. Joanne Scott goes further, noting changes ongoing in the United States that have already been affected by both the REACH requirements and the model of comprehensive chemicals policy reform that the enactment of REACH provides. See "From Brussels with Love: The Transatlantic Travels of European Law and the Chemistry of Regulatory Attraction," *American Journal of Comparative Law*, 57: 897, pp. 897–942. See also Harvey Black, "Chemical Reaction: The U.S. Response to REACH," *Environmental Health Perspectives*, 116:3, March 2008, pp. A124–A127.

35. Selin and VanDeveer, 2005, p. 14.

36. David Vogel, "Trading Up and Governing Across: Transnational Governance and Environmental Protection," *Journal of European Public Policy*, 4:4, December 1997, pp. 556–571. Quote is at p. 563.

37. For a global review see United National Environment Program (UNEP), Chemicals Bureau, *Global Chemicals Outlook: Towards Sound Management of Chemicals*, Geneva, 2013, see www.unep.org/hazardoussubstances/Portals/9/Mainstreaming/GCO/The%20Global%20Chemical%20Outlook_Full%20report_15Feb2013.pdf, accessed March 22, 2013.

38. Mark Schapiro predicts China's rapidly developing new laws will mean that "the EPA will have little power…to prevent the importation of substances into the United States from China that the new European Chemicals Agency refuses to 'authorize' for sale in Europe"; see Schapiro, 2007, p. 169.

Chapter 5.   Reframing the Chemicals Problem

1. The concept of a policy frame is drawn from the work of Donald Shon and Martin Rein and refers to the "underlying structure of belief, perception, and appreciation" that informs political action or professional practice. See *Frame Reflection: Toward a Resolution of Intractable Policy Controversies* (New York: Basic Books, 1994). The definition is at p. 23. A policy framework is merely the structural embodiment of such a frame.

2. Richard Judson, Ann Richard, David J. Dix, Keith Houck, Matthew Martin, Robert Kavlock, Vicki Dellarco, Tala Henry, Todd Holderman, Philip Sayre, Shirlee Tan, Thomas Carpenter, and Edwin Smith, "The Toxicity Data Landscape for Environmental Chemicals," *Environmental Health Perspectives*, 117(5), May 2009, pp. 685–695.

3. Reports were received on 6,200 individual chemical substances of which 4,800 were manufactured in the United States. Of the 3,100 chemicals imported, 43 percent were only imported. See U.S. Environmental Protection Agency, *2006 Inventory Update Reporting: Data Summary*, Washington, DC, 2006, see http://www.epa.gov/oppt/cdr/pubs/2006_data_summary.pdf, accessed September 30, 2014.

4. See J. Siemiatycki, L. Richardson, K. Straif et al., "Listing Occupational Carcinogens," *Environmental Health Perspectives*, 112, 2004, pp 1447–1459; and United National Environment Program (UNEP),Chemicals Bureau, *Global Chemicals Outlook: Toward Sound Management of Chemicals*, Geneva, 2013, see http://www.unep.org/chemicalsandwaste/Portals/9/Mainstreaming/GCO/The%20Global%20Chemical%20Outlook_Full%20report_15Feb2013.pdf, accessed September 30, 2014.

5. David Rejeski, "The Molecular Economy," *The Environmental Forum*, 27(1), 2010, pp. 36–41.

6. UNEP, 2013.

7. Geiser, 2001, p. 210.

8. Carson, 1962, pp. 155–184.

Chapter 6.   Understanding the Chemical Economy

1. This and other examples of community–facility "fenceline" struggles in Louisiana are described in Barbara Allen, *Uneasy Alchemy: Citizens and Experts in Louisiana's Chemical Corridor Disputes* (Cambridge, MA: MIT Press, 2003).

2. American Chemistry Council (ACC), *2011 Guide to the Business of Chemistry* (Washington, DC, 2011); and Innovest, *Overview of the Chemicals Industry*, unpublished report, March 2007, p. 28, see http://www.precaution.org/lib/07/innovest.pdf, accessed December 12, 2011; and James Heintz and Robert Pollin, *The Economic Benefits of a Green Chemical Industry in the United States*, Political Economy Research Institute, University of Massachusetts Amherst, May, 2011, see http://www.peri.umass.edu/fileadmin/pdf/other_publication_types/green_economics/Green_Chemistry_Report_FINAL.pdf, accessed October 3, 2014. The chemical manufacturing industry is conventionally defined as SIC code 28 or NAICS code 325.

3. The American Chemistry Council lists four categories by conflating agricultural chemicals and pharmaceuticals into "life science" chemicals see ACC, 2011.

4. In 2011, the industry shipped $720 billion in products; $180 billion were internal industry sales and about $540 billion were in final sales. See American Chemistry Council (ACC), *2012 Guide to the Business of Chemistry* (Washington, DC, 2012) p. 5.

5. See Heintz and Pollin, 2011; and U.S. Environmental Protection Agency, *2008 Sector Performance Report* (Washington, DC, 2008), see www.epa.gov/sectors/pdf/2008/2008-sector-report-508-full.pdf, accessed February 4, 2011.

6. Innovest, 2007, p. 22.

7. Innovest, 2007, pp. 33 and 35.

8. ICIS, *Top 100 Chemical Companies* (New York, 2013), see http://img.en25.com/Web/ICIS/FC0211_CHEM_201309.pdf, accessed January 10, 2014.

9. Peter H. Spitz, "The Chemical Industry at the Millennium," in Peter H. Spitz, ed., *The Chemical Industry at the Millennium: Maturity, Restructuring and*

Globalization (Philadelphia, PA: Chemical Heritage Press, 2003) pp. 311–341; and Frank Esposito, "Dow Chemical to Separate Chlorine Business," *Plastics News,* December 2, 2013, see http://www.plasticsnews.com/article/20131202/NEWS/131209999, accessed October 1, 2014.

10. Heintz and Pollin, 2011, p. 26.

11. ACC, 2011, p. 98.

12. Adenike Adeyeye, James Barrett, Jordan Diamond, Lisa Goldman, John Pendergrass, and Daniel Schrammm, *Estimating U.S. Government Subsidies to Energy Sources, 2002–2008,* "Washington, DC: Environmental Law Institute, September 2009, see http://www.eli.org/sites/default/files/eli-pubs/d19_07.pdf, accessed October 2, 2014.

13. These production chains are laid out in more detail in ACC, 2011, pp. 22–31.

14. See Martin Krayer von Krauss and Poul Harremoes, "MTBE in Petrol as a Substitute for Lead," in Poul Herremoes et al., eds., *Late Lessons from Early Warnings: The Precautionary Principle, 1896–2000* (Copenhagen: European Environment Agency, 2001) pp. 110–125; and Richard Johnson, James Pankow, David Bender, Curtis Price, and John Zogorski, "MTBE: To What Extent Will Past Releases Contaminate Community Water Supplies," *Environmental Science and Technology,* 34(9), May 1, 2000, pp. 210A–217A.

15. Forty-six of the top fifty high-production-volume chemicals in 1977 were still in the top fifty in 1993 and roughly in the same order. See Organization for Economic Cooperation and Development (OECD), *OECD Environmental Outlook for the Chemicals Industry* (Paris, 2001) p. 24, see http://www.oecd.org/env/ehs/2375538.pdf, accessed February 20, 2014.

16. United National Environment Program, Chemicals Bureau, *Global Chemicals Outlook, Trends and Indicators* (Rachel Massey and Molly Jacobs), unpublished draft, Geneva, November 2011.

17. U.S. Energy Information Administration (EIA), *Annual Energy Review, 2011* (Washington, DC, 2011), see http://www.eia.gov/totalenergy/data/annual/pdf/aer.pdf, accessed January 10, 2014, p. 47; and ACC, 2011, p. 95 for the percentages.

18. U.S. EIA, 2011; and ACC, 2011, pp. 99–101.

19. American Chemical Society, et al., *Technology Vision 2020-The U.S. Chemical Industry* (Washington, DC, 1996), see http://energy.gov/sites/prod/files/2013/11/f4/chem_vision.pdf, accessed September 30, 2014; ACC, 2011, p. 101; and Innovest, 2007, p. 42.

20. ACC, 2011, p. 104.

21. U.S. Environmental Protection Agency, *Toxics Release Inventory, 2012 TRI National Analysis Overview,* see http://www2.epa.gov/toxics-release-inventory-tri-program/2012-tri-national-analysis-overview, accessed September16, 2014. The 2005 data is from U.S. Environmental Protection Agency, *2008 Sector Performance Report* (Washington, DC, 2008), see http://www.epa.gov/sectors/pdf/2008/2008-sector-report-508-full.pdf, accessed September 16, 2014.

22. U.S. Census Bureau, Census of Manufacture, *The 2012 Statistical Abstract*, see https://www.census.gov/compendia/statab/cats/manufactures.html, accessed October 1, 2014.

23. ACC, 2011, p. 5.

24. ACC, 2011, pp. 2–3.

25. U.S. Department of Agriculture, Economic Research Service, website, *Fertilizer Use and Price*, see www.ers.usda.gov/data-products/fertilizer-use-and-price. aspx#26720, accessed, October 1, 2014; and U.S. EPA, *Pesticide Industry Sales and Usage, 2006 and 2007 Market Estimates* (Washington, DC, 2011).

26. World Trade Organization, *International Trade Statistics, United States Profile*, see http://stat.wto.org/TariffProfile/WSDBTariffPFView.aspx?Language =E&Country=US, accessed October, 2, 2014.

27. Raymond M. Wolfe, *Business R&D Performed in the United States Cost $291 Billion in 2008 and $282 Billion in 2009* (Washington, DC: National Science Foundation, NSF 12–309, March 2012), see www.nsf.gov/statistics/infbrief/ nsf12309/, accessed February 12, 2012.

28. The 2011 figure for hazardous waste comes from the U.S Environmental Protection Agency, Office of Solid Waste, *The National Biennial RCRA Hazardous Waste Report: Based on 2011 Data* (Washington, DC, no date), see www.epa. gov/epawaste/inforesources/data/br11/national11.pdf, accessed February 14, 2014; and the on- and off-site release data come from the U.S. Environmental Protection Agency, Toxics Release Inventory, *Industry Sector Profile: The Chemical Industry* (Washington, DC, 2012), see http://www2.epa.gov/toxics-release-inventory-tri-program/2011-tri-national-analysis-overview, accessed October 1, 2014.

29. U.S. Environmental Protection Agency, Office of Solid Waste and Emergency Response, *Municipal Solid Waste Generation, Recycling, and Disposal in the United States: Facts and Figures for 2010* (Washington, DC, 2011), see www.epa. gov/epawaste/nonhaz/municipal/pubs/msw_2010_rev_factsheet.pdf, accessed February 19, 2012.

30. This estimate is based on a 1985 study by the EPA. The number represents only that share of the nonhazardous industrial waste attributable to manufacturing. See U.S. Congress, Office of Technology Assessment, *Managing Industrial Solid Wastes from Manufacturing, Mining, Oil and Gas Production, and Utility Coal Production* (Washington, DC: U.S. Government Printing Office, 1992) p. 9, see http://ota-cdn.fas.org/reports/9225.pdf, accessed October 1, 2014.

Chapter 7.   Driving the Chemical Market

1. See Gordon R. Foxall, *Understanding Consumer Choice* (New York: Palgrave Macmillan, 2005); and John Thorgerson, "Promoting Green Consumer Behavior with Eco-labels," in Thomas Dietz and Paul Stern, eds., *New Tools for Environmental Protection: Education, Information and Voluntary Measures* (Washington, DC: National Academy Press, 2002) pp. 83–104.

2. Scott Bearse, Peter Capozucca, Laura Favret, and Brian Lynch, *Finding the Green in Today's Shoppers: Sustainability Trends and New Shopper Insights* (Washington, DC: Deloitte/Grocery Manufacturers Association, 2009), see www.deloitte.com/assets/Dcom-Shared%20Assets/Documents/US_CP_GMADeloitte-GreenShopperStudy_2009.pdf, accessed April 17, 2011.

3. *Ecolabel Index: Who's Deciding What's Green*, see www.ecolabelindex.com, accessed January 8, 2013.

4. U.S. Federal Trade Commission, *Guides for the Use of Environmental Marketing Claims*, see http://www.ftc.gov/enforcement/rules/rulemaking-regulatory-reform-proceedings/guides-use-environmental-marketing-claims, accessed October 1, 2014. 5. International Institute for Sustainable Development, *The ISO 14020 Series*, see https://www.iisd.org/business/markets/eco_label_iso14020.aspx, accessed October 2, 2014.

6. An organic label means that at least 95 percent of the products meets organic standards, with each organically produced ingredient identified on the label. Organic food standards are set by the National Organic Program established by the Organic Foods Production Act of 1990 (21 PL 101–624).

7. For instance, the standard for commercial adhesives sets the LD50 at or above 2,000 ppm of vapor gases and 20mg/lL of mists, dusts, or fumes. See Green Seal, http://www.greenseal.org/, accessed October 2, 2014.

8. European Commission, *European Union Ecolabel*, see http://ec.europa.eu/environment/ecolabel/, accessed February 12, 2010; and European Commission, *Green Public Procurement*, see http://ec.europa.eu/environment/gpp/pdf/brochure.pdf, accessed February 12, 2013.

9. Tera Choice, *The Sins of Greenwashing: Home and Family Edition,* 2010, see http://sinsofgreenwashing.org/index35c6.pdf, accessed April 8, 2011; and U.S. Federal Trade Commission *Guides for the Use of Environmental Marketing Claims*, see http://www.ftc.gov/enforcement/rules/rulemaking-regulatory-reform-proceedings/guides-use-environmental-marketing-claims, accessed October 1, 2014.

10. U.S. Environmental Protection Agency, Design for Environment Program, *Safer Product Labeling Program*, see www.epa.gov/dfe/pubs/projects/formulat/saferproductlabeling.htm, accessed January 8, 2013.

11. Abhijit Banergee and Barry Solomon, "Eco-Labeling for Energy Efficiency and Sustainability: A Meta-Evaluation of U.S. Programs," *Energy Policy*, 31, 2003, pp. 109–123.

12. Rita Schenck, *The Outlook and Opportunity for Type III Environmental Product Declarations in the United States of America*, Vashon, WA: Institute for Environmental Research and Education, September 23, 2009, see http://lcacenter.org/pdf/Outlook-for-Type-III-Ecolabels-in-the-USA.pdf, accessed October 2, 2014.

13. Health Product Declaration Collaborative, see http://hpdcollaborative.org/, accessed October 2, 2014.

14. For *Skin Deep*, see http://www.ewg.org/skindeep/, accessed October 1, 2014.

15. Dara O'Rouke, Director, *GoodGuide*, personal interview, February 27, 2009. For *GoodGuide*, see www.goodguide.com/, accessed April 5, 2011.

16. Daniel Goleman, *Ecological Intelligence: How Knowing the Hidden Impacts of What We Buy Can Change Everything* (New York: Broadway Books, 2009) pp. 79–80.

17. Quote is from Goleman, 2009, p. 84.

18. For much of this section, see Yve Torre, *Best Practices in Product Chemicals Management in the Retail Industry*, Green Chemistry and Commerce Council, December, Lowell Center for Sustainable Production, University of Massachusetts Lowell, 2009, see http://greenchemistryandcommerce.org/downloads/uml-rpt-bestprac1209.pdf, accessed April 5, 2011.

19. Tim Griener, Mark Rossi, Beverley Thorpe, and Bob Kerr, *Healthy Business Strategies for Transitioning the Toxic Chemical Economy* (Clean Production Action, 2006) pp. 25–29, see http://dgcommunications.com/documents/dg-CPA_Bus_Report.pdf, accessed April 4, 2011.

20. Kara Sissell, "Retailers and States Take the Lead," *Chemical Week*, April 14–21, 2008, pp. 26–31.

21. Home Depot, *Wood Purchasing Policy*, see https://corporate.homedepot.com/corporateresponsibility/environment/woodpurchasing/Pages/default.aspx, accessed October 2, 2014.

22. Roger McFadden, Senior Scientist, Staples, personal interview, October 6, 2010. For a description of the SPDS, see Torre, 2009, pp. 26–27.

23. In 2013, Walmart released its own *Policy on Sustainable Chemistry in Consumables*, see http://az204679.vo.msecnd.net/media/documents/wmt-chemical-policy_130234693942816792.pdf, accessed November 16, 2013.

24. Target, *Introducing the Target Sustainable Product Standard*, see https://corporate.target.com/discover/article/introducing-the-Target-Sustainable-Product-Standard, accessed January 4, 2014.

25. Richard Dennison, Environmental Defense Fund, personal interview, June 2, 2010.

26. U.S. Office of the President, *Executive Order 13514, Federal Leadership in Environmental, Energy and Economic Performance* (Washington, DC, October 5, 2009).

27. For the GSA program, see U.S. Government Services Administration, *Environmental Products Overview*, www.gsa.gov/enviro, accessed April 17, 2011. See also U.S. Department of Defense, *Green Procurement Strategy* (Washington, DC, 2008), see http://www.gsa.gov/portal/content/104543?utm_source=FAS&utm_medium=print-radio&utm_term=enviro&utm_campaign=shortcuts, accessed October 2, 2014.

28. Electronic Product Environmental Assessment Tool, see http://www.epeat.net/, accessed December 12, 2011.

29. Responsible Purchasing Network, see http://www.responsiblepurchasing.org/about/index.php, accessed December 12, 2011.

30. Markowitz and Rosner, 2002, p. 117.

31. Robert Gottlieb, *Forcing the Spring: The Transformation of the American Environmental Movement* (Washington, DC: Island Press, 1993) pp. 240–250. The early struggle to restrict agricultural uses of pesticides can be found in David Weir and Mark Schapiro, *Circle of Poison: Pesticides and People in a Hungry World* (San Francisco, CA: Institute for Food and Development Policy, 1981).

32. Leslie Byster and Ted Smith, "From Grass Roots to Global: The Silicon Valley Toxic Coalition's Milestones in Building a Movement for Corporate Accountability and Sustainability in the High Tech Sector," in Ted Smith, David A. Sonenfeld, and David Naguib Pellow, eds., *Challenging the Chip: Labor Rights and Environmental Justice in the Global Electronics Industry* (Philadelphia: Temple University Press, 2006) pp. 111–119.

33. Health Care without Harm, *The Global Movement for Mercury-Free Health Care*, Arlington, VA. October 15, 2007, p. 24, see https://noharm-global.org/sites/default/files/documents-files/746/Global_Mvmt_Mercury-Free.pdf, accessed October 2, 2014.

34. Health Care without Harm, *Green Guide for Healthcare*, Version 2.1, see www.gghc.org, accessed March 22, 2013.

35. Campaign for Safe Cosmetics, *Market Shift: The Story of the Safe Cosmetics Compact and the Growing Demand for Safer Products*, see www.safecosmetics.org/downloads/MarketShift_CSC_June15_2012.pdf, accessed March 22, 2013.

36. For a positive review of this emerging force, see Paul Hawkin, *Blessed Unrest: How the Largest Movement in the World Came into Being and Why No One Saw It Coming* (New York: Viking, 2007). For an assessment of the growing international influence of these organizations, see Michel M. Betsill and Elisabeth Corell, *NGO Diplomacy: The Influence of Nongovernmental Organizations in International Environmental Negotiations* (Cambridge: MIT Press, 2008).

37. Richard Dennison, Senior Scientist, Environmental Defense Fund, personal interview, June 2, 2010.

Chapter 8. Transforming the Chemical Industry

1. Christian Baumgaertel, "World Business Briefing, Europe: Nike Shirts Called Not Hazardous," *New York Times*, January 8, 2000.

2. See Milton Friedman, "The Social Responsibility of Business Is to Increase Its Profits," *New York Times Magazine*, September 13, 1970.

3. American Chemistry Council, *Responsible Care*, see http://responsiblecare.americanchemistry.com, accessed March 22, 2013.

4. International Council of Chemical Associations, *Global Product Strategy*, see http://www.icca-chem.org/en/Home/Global-Product-Strategy/, accessed December 12, 2013.

5. This information is drawn from a 2009 case study completed for the Green Chemistry and Commerce Council. Monica Becker, *Considered Chemistry at*

Nike: Creating Safer Products through Evaluation and Restriction of Hazardous Chemicals, unpublished case study, Green Chemistry and Commerce Council, 2009, see http://www.greenchemistryandcommerce.org/downloads/Nike_final.pdf, accessed September 30, 2014

6. John Frazier, Senior Director of Chemical Innovation, Nike, personal interview, May 10, 2012.

7. Lowell Center for Sustainable Production and the Green Chemistry and Commerce Council, *An Analysis of Restricted Substance Lists (RSLs) and Their Implication for Green Chemistry and Design for Environment*, unpublished paper, November 2008, see http://www.greenchemistryandcommerce.org/downloads/RSLAnalysisandList_000.pdf, accessed September 30, 2014.

8. This information is drawn from a 2009 case study completed for the Green Chemistry and Commerce Council. Monica Becker, *S.C. Johnson Is Transforming Its Supply Chain to Create Products That Are Better for the Environment*, unpublished case study, Green Chemistry and Commerce Council, see http://www.greenchemistryandcommerce.org/downloads/SCJ_final.pdf, accessed September 30, 2014.

9. H. A. Holder, P. H. Mazurkiewicz, C. D. Robertson, and C. A. Wray, "Hewlett Packard's Use of the GreenScreen™ for Safer Chemicals," in *Issues in Environmental Science and Technology: Chemical Alternatives Assessment* (Royal Society of Chemistry, 2013) pp. 157–176.

10. Hewlett Packard, *HP 2012 Global Citizenship Report*, 2012, p. 41, see http://h20195.www2.hp.com/V2/GetPDF.aspx/c03742928.pdf, accessed January 30, 2014. See also Monica Becker, *Managing Chemicals of Concern and Designing Safer Products at Hewlett-Packard*, unpublished case study, Green Chemistry and Commerce Council, 2009, see http://www.greenchemistryandcommerce.org/downloads/HP_final.pdf, accessed February 12, 2010.

11. Hewlett Packard, *HP 2012 Global Citizenship Report*, 2012, p. 41.

12. Lowell Center for Sustainable Production and the Green Chemistry and Commerce Council, *Gathering Chemical Information and Advancing Safer Chemistry in Complex Supply Chains*, 2009, see http://www.greenchemistryandcommerce.org/downloads/summaryreport_000.pdf, accessed March 15, 2011.

13. Rachel Massey, Janet Hutchinson, Monica Becker, and Joel Tickner, *Toxic Substances in Articles: The Need for Information* (Copenhagen: Nordic Council of Minister, 2008) pp. 58–60, see http://www.norden.org/en/publications/publikationer/2008-596/at_download/publicationfile, accessed February 20, 2010.

14. Outdoor Industry Association, *Chemicals Management Framework*, 2013, see http://dev.outdoorindustry.org/responsibility/chemicals/cmpilot.html accessed November 6, 2013. See also Corinna Kester and Dana Ledyard, *The Sustainable Apparel Coalition: A Case Study of a Successful Industry Collaboration* (Berkeley: Center for Responsible Business, University of California Berkeley, 2012), see http://responsiblebusiness.haas.berkeley.edu/CRB_SustainableApparelCaseStudy_FINAL.pdf, accessed November 6, 2013.

15. Roadmap to Zero Discharge of Hazardous Chemicals, see http://www. roadmaptozero.com, accessed November 7, 2013.

16. OECD, 2001, p. 34–35.

17. G. P. J. Dukema, J. Grievik, and M. P. C. Weijnen, "Functional Modeling for a Sustainable Petrochemical Industry," *Transaction of the Institution of Chemical Engineers*, 81, 2003, pp. 331–339.

18. European Commission, *High Level Group on the Competitiveness of the European Chemicals Industry: Final Report*, Brussels, 2009, see http://www.cefic. org/Documents/PolicyCentre/HLG-Chemical-Final-report-2009.pdf, accessed February 7, 2012.

19. See Jean-Claude Charpentier, "Four Main Objectives for the Future of Chemical and Process Engineering Mainly Concerned by the Science and Technology of New Material Production," *Chemical Engineering Journal*, 107, 2005, pp. 3–17; and A. J. A. Stankiewicz and A. Moulijn, "Process Intensification: Transforming Chemical Engineering," *Chemical Engineering Progress*, 96(1), 2000, pp. 22–34.

20. Sigurd Buchholz, "Future Manufacturing Approaches in the Chemical and Pharmaceutical Industry," *Chemical Engineering and Processing*, 49, 2010, pp. 993–995.

21. Peter H. Spitz, "The Chemical Industry and the Environment," in Spitz, 2003, pp. 206–243, p. 224 in particular.

22. Thomas Jakl and Petra Schwager, eds., *Chemical Leasing Goes Global: Selling Services Instead of Barrels* (New York: Springer Wien, 2008), see www.chemicalstrategies.org, accessed March 14, 2011.

23. James H. Clark et al., "Green Chemistry and the Biorefinery: A Partnership for a Sustainable Future," *Green Chemistry*, 8, 2006, pp. 853–860.

24. James H. Clark and Fabian E. I. Deswarte, "The Biorefinery Concept—An Integrated Approach," in James H. Clark and Fabian E. I. Deswarte, esds., *Introduction to Chemicals from Biomass* (New York: Wiley, 2008) pp. 1–20, see p. 7.

25. Clark and Deswarte, 2008, p. 8.

26. A. J. Ragauskas et al., "The Path Forward for Biofuels and Biomaterials," *Science*, 311, January 27, 2006, pp. 484–489.

27. Clark and Deswarte, 2008, pp. 14–17.

28. ACC, 2011.

29. International Organization for Standardization, *ISO 14000-Environmental Management*, see http://www.iso.org/iso/home/standards/management-standards/iso14000.htm, accessed March 14, 2011.

30. See citations from Toshi H. Arimura, Nicole Darnall, and Hajime Katayama, *Is ISO 14001 a Gateway to More Advanced Voluntary Action?*, Resources for the Future, Washington, DC, March 2009, see http://www.rff.org/documents/rff-dp-09-05.pdf, accessed March 15, 2011.

31. U.S. Green Building Council, LEED website, see http://usgbc.org/leed, accessed October 10, 2013.

32. Living Building Challenge, see http://living-future.org/lbc, accessed October 10, 2013.

33. See Clean Production Action, see www.cleanproduction.org, accessed October 10, 2013.

34. See Healthy Building Network, *About Pharos*, see www.pharosproject.net, accessed February 26, 2010.

35. Mark Rossi, Co-Director, Clean Production Action, personal interview, December 12, 2011.

36. Social Investment Forum Foundation, *Report on Socially Responsible Investing Trends in the United States* (New York, 2012), see http://www.ussif.org/files/Publications/12_Trends_Exec_Summary.pdf, accessed September 30, 2014.

37. ChemSec, *Chemicals Criteria Catalogue*, see http://www.chemsec.org/images/stories/2012/chemsec/Chemicals_Criteria_Catalogue_Executive_Summary.pdf, accessed March 14, 2011.

38. Richard Liroff, presentation, Green Chemistry and Commerce Council Conference, Houston, TX, May 12, 2009.

39. The term *sustainability* here references the definition put forward in the World Commission on Environment and Development, *Our Common Future* (New York: Oxford University Press, 1987).

40. William F. Carroll, Jr., and Douglas J. Raber, *The Chemistry Enterprise in 2015*, American Chemicals Society, Washington, DC, 2005, see http://www.acs.org/content/dam/acsorg/membership/acs/welcoming/industry/the-chemistry-enterprise-2015.pdf, accessed January 11, 2012.

41. Cynthia Challener, "Sustainable Development at a Crossroads," *Chemical Market Reporter*, 260, July 16, 2001, pp. Fr3–Fr4.

42. American Chemical Society, et al., *Technology Vision 2020-The U.S. Chemical Industry* (Washington, DC, 1996), see http://energy.gov/sites/prod/files/2013/11/f4/chem_vision.pdf; Council of Chemical Research et al., *Vision 2020: Chemical Industry of The Future Technology—Roadmap for Materials* (New York, 2000), see http://all-experts.com/assets/roadmaps/518__materials_tech_roadmap.pdf ; American Institute of Chemical Engineers, Center for Waste Reduction Technologies, *Vision 2020: Reaction Engineering Roadmap* (New York, 2001), see http://all-experts.com/assets/roadmaps/482__reaction_roadmap.pdf, all accessed September 30, 2014.

43. National Research Council, *Sustainability in the Chemicals Industry: Grand Challenges and Research Needs* (Washington, DC: National Academies Press, 2005), see http://digilib.bppt.go.id/sampul/Sustainability_in_the_Chemical_Industry_Grand_Challenges_and_Research_Needs_A_Workshop_Report.pdf, accessed March 7, 2011.

44. Michael Porter and Claas van der Linde, "Toward a New Conception of the Environment Competitiveness Relationship," *Journal of Economic Perspective*, 9(4), 1995, pp. 97–118.

Chapter 9. Designing Greener Chemistry

1. U.S. Environmental Protection Agency, Presidential Green Chemistry Challenge Awards, *Award Recipients, 1996–2013* (Washington, DC), see http://www2. epa.gov/sites/production/files/documents/award_recipients_1996_2012.pdf, accessed March 13, 2013.

2. For an early review, see J. A. Cano-Ruiz and G. J. McRae, "Environmentally Conscious Chemical Process Design," *Annual Review of Energy and Environment*, 23, 1998, pp. 499–536.

3. E. J. Woodhouse and S. Breyman, "Green Chemistry as Social Movement?" *Science, Technology and Human Values*, 30, 2005, pp. 199–222.

4. See Kenneth Hancock and Margaret Cavanaugh, "Environmentally Benign Chemical Synthesis and Processing for the Economy and the Environment," in P. T. Anastas and C. A. Farris, eds., *Benign by Design: Alternative Synthetic Pathways for Pollution Prevention* (Washington, DC: American Chemical Society, 1994) pp. 23–30; Ivan Amoto, "The Slow Birth of Green Chemistry," *Science*, 259, March 12, 1993, pp. 1538–1541; and Paul Anastas and Mary Kirchhoff, "Origins, Current Status and Future Challenges of Green Chemistry," *Accounts of Chemical Research*, 35(9), 2002, pp. 686–694.

5. This big vision is explored in John Warner, Amy Cannon, and Kevin Dye, "Green Chemistry," *Environmental Impact Assessment Review*, 24, 2004, pp. 775–799.

6. Paul Anastas and John Warner, *Green Chemistry: Theory and Practice* (New York: Oxford University Press, 1998) p. 11.

7. The principles have been slightly adapted since the publication of the Anastas and Warner book. See U.S. Environmental Protection Agency, "12 Principles of Green Chemistry," see www.epa.gov/sciencematters/june2011/principles.htm, accessed March 13, 2013.

8. Paul T. Anastas and Tracy C. Williamson, "Frontiers in Green Chemistry," in Paul T. Anastas and Tracy C. Williamson, eds., *Green Chemistry: Frontiers in Benign Chemical Syntheses and Processes* (New York: Oxford University Press, 1998) p. 10.

9. Jesper Sjostrom calls green chemistry a "meta-discipline" and notes two classifications types of study: "green chemistry activities" and "green chemistry policy and knowledge areas." See "Green Chemistry in Perspective—Models of GC Activities and GC Policy and Knowledge Areas," *Green Chemistry*, 8, 2006, pp. 130–137.

10. M. Lancaster, *Green Chemistry: An Introductory Text* (Cambridge, UK: Royal Society of Chemistry, 2002); V. K. Ahluwalia and M. Kidwai, *New Trends in Green Chemistry*, Second Edition (New Delhi, India: Anamaya, 2006); and Francesca M. Kerton, "Green Chemical Technologies," in Clark and Deswarte, 2008, pp. 47–76.

11. Green engineering is respectfully presented in David Allen and David Shonnard, *Green Engineering: Environmentally Conscious Design of Chemical Processes* (Upper Saddle River, NJ: Prentice Hall, 2002).

12. Scott Sieburth, "Isosteric Replacement of Carbon with Silicon in the Design of Safer Chemicals," in Stephen C. DeVito and Roger L. Garrett, eds., *Designing Safer Chemicals: Green Chemistry for Pollution Prevention* (Washington, DC: American Chemical Society, 1996) pp. 74–83.

13. For more details, see Alexei Lapkin and David Constable, *Green Chemistry Metrics: Measuring and Monitoring Sustainable Processes* (New York: J. Wiley, 2008).

14. G. Centi, P. Ciambelli, S. Perathoner, and P. Russo, "Environmental Catalysis: Trends and Outlooks," *Catalysis Today*, 1–(13), 2002, p. 2691.

15. See Centi, Ciabelli, Perathoner, and Russo, 2002, pp. 1–13.

16. See Jurgen Metzger, "An Organic Reactions without Organic Solvents and Oils and Fats as Renewable Raw Materials for the Chemical Industry," *Chemosphere*, 43, 2001, pp. 83–87.

17. C. J. Li, "Organic Reactions in Aqueous Media with a Focus on Carbon-Carbon Bond Formations: A Decade Update," *Chemical Reviews*, 105, 2005, pp. 3095–3165.

18. M. B. Nuchter, W. Ondruschka, W. Bonrath, and A. Gum, "Microwave-Assisted Synthesis—A Critical Technology Overview," *Green Chemistry*, 6, 2004, pp. 128–141.

19. Trevor Kletz, "Inherently Safer Plants—The Concept, Its Scope and Benefits," *Loss Prevention Bulletin*, 51, June 1983, pp. 1–8. See also Robert E. Bollinger, David G. Clark, Arthur M. Dowell, Roger M. Ewbank, Dennis C. Hendershot, William K. Lutz, Steven I. Meszaros, Donald E. Park, and Everett D. Wixom (Daniel A. Crowl, ed.), *Inherently Safer Chemical Processes: A Life Cycle Approach* (New York: Center for Chemical Process Safety, 1996).

20. Mark Finlay, "Old Efforts at New Uses: A Brief History of Chemurgy and the American Search for Biobased Materials," *Journal of Industrial Ecology*, 7(3–4), 2004, pp. 33–46.

21. See Robert C. Brown, *Biorenewable Resources: Engineering New Products from Agriculture* (Ames, IA: Iowa State Press, 2003).

22. NatureWorks, LLC, see www.natureworksllc.com/, accessed March 12, 2011; and Maine Technology Institute, *Bioplastics Cluster Looks to Maine Potatoes*, see www.mainetechnology.org/results/success-stories/bioplastics-cluster-looks-to-maine-potatoes-and-forests, accessed March 12, 2011.

23. Biomass Research and Development Initiative, *Roadmap for BioEnergy and Biobased Products in the United States*, see http://www1.eere.energy.gov/bioenergy/pdfs/obp_roadmapv2_web.pdf, accessed September 30,2014. ; and Marvin Duncan, "Federal Initiatives to Support Biomass Research and Development," *Journal of Industrial Ecology*, 7(3–4), 2004, pp. 193–201.

24. David B. Turley, "The Chemical Value of Biomass," in Clark and Deswarte, 2008, pp. 21–46.

25. P. A. Matson, W. J. Parton, A. G. Power, and M. J. Smith, "Agricultural Intensification and Ecological Properties," *Science*, 277, July 25, 1997, pp. 504–509.

26. David Pimentel and Wen Dazhong, "Technology Changes in Energy Use in U. S. Agricultural Production," in Ronals Carroll, John H. Vandermeer, and Peter M. Rosset, eds., *Aroecology* (New York: McGraw-Hill, 1990) pp. 150–155; and Sandra Postel, *Water for Agriculture: Facing the Limits*, Worldwatch Paper No. 93 (Washington, DC: Worldwatch Institute, 1989).

27. As an example, see E. S. Stevens, *Green Plastics: An Introduction to the New Science of Biodegradable Plastics* (Princeton, NJ: Princeton University Press, 2002).

28. Sustainable Materials Collaborative, *Biomaterials: Essential for the Next Generation of Products*, see www.sustainablebiomaterials.org/, accessed March 12, 2011.

29. Janine M. Benyus has written a nontechnical and quite readable review of these investigations in *Biomimicy: Innovation Inspired by Nature* (New York: William Morrow, 1997).

30. An element-by-element description in abbreviated but quite readable form can be found in John Emsley, *Natures Building Blocks: An A-Z Guide to the Elements* (New York: Oxford University Press, 2001).

31. Berkeley (Buzz) Cue, BWC Pharma Consulting, personal interview, January 31, 2011. For a good review of work at GlaxoSmithKline, see Alan Curzons, David Constable, David Mortimer, and Virginia Cunningham, "So You Think Your Process is Green? Using Principles of Sustainability to Determine what is 'Green': A Corporate Perspective," *Green Chemistry*, 3, 2001, pp. 1–6.

32. U.S. Environmental Protection Agency, *The Presidential Green Chemistry Challenge, Awards Recipients, 1996–2009*, 2009, pp. 62–63 and 128–129.

33. See Jeffrey S. Plotkin, "Petrochemical Technology Development," in Peter H. Spitz, ed., *The Chemical Industry at the Milennium: Maturity, Restructuring and Globalization* (Philadelphia: Chemical Heritage Press, 2003) pp. 51–84.

34. Th. F. O'Brien, T. V. Bommaraju, and F. Hine, *Handbook of Chlor-Alkali Technology* (New York: Springer, 2005).

35. Islvan T. Horvath and Paul T. Anastas, "Innovations in Green Chemistry," *Chemical Reviews*, 107(6), 2007, pp. 2169–2173.

36. U.S. Department of Energy, *Top Value Added Chemicals from Biomass: Volume 1: Results of Screening for Potential Candidates from Sugars and Syngas* (Washington, DC, August 2004), see http://www1.eere.energy.gov/bioenergy/pdfs/35523.pdf, accessed January 14, 2014.

37. Paul Anastas, Director, Center for Green Chemistry and Engineering, Yale University, personal interview, January 23, 2014.

38. Thomas Kuhn argued that science developed through "episodic upheavals" rather than the "development-by-accumulation" model that was then conventionally accepted. See Thomas Kuhn, *The Structure of Scientific Revolutions* (Chicago: University of Chicago Press, 1962).

39. Stephen K. Ritter, "Teaching Green: Initiative Encourages Faster Uptake of Green Chemistry and Toxicology in the Undergraduate Curriculum," *Chemical and Engineering News*, 90(40), October 1, 2012, 64–65.

40. John Warner, Director, Warner-Babcock Institute, personal interview, November 10, 2011.

Chapter 10.   Characterizing and Prioritizing Chemicals

1. Nicholas Ashford, C. W. Ryan, and Charles Caldart, "A Hard Look at Federal Regulation of Formaldehyde: A Departure from Reasoned Decision Making," *Harvard Law Review*, 297, 1983.

2. See National Research Council, *Review of the Environmental Protection Agency's Draft IRIS Assessment of Formaldehyde* (Washington, DC: National Academy Press, 2011). In 2010, Congress passed a law, the Formaldehyde Standards for Wood Products Act, which finally set national standards at least for formaldehyde emissions from composite wood products.

3. Chemical Abstract Service, see http://www.cas.org/, accessed December 30, 2014.

4. The National Institute of Occupational Safety and Health developed the RTECS system, which today includes some 153,000 substances, and SMILES is a linear code for unambiguously specifying the structure of chemical molecules.

5. A recent survey of 22,000 chemicals found 610 substances likely to be persistent and bioaccumulative in the environment. About 100 have been measured in the environment, and another forty-seven have appeared in monitoring data. See Philip H. Howard and Derek C. G. Muir, "Identifying New Persistent and Bioaccumulatve Organics among Chemicals in Commerce," *Environmental Science and Technology*, 44(7), 2010, pp. 2277–2285.

6. California Office of Environmental Health Hazard Assessment, *Green Chemistry Hazard Traits, End Points and Other Relevant Data* (Sacramento, CA, August 10, 2010), see http://oehha.ca.gov/multimedia/green/pdf/081110prereghazard.pdf, accessed June 20, 2011.

7. European Parliament and Council of the European Union, *REACH Regulation*, Annex I, Regulation (EC) No. 1907/2006, Brussels, December 18, 2006.

8. Health Canada, *Categorization of Substances on the Domestic Substances List*, see http://www.hc-sc.gc.ca/ewh-semt/contaminants/existsub/categor/index-eng.php, accessed February 4, 2011.

9. The FDA maintains a list of substances "generally regarded as safe" for identifying safer food additives. However, Sarah Vogel suggests that a number of chemicals on the list were "grandfathered in" without adequate testing. See Vogel, 2013, p. 36.

10. Clean Production Action, *Green Screen for Safer Chemicals*, see http://www.greenscreenchemicals.org/, accessed October 1, 2014.

11. United Nations Economic and Social Council, *A Guide to the Globally Harmonized System of Classification and Labeling of Chemicals* (GHS), see www.unece.org/fileadmin/DAM/trans/danger/publi/ghs/ghs_rev04/English/ST-SG-AC10-30-Rev4e.pdf, accessed March 7, 2013.

12. Risk phases were first laid out in Annex III of the European Union, Directive 67/548/EEC, *Nature of Special Risks Attributed to Dangerous Substances and Preparations*, and updated in Directive 2001/59/EC. They will be used along with the GHS hazard statements until 2016.

13. World Health Organization, *The WHO Recommended Classification of Pesticides by Hazard* (Geneva, 2009), see http://www.who.int/ipcs/publications/pesticides_hazard_2009.pdf, accessed January 4, 2014.

14. European Chemicals Agency, see www.echa.europa.eu/regulations/clp/classification, accessed, January, 30, 2013.

15. European Chemicals Agency, "ECHA Receives 3.1 Million Classification and Labeling Notifications," Press Release, Helsinki, January 4, 2011, see http://echa.europa.eu/documents/10162/13585/pr_11_01_clp_deadline_20110104_en.pdf, accessed October 1, 2014.

16. See Siemiatycki, Richardson, Straif, et al., 2004; and UNEP, *Global Chemicals Outlook*, 2013.

17. The ECHA list only includes substances classified as known (group 1) human carcinogens or reproductive toxins. European Chemicals Agency, *CMR Substances from Annex VI of the CLP Regulation Registered under REACH and/or Notified under CPL: A First Screening-Report 2012*, see www.echa.europa.eu/documents/10162/13562/cmr_report_en.pdf; and OSPAR Convention, *The OSPAR List of Substances of Possible Concern*, see www.ospar.org/content/content.asp?menu=00950304450000_000000_000000, both accessed March 7, 2013. A more recent analysis identifies some 600 PBTs; see Howard and Muir, 2010.

18. California Office of Environmental Health Hazard Assessment, *Proposition 65 List of Chemicals*, see www.oehha.ca.gov/prop65/prop65_list/newlist.html; European Chemicals Agency, *Candidate List of Substances of Very High Concern for Authorization*, Updated December 12, 2012, see www.echa.europa.eu/candidate-list-table; and International Chemicals Secretariat, *Substitute it Now List*, see http://www.chemsec.org/what-we-do/sin-list, accessed October 1, 2014.

19. Alex Stone and Damon Delistraty, "Sources of Toxicity and Exposure Information for Identifying Chemicals of High Concern to Children," *Environmental Impact Assessment Review*, 30, 2010, pp. 380–387.

20. European Chemicals Agency, *General Approach for Substances of Very High Concern (SVHC) for inclusion in the List of Substances Subject to Authorization* (Helsinki, May 28, 2010), see http://echa.europa.eu/documents/10162/13640/axiv_prioritysetting_general_approach_20100701_en.pdf, accessed March 22, 2013.

21. U.S. Environmental Protection Agency, Office of Pollution Prevention and Toxics, *TSCA Work Plan Chemicals: Methods Document* (Washington, DC, Feb-

ruary 2012), see www.epa.gov/oppt/existingchemicals/pubs/wpmethods.pdf, accessed March 7, 2013.

22. National Research Council, Committee on Improving Risk Analysis Approaches Used by the U.S. EPA, *Science and Decisions: Advancing Risk Assessment* (Washington, DC: National Academy Press, 2009); and National Research Council, Committee on the Health Risks of Phthalates, *Phthalates and Cumulative Risk Assessment: The Tasks Ahead* (Washington, DC: National Academy Press, 2008).

23. See Robert D. Bullard, *Unequal Protection: Environmental Justice and Communities of Color* (San Francisco, Sierra Club Books, 1994); and David Naguib Pellow, *Resisting Global Toxics: Transnational Movements for Environmental Justice* (Cambridge, MA: MIT Press, 2007).

Chapter 11.   Generating Chemical Information

1. Katy Wolfe, *Tert-Butyl Acetate: Safer Alternatives in Cleaning and Thinning Applications*, report prepared by the Institute for Research and Technical Assistance for U.S. Environmental Protection Agency, March, 2007, see http://www.irta.us/reports/TBAC%20Report.pdf.pdf, accessed October 1, 2014.

2. Joseph H. Guth, Richard A. Dennison, and Jennifer Sass, "Require Comprehensive Safety Data for all Chemicals," *New Solutions*, 17(3), 2007, pp. 233–258.

3. National Library of Medicine, *Toxicology Data Network*, see http://toxnet.nlm.nih.gov/, accessed March 7, 2013.

4. European Chemicals Agency, *Information on Chemicals,* see http://echa.europa.eu/information-on-chemicals; and Organization for Economic Cooperation and Development, *The Global Portal to Information on Chemical Substances*, see http://www.echemportal.org/echemportal/index?pageID=0&request_locale=en, both accessed October 1, 2014.

5. U.S. Environmental Protection Agency, Computational Toxicology Research Program, *ACToR*, see www.epa.gov/ncct/actor, accessed January 14, 2011; for a fuller description see Judson et al., 2009.

6. U.S. Environmental Protection Agency, *Analog Identification Methodology (AIM)*, see www.epa.gov/oppt/sf/tools/aim.htm, accessed March 7, 2013.

7. U.S. Environmental Protection Agency, *PBT Profiler*, see www.pbtprofiler.net, accessed March, 7, 2013.

8. U.S. Environmental Protection Agency, Sustainable Futures, *OncoLogic: A Computer System to Evaluate the Carcinogenic Potential of Chemicals*; and *Ecological Structure Activity Relationships (ECOSAR)*, see www.epa.gov/oppt/sf/pubs/oncologic.htm and www.epa.gov/oppt/newchems/tools/21ecosar.htm, accessed March 7, 2013.

9. Environment Canada, *Existing Substances Evaluation*, see http://www.ec.gc.ca/ese-ees/, accessed October 1, 2014.

10. U.S. Environmental Protection Agency, Office of Chemical Safety and Pollution Prevention, *Harmonized Test Guidelines—Master List*, see www.epa.gov/ocspp/pdfs/OCSPP-TestGuidelines_MasterList.pdf, accessed March, 7, 2013. For the struggles over endocrine disruption see Langston, 2010, and Vogel, 2013.

11. National Research Council, Committee on Toxicity Testing and Assessment of Environmental Agents, *Toxicity Testing in the 21st Century: A Vision and a Strategy* (Washington, DC: National Academy Press, 2007).

12. U.S. Environmental Protection Agency, *Toxicity Forecaster (ToxCast)*, see www.epa.gov/ncct/download_files/factsheets/toxcast_12-13-2010.pdf, accessed February 11, 2011; R. S. Judson, K. A. Houck, R. J. Kavlock, T. B. Knudsen, M. T. Martin, et al., "*In vitro* Screening of Environmental Chemicals for Targeted Testing Prioritization: The ToxCast Project," *Environmental Health Perspectives*, 118, 2010, pp. 485–492; and Erik Stokstad, "Putting Chemicals on a Path to Better Risk Assessment," *Science*, 325, August 7, 2009, pp. 694–695.

13. The OECD provides a periodic assessment on global chemical production and use, but the data are quite generalized. Private surveys such as Stanford Research Institute's *Chemicals Economics Handbook* provide more detailed information but from quite uneven sources. For a general overview, see United Nations Environment Program's *Global Chemicals Outlook*, 2013.

14. U.S. Environmental Protection Agency, *2012 Chemical Data Reporting Results*, see www.epa.gov/cdr/pubs/guidance/cdr_factsheets.html, accessed November 20, 2013.

15. California Department of Pesticide Regulation, *California Pesticide Information Portal*, see http://www.cdpr.ca.gov/docs/pur/purmain.htm, accessed February 25, 2013.

16. See Interstate Chemicals Clearinghouse, *U.S. State-Level Chemical Policy Database*, see www.theic2.org/chemical-policy, accessed September 30, 2014.

17. Green Chemistry and Commerce Council, *Meeting Customer Needs for Chemical Data: A Guidance Document for Suppliers*, February 2011, see www.greenchemistryandcommerce.org/downloads/GC3_guidance_final_031011.pdf, accessed March 15, 2013.

18. Massey, Hutchinson, Becker, and Tickner, 2008.

19. GreenBlue, *CleanGredients*, see www.cleangredients.org, accessed February 26, 2010.

20. California Department of Public Health, *California Safe Cosmetics Program*, see www.cdph.ca.gov/programs/cosmetics/Pages/default.aspx, accessed February 25, 2013.

21. Northeast Waste Management Association, Interstate Mercury Education and Research Clearinghouse, see www.newmoa.org/prevention/mercury/imerc.cfm, accessed January 14, 2011.

22. Torre, 2009, p. 11.

23. United Nations Environment Program, Chemicals Bureau, *Chemicals in Products Project*, see http://www.unep.org/chemicalsandwaste/UNEPsWork/

ChemicalsinProductsproject/tabid/56141/Default.aspx, accessed December 20, 2013.

24. U.S. Environmental Protection Agency, *Toxics Release Inventory Program*, see http://www2.epa.gov/toxics-release-inventory-tri-program/, accessed October 1, 2014. A searchable database derived from the national PRTRs is maintained by the Organization for Economic Cooperation and Development, *Center for PTR Data*, see www.oecd.org/env_prtr_data/, accessed February 10, 2011.

25. U.S. Environmental Protection Agency, *Ground Water and Drinking Water*, see water.epa.gov/drink/index.cfm; and California Air Resources Board, see www.arb.ca.gov/homepage.htm, both accessed January 14, 2011.

26. U.S. Centers for Disease Control, *Fourth National Report on Human Exposure to Environmental Chemicals* (Atlanta, 2009).

27. U.S. Environmental Protection Agency, Office of Pollution Prevention and Toxics, *Exposure Assessment Tools and Methods*, see www.epa.gov/oppt/exposure/, accessed March 7, 2013.

28. SubsPort, *Moving Toward Safer Alternatives*, see www.subsport.eu/, accessed March 28, 2011.

29. The paragraph 15(c) of the SAICM text is clear on this: "In the context of this paragraph, information on chemicals relating to the health and safety of humans and the environment should not be regarded as confidential." See *Overarching Policy Statement*, p. 15, in United Nations Environment Program, *SAICM*, 2006.

30. Pat Rizzuto, "Chemical Manufacturers Made Far Fewer Confidentiality Claims in 2010," *Chemical Regulation Reporter*, December 23, 2013, see http://www.bna.com/chemical-manufacturers-made-n17179880950/, accessed December 27, 2013.

31. Chemical Scorecard, see http://scorecard.goodguide.com/chemical-profiles/, accessed February 6, 2013.

32. European Trade Union Institute, see www.istas.net/risctox/en, accessed February 6, 2013.

33. Charlotte Brody, BlueGreen Alliance, personal interview, January 16, 2014; and BlueGreen Alliance, see www.chemhat.org, accessed February 6, 2013.

34. U.S. Environmental Protection Agency, Office of Chemical Safety and Pollution Prevention, *ChemView*, see http://www.epa.gov/chemview/, accessed January 10, 2014.

35. This argument was put forth early by Dorthy Nelkin in *Controversy: The Politics of Technical Decisions* (London: Sage, 1979) and carried on by Sheila Jasanoff, "Beyond Epistemology: Relativism and Engagement in the Politics of Science," *Social Studies of Science*, 26(2), 1997, pp. 393–418; and Brian Wynn, "Knowledge in Context," *Science, Technology and Human Values*, 16(1), 1991, pp. 11–121.

36. Daniel Sarewitz and David Kriebel, *The Sustainable Solutions Agenda*, Consortium for Science, Policy and Outcomes, Arizona State University and the Lowell Center for Sustainable Production (Lowell, MA: University of Massachusetts

Lowell, 2010) p. 7, see www.sustainableproduction.org/downloads/SSABooklet. pdf, accessed February 12, 2011.

37. This approach has been described as "sustainability science." See R. Kates, W. Clark, R. Corell, J. Hall, C. Jaeger et al., "Sustainability Science," *Science*, 292(5517), 2001, pp. 641–642; and Daniel Faber, *Eco-Pragmatism: Making Sensible Environmental Decisions in an Uncertain World* (Chicago: University of Chicago Press, 1999).

Chapter 12.   Substituting Safer Chemicals

1. Richard Doherty, "A History of the Production and Use of Carbon Tetrachloride, Tetrachloroethylene, Trichloroethylene and 1,1,1-Trichloroethane in the United States: Part 2—Trichloroethylene and 1,1,1-Trichloroethane," *Environmental Forensics*, 1(2), 2000, pp. 83–93.

2. These included methylene chloride, trichloroethylene, lead, organo-tin compounds, chlorinated paraffins, phthalates, nonylphenolethoxylates, and brominated flame retardant. See Swedish National Chemicals Inspectorate (KEMI), *Risk Reduction of Chemicals: A Government Commission Report* (Solna, Sweden, 1991).

3. United Nations, *Consolidated List of Products Whose Consumption and/or Sale Have Been Banned, Withdrawn, Severely Restricted or Not Approved by Governments, Second Issue* (ST/ESA192) (New York, 1987).

4. Linda Baker, "The Hole in the Sky: Think the Ozone Layer Is Yesterday's Issue? Think Again," *E-Magazine*, October 31, 2000.

5. Beth Rosenberg, "The Story of the Alar Ban: Politics and Unforeseen Consequences," *New Solutions*, 6(2), 1996, pp. 34–50.

6. Swedish National Chemicals Inspectorate, *Observation List: Examples of Substances Requiring Particular Attention, Revised Edition* (Solna, Sweden, 1998), see www.chemicalspolicy.org/downloads/Swedish%20Obs%20List.pdf, accessed February 2, 2011.

7. Janet Raloff, "Danger on Deck," *Science News*, 165(5), January 21, 2004, p. 74.

8. United Nations Environment Program, *Stockholm Convention on Persistent Organic Pollutants, As Amended in 2009* (Stockholm, 2009), see http:// www.env.go.jp/chemi/pops/treaty/treaty_en2009.pdf, accessed January 14, 2014.

9. This is also called the "product choice principle." See Swedish National Chemicals Inspectorate, *The Substitution Principle, a Report from the Swedish Chemicals Agency*, Report NR-8/07 (Solna, Sweden, 2007) p. 5, see http://www.kemi.se/ Documents/Publikationer/Trycksaker/Rapporter/Report8_07_The_Substitution_ Principle.pdf, accessed October 30, 2014.

10. For further examples, see David Allen and Kristen Sinclair Rosselot, *Pollution Prevention for Chemical Processes* (New York: John Wiley, 1997).

11. U.S. Environmental Protection Agency, Office of Wastewater Management, *Environmentally Acceptable Lubricants* (Washington, DC, November 2011); and U.S. Environmental Protection Agency, Design for Environment Program, *Safer Ingredient List for Use in DfE Labeled Products*, see www.epa.gov/dfe/saferingredients.htm, accessed January 8, 2013.

12. Lowell Center for Sustainable Production, *Decabromodiphenylether: An Investigation of Non-Halogen Substitutes to Electronic Enclosures and Textile Applications* (prepared by Pure Strategies, Inc.) (Lowell, MA, University of Massachusetts Lowell, April 2005).

13. Lowell Center for Sustainable Production, *Phthalates and Their Alternatives: Health and Environmental Concerns* (Lowell, MA: University of Massachusetts Lowell, 2011), see http://www.sustainableproduction.org/downloads/PhthalateAlternatives-January2011.pdf, accessed December 11, 2013.

14. Kim Alfonsi, Juan Colberg, Peter. J. Dunn, Thomas Fevig, Sandra Jennings, Timothy A. Johnson, H. Peter Kleine, Craig Knight, Mark A. Nagy, David A. Perry, and Mark Stefaniak, "Green Chemistry Tools to Influence a Medicinal Chemistry and Research Chemistry Based Organisation," *Green Chemistry*, 2008, 10, pp. 31–36.

15. See U.S. Environmental Protection Agency, Office of Chemical Safety and Pollution Prevention, *Existing Chemical Action Plans*, see http://www.epa.gov/oppt/existingchemicals/pubs/ecactionpln.html#posted, accessed February 16, 2012.

16. For a more detailed review of the process, see Massachusetts Department of Environmental Protection, *Toxics Use Reduction Planning and Plan Update Guidance* (Boston, MA, Dcember 2009), see http://www.mass.gov/eea/docs/dep/toxics/laws/planguid.pdf, accessed October 1, 2014.

17. Mary O'Brien was early in advocating this approach common in environmental impact assessment as a tool for assessing alternatives to hazardous chemicals. See O'Brien, 2000, During the 1990s, Nicholas Ashford at MIT developed the approach in his Technology Options Analysis tool. See Nicholas Ashford, "An Innovation-Based Strategy for the Environment and Workplace," in A. M. Finkel and D. Golding, eds., *Worst Things First? The Debate over Risk Based National Priorities* (Washington, DC: Resources for the Future, 1994) pp. 275–314.

18. Brandon Kuczenski and Roland Geyer have described the relationship between multiple criteria decision analysis and alternatives assessment in a special report for the California Department of Toxics Substances Control. See *Chemical Alternatives Analysis: Methods Models and Tools*, unpublished, Bren School of Environmental Science and Management, University of California, Santa Barbara, 2010, see http://www.dtsc.ca.gov/PollutionPrevention/GreenChemistryInitiative/upload/08-T3629-AA-Report-Final-Aug-24-2010.pdf, accessed February 22, 2011. For an example of an alternatives assessment framework using multi-criteria decision analysis see T. F. Malloy, P. J. Sinsheimer, A. Blake and I. Linkov, *Developing Regulatory Alternatives Analysis Methodologies for the California Green Chemistry Initiative*, Sustainable Technology and Policy Program, University of California, Los Angeles, CA, October 20, 2011, see http://www.stpp.ucla.

edu/sites/default/files/Final%20AA%20Report.final%20rev.pdf, accessed October 1, 2014.

19. See Martin Charter and Ursala Tischner, *Sustainable Solutions: Developing Products and Services for the Future* (Sheffiled, UK: Greenleaf Publishing, 2001); and Nathan Shedroff, *Design Is the Problem: The Future of Design Must Be Sustainable* (Brooklyn, NY: Rosenfeld Media, 2009).

20. Mark Rossi, Joel Tickner, and Ken Geiser, *Alternatives Assessment Framework of the Lowell Center for Sustainable Production* (Lowell, MA: Lowell Center for Sustainable Production, University of Massachusetts Lowell, 2006), see www.sustainableproduction.org/downloads/FinalAltsAssess06_000.pdf, accessed March 7, 2013.

21. Massachusetts Toxics Use Reduction Institute, *Five Chemicals Alternatives Assessment Study* (Lowell, MA: University of Massachusetts Lowell, June 2006), see http://www.turi.org/TURI_Publications/TURI_Methods_Policy_Reports/Five_Chemicals_Alternatives_Assessment_Study._2006, accessed October 1, 2014.

22. Michael Ellenbecker, Director, Massachusetts Toxics Use Reduction Institute, personal interview, November 12, 2010.

23. Today, this seven-step process is formalized in an interactive "Alternatives Assessment Wiki" that is maintained at the Interstate Chemicals Clearinghouse, see www.ic2saferalternatives.org/, accessed March 10, 2011.

24. For a review of these frameworks see Organization for Economic Cooperation and Development, Environment Directorate, *Current Landscape of Alternatives Assessment Practice: A Meta Review*, (Paris, 2013), see http://www.oecd.org/officialdocuments/publicdisplaydocumentpdf/?cote=ENV/JM/MONO(2013)24&docLanguage=En/, accessed October 1, 2014.

25. For a good review of these tools and more, see Sally Edwards, Mark Rossi, and Pam Civie, *Alternatives Assessment for Toxics Use Reduction: A Survey of Methods and Tools, Methods and Policy Report No. 23* (Lowell, MA: Massachusetts Toxics Use Reduction Institute, 2005).

25 U.S. Environmental Protection Agency, Office of Pollution Prevention and Toxics, *Design for Environment Program Alternatives Assessment Criteria for Hazard Evaluation* (Washington, DC, November 2010), see http://www.epa.gov/dfe/alternatives_assessment_criteria_for_hazard_eval.pdf, accessed March 7, 2011; and Emma T. Lavoie, Lauren G. Heine, Helen Holder, Mark S. Rossi, Robert E. Lee, Emily A. Connor, Melanie A. Veabel, David M. Difiore, and Clive L. Davies, "Chemical Alternatives Assessment: Enabling Substitution to Safer Chemicals," *Environmental Science and Technology*, 44, 2010, pp. 9244–9249.

27. Holder, Mazurkiewicz, Robertson, and Wray, 2013, pp. 157–175.

28. U.S. Environmental Protection Agency, *Cleaner Technologies Substitution Assessment*, EPA 774R-95–002 (Washington, DC, December 1996).

29. For a good text on environmental cost accounting, see Stefan Schaltegger and Roger Burritt, *Contemporary Environmental Accounting: Issues, Concepts and Practice* (Sheffield, UK: Greenleaf, 2000).

30. U.S. Environmental Protection Agency, *Guidelines for Exposure Assessment*, see http://cfpub.epa.gov/ncea/cfm/recordisplay.cfm?deid=15263, accessed January 12, 2014.

31. P. H. Brunner and H. Rechberger, *Practical Handbook for Material Flow Analysis* (New York: Lewis, 2004).

32. James Fava, Frank Consoli, Richard Denison, Kenneth Dickson, Tim Mohin, and Bruce Vigon, *Conceptual Framework for Life Cycle Assessment: A Code of Practice* (Pensacola, FL: Society of Environmental Toxicology and Chemistry, 1993).

33. Thomas Graedel, *Streamlined Life Cycle Assessment* (Englewood Cliffs, NJ: Prentice-Hall, 1998).

34. U.S. Department of Commerce, NOAA Inter-organizational Committee on Principles and Guidelines for Social Impact Assessment, *Guidelines and Principles for Social Impact Assessment*, Tech Memo MNFS-F/SPO-16, 1993, reprinted in *Impact Assessment*, 12(2), 1994, pp. 107–152.

35. C. Benoit, G. A. Norris, S. Valdivia, S. Ciroth, A. Moberg, U. Bos, S. Prakash, C. Ugaya, and T. Beck, "The Guidelines for Social Life Cycle Assessment of Products: Just in Time," *International Journal of Life Cycle Assessment*, 15, 2000, pp. 156–163.

36. United Nations Environment Program, Persistent Organic Pollutants Review Committee, *Substitution and Alternatives, Addendium: General Guidance on Consideration Related to Alternatives and Substitutes for Persistent Organic Pollutants*, Geneva, October 12–16, 2009.

37. Debbie Raphael, Past-Director, California Department of Toxic Substances Control, personal interview, January 20, 2014. See also California Assembly, *Assembly Bill 1879*, see www.chemicalspolicy.org/legislationdocs/California/CA_AB1879.pdf, accessed February 22, 2011.

38. National Academy of Sciences, Board of Chemical Sciences and Technology, *A Framework to Guide Selection of Chemical Alternatives* (Washington, DC: National Academy Press, October, 2014).

39. Joel Tickner, Associate Professor, University of Massachusetts Lowell, personal interview, January 31, 2011.

Chapter 13.   Developing Safer Alternatives

1. U.S. Environmental Protection Agency, Presidential Green Chemistry Challenge, *2007 Greener Synthetic Pathways Award*, see http://www2.epa.gov/green-chemistry/2007-greener-synthetic-pathways-award/, accessed January 7, 2014.

2. U.S. Environmental Protection Agency, Presidential Green Chemistry Challenge, *2002 Designing Greener Chemicals Award*, see http://www2.epa.gov/green-chemistry/2002-designing-greener-chemicals-award/, accessed January 7, 2014.

3. Café Foundation, *The 2011 Green Flight Challenge Sponsored by Google*, see http://cafefoundation.org/v2/gfc_main.php, accessed December 13, 2013.

4. U.S. Environmental Protection Agency, *Presidential Green Chemistry Challenge Winners*, see www2.epa.gov/green-chemistry/presidential-green-chemistry-challenge-winners/, accessed December 12, 2013.

5. U.S. EPA, *Information About the U.S. Presidential Green Chemistry Challenge*, see http://www2.epa.gov/green-chemistry/information-about-presidential-green-chemistry-challenge/, accessed December 12, 2013.

6. See Woodrow Wilson Institute, *Project on Emerging Nanotechnologies*, see www.nanotechproject.org, accessed September 30, 2014.

7. See C. W. Lam, J. T. James, R. McCluskey, S. Arepalli, and R. L. Hunter, "A Review of Carbon Nanotube Toxicity and Assessment of Potential Occupational and Environmental Health Risks," *Critical Reviews in Toxicology*, 36, 2006, pp. 189–217; and G. Oberdorster, E. Oberdorster, and J. Oberdorster, "Nanotoxicology: An Emerging Discipline Evolving from Studies of Ultra Fine Particles," *Environmental Health Perspectives*, 113, 2005, pp. 823–839.

8. European Parliament, Scientific Technology Options Analysis, *The Role of Nanotechnology in Substituting Hazardous Chemicals* (Brussels, 2006), see https://www.itas.kit.edu/downloads/etag_fied07a.pdf, accessed September 30, 2014.

9. For the adaptation of the Twelve Principles, see J. A. Dahl, B. L. S. Maddux, James E. Hutchinson, "Toward Greener Nanosynthesis," *Chemical Reviews*, 107, 2007, pp. 2228–2269. See also Martin J. Mulvihill, Evan S. Beach, Julie B. Zimmerman, and Paul T. Anastas, "Green Chemistry and Green Engineering: A Framework for Sustainable Technology Development," *Annual Review of Environment and Resources*, 36, 2011, pp. 271–293. Additional research on the safety and effects of nanotechnologies is ongoing at university centers such as Rice University's Center for Biological and Environmental Nanotechnology and collaborating centers at Duke University and the University of California, Los Angeles; see Rachel Petkewich, "Probing Hazards of Nanomaterials," *Chemical and Engineering News*, 86(42), October 20, 2008, pp. 53–56.

10. See D. J. C. Constable, A. D. Cuzons, L. M. F. dos Santos, G. R. Green, R. E. Hannah et al., "Green Chemistry Measures for Process Research and Development," *Green Chemistry*, 3, 2001, pp. 7–9.

11. U.S. Environmental Protection Agency, U.S. EPA and Other Organization Tools for Chemicals Assessment and Design, see http://www.chemicalspolicy.org/downloads/EPAandOtherResources.pdf, accessed September 30, 2014.

12. U.S. Environmental Protection Agency, *Sustainable Futures*, see http://www.epa.gov/oppt/sf/, accessed February 19, 2012.

13. U.S. Environmental Protection Agency, *Toxicological Priority Index (ToxPi)*, see http://www.epa.gov/ncct/ToxPi/, accessed October 5, 2014.

14. Stephen Ritter, "Designing Away Endocrine Disruption," *Chemical and Engineering News*, 90(51), December 17, 2012, pp. 33–34.

15. David Pimentel and Rajinder Peshin, eds., *Integrated Pest Management: Pesticide Problems, Volume 3* (New York: Springer, 2013).

16. There have been two efforts to establish principles of green engineering that are parallel to the principles of green chemistry. See U.S. Environmental Protection Agency, *Principles of Green Engineering*, see http://www.epa.gov/oppt/greenengineering/pubs/whats_ge.html, accessed December 12, 2012; and P. T. Anastas and J. B. Zimmerman, "Design through the Twelve Principles of Green Engineering," *Environmental Science and Technology*, 37(5), 2003, pp. 94A–101A.

17. Everett Rogers, *Diffusion of Innovations, Fourth Edition* (New York: Free Press, 1983).

18. James S. Coleman, *Medical Innovation: A Diffusion Study* (New York: Bobbs-Merrill, 1966); and Bryce Ryan and Neil C. Gross, "The Diffusion of Hybrid Seed Corn in Two Iowa Communities," *Rural Sociology*, 8, 1943, pp. 15–24.

19. Roger McFadden, Senior Scientist, Staples, personal interview, October 6, 2010 .

20. For similar recommendations, see Julie B. Manley, Paul Anastas, and Berkeley W. Cue, "Frontiers in Green Chemistry: Meeting the Grand Challenges for Sustainability in R&D and Manufacturing," *Journal of Cleaner Production*, 16, 2008, pp. 743–750; Mulvihill, Beach, Zimmerman, and Anastas, 2011; and Green Chemistry and Commerce Council and National Pollution Prevention Roundtable, *Growing the Green Economy through Green Chemistry and Design for the Environment* (Lowell, MA: Lowell Center for Sustainable Production, University of Massachusetts Lowell, 2009), see http://www.p2.org/wp-content/uploads/growing-the-green-economy.pdf, accessed December 11, 2012.

21. See Heintz and Pollin, 2011; and Navigant Research, "Green Chemical Industry to Soar to $98.5 by 2020," June 20, 2011, see www.navigantresearch.com/newsroom/green-chemical-industry-to-soar-to-98-5-billion-by-2020/, accessed December 12, 2013.

22. Amory B. Lovins et al., *Winning the Oil End Game: Innovation for Profits, Jobs and Secutiry* (Boulder, CO: Rocky Mountain Institute, 2005).

23. American Chemistry Council, "Shale Gas and New Petrochemical Investment: Benefits for the Economy, Jobs and US Manufacturing," March 2011, see http://chemistrytoenergy.com/sites/chemistrytoenergy.com/files/ACC-Shale-Report.pdf, accessed September 30, 2014.

Chapter 14.   Drafting Safer Chemical Policies

1. United Nations Conference on Environment and Development, *Agenda 21*, Rio de Janeiro, 1992, see http://sustainabledevelopment.un.org/content/documents/Agenda21.pdf, accessed March 10, 2013.

2. United Nations Environment Program, *Plan of Implementation of the World Summit on Sustainable Development*, Johannesburg, 2002, Section 23, see www.

unmillenniumproject.org/documents/131302_wssd_report_reissued.pdf, accessed March 10, 2013.

3. For a good review of these policy options directed at the states, see Lowell Center for Sustainable Production, *Options for State Chemicals Policy Reform: A Resource Guide* (Lowell: University of Massachusetts Lowell, January 2008), see http://www.chemicalspolicy.org/downloads/OptionsExecutiveSummary.pdf, accessed December 11, 2011.

4. The Department of Health and Human Service's *Healthy People 2020* provides a useful example. This national plan has 1,200 objectives organized around forty-two topic areas. See U.S. Department of Health and Human Service, *Healthy People, 2020*, 2010, see http://www.cdc.gov/nchs/healthy_people/hp2020.htm, accessed January 20, 2014.

5. U.S. Environmental Protection Agency, *Reinventing Environmental Protection, 1998 Annual Report* (Washington, DC, 1998).

6. European Commission, *A European Union Strategy for Sustainable Development* (Luxemburg: Office of Official Publications of the European Commission, 2002) p. 35.

7. Swedish Parliament, Government Bill 1997/98:145, 1997.

8. *Clean Air Act Amendments of 1990*, 104 Stat. 2468, Section 612.

9. See Nicholas Ashford, "Understanding Technological Responses of Industrial Firms to Environmental Problems: Implications for Government Policy," in Kurt Fischer and Johann Schot, eds., *Environmental Strategies for Industry* (Washington, DC: Island Press, 1993) pp. 277–307.

10. Such incentives are similar to what Richard H. Thaler and Cass R. Sunstein call "nudges." See *Nudge: Improving Decisions about Health, Wealth and Happiness*, rev. ed. (New York: Penguin, 2009).

11. Annika Gottberg, Joe Morris, Simon Pollard, Ceceilia Mark-Herbert, and Matthew Cook, "Producer Responsibility, Waste Minimization and WEEE Directive: Case Studies in Eco-Design from the European Lighting Sector," *Science of the Total Environment*, 359(1–13), April 15, 2006, pp. 38–56.

12. Robert Socolow and Valerie Thomas, "The Industrial Ecology of Lead and Electric Vehicles," *Journal of Industrial Ecology*, 1(1), January 1997, pp. 30–36.

13. See Neil Hawke, *Environmental Policy: Implementation and Enforcement* (Hempshire, UK: Ashgate, 2008) pp. 217–218.

14. U.S. Environmental Protection Agency, *Cap and Trade Acid Rain Program Results*, see www.epa.gov/capandtrade/documents/ctresults.pdf, accessed February 16, 2012; G. Chan, R. Stavins, R. Stowe, and R. Sweeney, "The SO_2 Allowance-Trading System and the Clean Air Act Amendments of 1990: Reflections on 20 Years of Policy Innovation," *National Tax Journal*, 65, 2012, pp. 419–452; and Daniel A. Faber, *Emissions Trading and Environmental Justice*, Center for Law, Energy and the Environment, University of California, August 2011, see www.law.berkeley.edu/files/Emissions_Trading_and_Social_Justice.pdf, accessed October 22, 2013.

15. See Massachusetts Toxics use Reduction Institute, *Cleaning Laboratory*, see http://www.turi.org/Our_Work/Cleaning_Laboratory/, accessed February 16, 2012.

16. U.S. Environmental Protection Agency, *Verified Technologies*, see http://www. epa.gov/etv/verifiedtechnologies.html, accessed December 12, 2012.

17. National Research Council, Committee on Human Biomonitoring for Environmental Toxicants, *Human Biomonitoring for Environmental Chemicals* (Washington, DC: National Academy Press, 2006).

Chapter 15.   Reconstructing Government Capacity

1. U.S. Government Accounting Office, *Workplace Safety and Health: Multiple Challenges Lengthen OSHA's Standard Setting*, April 2012, see www.gao.gov/assets/590/589825.pdf, accessed March 12, 2013.

2. Robert Innes and Abdoul Sam, "Voluntary Pollution Reductions and the Enforcement of Environmental Law: An Empirical Study of the 33/50 Program," *Journal of Law and Economics*, 51(2), 2004, pp. 271–296; and Martina Vidovic and Neha Khanna, "Can Voluntary Pollution Prevention Programs Keep Their Promise," *Journal of Environmental Economics and Management*, 53(2), 2007, pp. 180–195.

3. Scott Hassell, Noreen Clancy, Nicholas Burger, Christopher Nelson, Rena Rudavsky, and Sarah Olmstead, *An Assessment of the U.S. Environmental Protection Agency's National Environmental Performance Track Program* (Santa Monica, CA: RAND Corporation, 2010), see www.rand.org/content/dam/rand/pubs/technical_reports/2010/RAND_TR732.pdf, accessed September 30, 2014.

4. U.S. Environmental Protection Agency, *Reinventing Environmental Protection* (Washington, DC, 1998).

5. Cary Coglianese and Laurie K. Allen, "Building Sector-Based Consensus: A Review of the U.S. EPA's Common Sense Initiative," in Theo de Bruijn and Vicky Noberg-Bohm, eds., *Industrial Transformation: Environmental Policy Innovation in the United States and Europe* (Cambridge, MA: MIT Press, 2005) pp. 65–92.

6. U.S. Environmental Protection Agency, Office of Reinvention, *The Common Sense Initiative: Lessons Learned* (Washington, DC, 1998).

7. Clive Davies, Director, U.S. EPA Design for Environment Program, personal interview, April 25, 2012. See also U.S. Environmental Protection Agency, *Design for the Environment*, see www.epa.gov/dfe/, accessed February 19, 2012.

8. Rie Tsutsumi, "The Nature of Voluntary Agreements in Japan—Functions of Environment and Pollution Control Agreements," *Journal of Cleaner Production*, 9, 2001, pp. 145–153; and Hans Bressers, Theo de Bruijn, Kris Lulof, and Lawrence J. O'Toole, Jr., "Negotiation-Based Policy Instruments and Performance: Dutch Covenants and Environmental Policy Outcomes," *Journal of Public Policy*, 31(2), 2011, pp. 187–208.

9. Interstate Chemicals Clearinghouse, *U.S. State Chemicals Policy Database*, 2014.

10. Lowell Center for Sustainable Production, *State Leadership in Formulating and Reforming Chemicals Policy: Actions Taken and Lessons Learned* (Lowell: University of Massachusetts Lowell, July 2009), see http://www.chemicalspolicy.org/downloads/StateLeadership.pdf, accessed December 15, 2012.

11. See Maine Legislature, *Act to Protect Children's Health and the Environment from Toxic Chemicals in Toys and Children's Products* (Maine Rev. Stat. Ann. Lit. 38–1691); and Washington's *Children's Safe Product Act* (HB 2647).

12. Interstate Chemicals Clearinghouse, see www.newmoa.org/prevention/ic2/, accessed February 19, 2012.

13. Mike Belliveau, Director Environmental Health Strategy Center, personal interview, December 10, 2012.

14. U.S. Environmental Protection Agency, Office of Inspector General, *EPA Needs to Comply with the Federal Insecticide, Fungicide, and Rodenticide Act and Improve Its Oversight of Exported Never-Registered Pesticides*, Report No. 10-P-0026, November 10, 2009, see http://www.epa.gov/oig/reports/2010/20091110-10-P-0026.pdf, accessed December 12, 2012.

15. Thomas J. Bollyky and Anu Bradford, "Getting to Yes on Transatlantic Trade," *Foreign Affairs*, Blog, July 10, 2013, see http://www.foreignaffairs.com/articles/139569/thomas-j-bollyky-and-anu-bradford/getting-to-yes-on-transatlantic-trade/, accessed September 30, 2014.

16. U.S. Environmental Protection Agency, *Reinventing Environmental Protection* (Washington, DC, 1998).

17. J. Clarence Davies, "Nanolessons for Revamping Government Oversight of Technology," *Issues in Science and Technology*, 26(1), Fall 2009, pp. 43–48.

18. Richard Dennison, Environmental Defense Fund, personal interview, June 2, 2010.

19. U.S. Senate, *S. 847, Safe Chemicals Act of 2011*, see www.govtrack.us/congress/bills/112/s847/text/, accessed September 30, 2014.

20. Another $129 million was allocated to pesticides in 2013. See U.S. Environmental Protection Agency, *FY 2013: Budget in Brief* 2012; and Schifano, Geiser and Tickner 2011.

21. U.S. Environmental Protection Agency, *CERCLA Overview*, see http://www.epa.gov/superfund/policy/cercla.htm, accessed December 2, 2013.

Chapter 16.   Solving the Chemicals Problem

1. Health Care for America Now, see http://healthcareforamericanow.org/about-us/mission-history/, accessed January 1, 2014.

2. Two different assessments of the national campaigns to pass health care reform and climate change legislation have been offered by Harvard University's Theda Skocpol and Yale University's Nathaniel Loewentheil. However, both agree that the success of the health care reform struggle was based on the mobilization of a broadly based grassroots constituency, whereas the failure to build a similar

popular movement was a significant factor in the lack of success of the climate protection campaign. See Theda Skocpol, *Naming the Problem: What It Will Take to Counter Extremism and Engage Americans in the Fight against Global Warming*, unpublished paper prepared for the Symposium on the Politics of America's Fight Against Global Warming, February 14, 2013, Harvard University, Cambridge, MA, see http://www.scholarsstrategynetwork.org/sites/default/files/skocpol_captrade_report_january_2013_0.pdf, accessed January 1, 2014; and Nathaniel Loewentheil in *Of Stasis and Movements: Climate Change in the 111th Congress* (New Haven, CT: Institution for Social and Policy Studies, Yale University, 2012), see http://isps.yale.edu/research/publications/isps12-020-0#.UsT3E1WA270/, accessed January 1, 2014.

3. See Gottlieb 1993; Davies 2013; and Phil Brown, *Toxic Exposures: Contested Illnesses and the Environmental Health Movement* (New York: Columbia University Press, 2007).

4. Some current organizations, including the Business-NGO Working Group on Safer Chemicals and Sustainable Materials (52 organizations), the Green Chemistry and Commerce Council (69 organizations), the Interstate Chemicals Clearinghouse (11 states), Coming Clean (200 organizations), Health Care without Harm (473 organizations), Safer Chemicals, Healthy Families (450 organizations), the Blue Green Alliance (15 trade unions and environmental organizations), the Green Chemistry Institute, and the annual Green Chemistry and Engineering Conference, provide useful vehicles.

5. The best title for this movement is not clear. The most appropriate title may not emerge for years.

6. The new environmentalism is described in books such as Tim Jackson, *Prosperity without Growth: Economics for a Finite Planet* (London: Earthscan, 2009); Bill McKibbin, *Deep Economy: The Wealth of Communities and the Durable Future* (New York: Henry Holt, 2007); and James Gustave Speth, *The Bridge at the Edge of the World: Capitalism, The Environment, and Crossing from Crisis to Sustainability* (New Haven, CT: Yale University Press, 2008).

7. There is a robust literature on the new economy. See Richard Sennett, *The Culture of the New Capitalism* (New Haven, CT: Yale University Press, 2006); Gar Alperovitz, *America beyond Capitalism: Reclaiming Our Wealth, Our Liberty and Our Democracy, Second Edition* (Takoma Park, MD: Democracy Collaborative Press, 2005); Peter Barnes, *Capitalism 2.0: A Guide to Reclaiming the Commons* (San Francisco: Berrett-Koehler, 2006); and David C. Korten, *The Great Turning: From Empire to Earth Community* (San Francisco: Berrett-Koehler, 2006).

8. See Poul Herremoes et al., eds, *Late Lessons from Early Warnings: The Precautionary Principle, 1896–2000* (Copenhagen: European Environment Agency, 2001); and *Late Lessons from Early Warnings: Science, Precaution and Innovation* (Copenhagen: European Environment Agency, 2013), see www.eea.europa.eu/publications/environmental_issue_report_2001_22, and www.eea.europa.eu/publications/late-lessons-2/, accessed March 7, 2013.

9. Michael Wilson, Daniel Chia, and Bryan Ehlers, *Green Chemistry in California: A Framework for Leadership in Chemical Policy and Innovation* (Berkeley: California Policy Research Center, University of California, Berkeley, 2006), Table K, see http://coeh.berkeley.edu/FINALgreenchemistryrpt.pdf, accessed January 14, 2014.

10. Philip Landrigan, Clyde B. Schechter, Jeffrey M. Lipton, Marianne C. Fahs, and Joel Schwartz, "Environmental Pollutants and Disease in American Children: Estimates of Morbidity, Mortality, and Costs for Lead Poisoning, Asthma, Cancer and Developmental Disabilities," *Environmental Health Perspectives*, 110(7), July 2002, pp. 721–728.

11. A. Pruss-Ustiun, C. Vickers, P. Haeflinger, and R. Bertollini, "Knowns and Unknowns on the Burden of Disease Due to Chemicals: A Systematic Review," *Environmental Health*, 10–9, 2011, see www.ehjournal.net/content/10/1/9/, accessed December 12, 2012.

12. Leo Trusande, Philip J. Landrigan, and Clyde Schechter, "Public Health and Economic Consequences of Methylmercury to the Developing Brain," *Environmental Health Perspectives*, 5, May 2005, pp. 509–596.

13. DHI Water and Environment, *The Impact of REACH on the Environment and Human Health* (Horsholm, Denmark, 2005) (ENV.C.3/SER/2004/0042r) p. 3, see http://catt.univ-pau.fr/live/digitalAssets/91/91688_Final_report_2005_09.pdf, accessed December 15, 2013.

Bibliography

Ackerman, Frank, and Lisa Hienzerling. *Priceless: On Knowing the Price of Everything and the Value of Nothing*. New York: New Press, 2004.

Adenike Adeyeye, James Barrett, Jordan Diamond, Lisa Goldman, John Pendergrass, and Daniel Schrammm, *Estimating U.S. Government Subsidies to Energy Sources, 2002–2008*. Washington, DC: Environmental Law Institute, September 2009. See http://www.eli.org/sites/default/files/eli-pubs/d19_07.pdf, accessed October 2, 2014.

Adler, Robert. "From Model Agency to Basket Case—Can the Product Safety Commission Be Redeemed." *Administrative Law Review* 41(1), 1989, pp. 61–129.

Aftaion, Fred. *A History of the International Chemical Industry: From the "Early Days" to 2000*. Philadelphia: Chemical Heritage Press, 2001.

Ahluwalia, V. K., and M. Kidwai. *New Trends in Green Chemistry*, Second Edition. New Delhi, India: Anamaya, 2006.

Alfonsi, Kim, Juan Colberg, Peter. J. Dunn, Thomas Fevig, Sandra Jennings, Timothy A. Johnson, H. Peter Kleine, Craig Knight, Mark A. Nagy, David A. Perry, and Mark Stefaniak. "Green Chemistry Tools to Influence a Medicinal Chemistry and Research Chemistry Based Organisation," *Green Chemistry*, 2008, 10, pp. 31–36

Allen, Barbara. *Uneasy Alchemy: Citizens and Experts in Louisiana's Chemical Corridor Disputes*. Cambridge, MA: MIT Press, 2003.

Allen, David, and David Shonnard. *Green Engineering: Environmentally Conscious Design of Chemical Processes*. Upper Saddle River, NJ: Prentice Hall, 2002.

Allen, David, and Kristen Sinclair Rosselot. *Pollution Prevention for Chemical Processes*. New York: John Wiley, 1997.

Alperovitz, Gar. *America beyond Capitalism: Reclaiming Our Wealth, Our Liberty and Our Democracy, Second Edition*. Takoma Park, MD: Democracy Collaborative Press, 2005.

American Chemical Society et al. *Technology Vision 2020-The U.S. Chemical Industry*. Washington, DC, 1996. See http://energy.gov/sites/prod/files/2013/11/f4/chem_vision.pdf, accessed September 30, 2014.

American Chemistry Council. *2011 Guide to the Business of Chemistry*. Washington, DC, 2011.

American Chemistry Council. *2012 Guide to the Business of Chemistry*. Washington, DC, 2012.

American Chemistry Council. *Responsible Care, Product Stewardship*. Washington, DC, 2007. See www.americanchemistry.com/s_responsiblecare, accessed February, 2010.

American Chemistry Council. "Shale Gas and New Petrochemical Investment: Benefits for the Economy, Jobs and US Manufacturing." Washington, DC, March 2011. See http://chemistrytoenergy.com/sites/chemistrytoenergy.com/files/ACC-Shale-Report.pdf, accessed September 30, 2014.

American Institute of Chemical Engineers, Center for Waste Reduction Technologies. *Vision 2020: Reaction Engineering Roadmap*. New York, 2001. See http://all-experts.com/assets/roadmaps/482__reaction_roadmap.pdf, accessed September 30, 2014.

Amoto, Ivan. "The Slow Birth of Green Chemistry." *Science*, 259, March 12, 1993, pp. 1538–1541.

Anastas, Paul, and Mary Kirchhoff. "Origins, Current Status and Future Challenges of Green Chemistry." *Accounts of Chemical Research*, 35(9), 2002, pp. 686–694.

Anastas, Paul, and John Warner. *Green Chemistry: Theory and Practice*. New York: Oxford University Press, 1998.

Anastas, Paul T., and Tracy C. Williamson. "Frontiers in Green Chemistry." In Paul T. Anastas and Tracy C. Williamson, eds., *Green Chemistry: Frontiers in Benign Chemical Syntheses and Processes*. New York: Oxford University Press, 1998.

Anastas, P. T., and J. B. Zimmerman. "Design through the Twelve Principles of Green Engineering." *Environmental Science and Technology*, 37(5), 2003, pp. 94A–101A.

Andrews, David, and Richard Wiles. *Off the Books: Industry's Secret Chemicals*. Washington, DC: Environmental Working Group, 2009. See http://www.ewg.org/sites/default/files/report/secret-chemicals.pdf, accessed February 3, 2011.

Applegate, John S. "Synthesizing TSCA and REACH: Practical Principles for Chemical Regulation Reform." *Ecology Law Quarterly*, 35, 2008, pp. 721–769.

Arimura, Toshi H., Nicole Darnall, and Hajime Katayama. *Is ISO 14001 a Gateway to More Advanced Voluntary Action?* Washington, DC: Resources for the Future, March 2009. See http://www.rff.org/documents/rff-dp-09-05.pdf, accessed March 15, 2011.

Arora, Ashish, and Nathan Rosenberg. "Chemicals: A U.S. Success Story." In Ashish Arora, Ralph Landeau, and Nathan Rosenberg, eds., *Chemicals and Long Term Economic Growth; Insights from the Chemical Industry*. New York: John Wiley, 1998, pp. 71–102.

Ashford, Nicholas. "Understanding Technological Responses of Industrial Firms to Environmental Problems: Implications for Government Policy." In Kurt Fischer and Johann Schot, eds., *Environmental Strategies for Industry*. Washington, DC: Island Press, 1993, pp. 277–307.

Ashford, Nicholas. "An Innovation-Based Strategy for the Environment and Workplace." In A. M. Finkel and D. Golding, eds., *Worst Things First? The Debate over Risk Based National Priorities*. Washington, DC: Resources for the Future, 1994, pp. 275–314.

Ashford, Nicholas, C. W. Ryan, and Charles Caldart. "A Hard Look at Federal Regulation of Formaldehyde: A Departure from Reasoned Decision Making." *Harvard Law Review*, 297, 1983.

Baker, Linda. "The Hole in the Sky: Think the Ozone Layer Is Yesterday's Issue? Think Again." *E-Magazine*, October 31, 2000.

Baker, Nena. *The Body Toxic: How the Hazardous Chemistry of Everyday Things Threatens Our Health and Well-being*. New York: North Point Press, 2008.

Banergee, Abhijit, and Barry Solomon. "Eco-Labeling for Energy Efficiency and Sustainability: A Meta-Evaluation of U.S. Programs." *Energy Policy*, 31, 2003, pp. 109–123.

Barkan, I. D. "Industry Invites Regulation: The Passage of the Pure Food and Drug Act of 1906." *American Journal of Public Health*, 75(1), 1985.

Barnes, Peter. *Capitalism 2.0: A Guide to Reclaiming the Commons*. San Francisco: Berrett-Koehler, 2006.

Baumgaertel, Christian. "World Business Briefing, Europe: Nike Shirts Called Not Hazardous." *New York Times*, January 15, 2000.

Bearse, Scott, Peter Capozucca, Laura Favret, and Brian Lynch. *Finding the Green in Today's Shoppers: Sustainability Trends and New Shopper Insights*. Washington, D.C.: Deloitte/Grocery Manufacturers Association, 2009. See www.deloitte.com/assets/Dcom-Shared%20Assets/Documents/US_CP_GMADeloitteGreenShopperStudy_2009.pdf, accessed April 17, 2011.

Becker, Monica. *Considered Chemistry at Nike: Creating Safer Products through Evaluation and Restriction of Hazardous Chemicals*. Unpublished case study, Green Chemistry and Commerce Council, 2009. See http://www.greenchemistryandcommerce.org/downloads/Nike_final.pdf., accessed September 30, 2014.

Becker, Monica. *Managing Chemicals of Concern and Designing Safer Products at Hewlett-Packard*. Unpublished case study, Green Chemistry and Commerce Council, 2009. See http://www.greenchemistryandcommerce.org/downloads/HP_final.pdf, accessed February 12, 2010.

Becker, Monica, *S.C. Johnson Is Transforming Its Supply Chain to Create Products That Are Better for the Environment*. Unpublished case study, Green Chemistry and Commerce Council. See http://www.greenchemistryandcommerce.org/downloads/SCJ_final.pdf, accessed September 30, 2014.

Benoit, C., G. A. Norris, S. Valdivia, S. Ciroth, A. Moberg, U. Bos, S. Prakash, C. Ugaya, and T. Beck. "The Guidelines for Social Life Cycle Assessment of Products: Just in Time." *International Journal of Life Cycle Assessment*, 15, 2000, pp. 156–163.

Benyus, Janine M. *Biomimicy: Innovation Inspired by Nature*. New York: William Morrow, 1997.

Betsill, Michel M., and Elisabeth Corell. *NGO Diplomacy: The Influence of Nongovernmental Organizations in International Environmental Negotiations*. Cambridge: MIT Press, 2008.

Biomass Research and Development Initiative. *Roadmap for BioEnergy and Biobased Products in the United States*. See http://www1.eere.energy.gov/bioenergy/pdfs/obp_roadmapv2_web.pdf, accessed September 30, 2014.

Black, Harvey. "Chemical Reaction: The U.S. Response to REACH." *Environmental Health Perspectives*, 116(3), March 2008, pp. A124–A127.

BlueGreen Alliance. *ChemHAT* website. See www.chemhat.org, accessed February 6, 2013.

Birnbaum, Linda S., and Daniele F. Staskal. "Brominated Flame Retardants: Cause for Concern?" *Environmental Health Perspectives*, 112(1), 2004, pp. 9–17.

Bollinger, Robert E., David G. Clark, Arthur M. Dowell, Roger M. Ewbank, Dennis C. Hendershot, William K. Lutz, Steven I. Meszaros, Donald E. Park, and Everett D. Wixom. *Inherently Safer Chemical Processes: A Life Cycle Approach*, Daniel A. Crowl, ed. New York: Center for Chemical Process Safety, 1996.

Bollyky, Thomas J. and Anu Bradford, "Getting to Yes on Transatlantic Trade," *Foreign Affairs*, Blog, July 10, 2013. See http://www.foreignaffairs.com/articles/139569/thomas-j-bollyky-and-anu-bradford/getting-to-yes-on-transatlantic-trade, accessed September 30, 2014.

Bressers, Hans, Theo de Bruijn, Kris Lulof, and Lawrence J. O'Toole, Jr. "Negotiation-Based Policy Instruments and Performance: Dutch Covenants and Environmental Policy Outcomes." *Journal of Public Policy*, 31(2), 2011, pp. 187–208.

Brickman, Ronald, Shiela Jasanoff, and Thomas Ilgen. *Controlling Chemicals, The Politics of Regulation in Europe and the United States*. Ithaca, NY: Cornell University Press, 1985.

Bronstein, Alvin C., et al. "2001, Annual Report of the American Association of Poison Control Center's National Poison Data System, 29th Annual Report." *Clinical Toxicology*, 50, 2012, pp. 911–1164, Table 16A, p. 933. See https://aapcc.s3.amazonaws.com/pdfs/annual_reports/2011_NPDS_Annual_Report.pdf, accessed March 22, 2012.

Brown, Phil. *Toxic Exposures: Contested Illnesses and the Environmental Health Movement*. New York: Columbia University Press, 2007.

Brown, Robert C. *Biorenewable Resources: Engineering New Products from Agriculture*. Ames, IA: Iowa State Press, 2003.

Brunner, P. H., and H. Rechberger. *Practical Handbook for Material Flow Analysis*. New York: Lewis, 2004.

Buchholz, Sigurd. "Future Manufacturing Approaches in the Chemical and Pharmaceutical Industry." *Chemical Engineering and Processing*, 49, 2010, pp. 993–995.

Bullard, Robert D. *Unequal Protection: Environmental Justice and Communities of Color.* San Francisco: Sierra Club Books, 1994.

Bullard, Robert D., Paul Mohai, Robin Saha, and Beverly Wright. *Toxic Waste and Race at Twenty: 1987–2007.* United Church of Christ, March 2007. See http://www.ucc.org/justice/pdfs/toxic20.pdf, accessed February 25, 2013.

Butte, W., and B. Heinzow. "Pollutants in House Dust as Indicators of Indoor Contamination." *Review of Environmental Contamination and Toxicology*, 175, 2002, pp. 1–46.

Byster, Leslie, and Ted Smith. "From Grass Roots to Global: The Silicon Valley Toxic Coalition's Milestones in Building a Movement for Corporate Accountability and Sustainability in the High Tech Sector." In Ted Smith, David A. Sonenfeld, and David Naguib Pellow, eds., *Challenging the Chip: Labor Rights and Environmental Justice in the Global Electronics Industry.* Philadelphia: Temple University Press, 2006, pp. 111–119.

Café Foundation. *The 2011 Green Flight Challenge Sponsored by Google.* See http://cafefoundation.org/v2/gfc_main.php, accessed December 13, 2013.

California Air Resources Board. See www.arb.ca.gov/homepage.htm, accessed January 14, 2011.

California Assembly. *Assembly Bill 1879.* See www.chemicalspolicy.org/legislationdocs/California/CA_AB1879.pdf, accessed February 22, 2011.

California Department of Pesticide Regulation, California Pesticide Information Portal. See http://www.cdpr.ca.gov/docs/pur/purmain.htm, accessed February 25, 2013.

California Department of Public Health. *California Safe Cosmetics Program.* See www.cdph.ca.gov/programs/cosmetics/Pages/default.aspx, accessed February 25, 2013.

California Office of Environmental Health Hazard Assessment. *Green Chemistry Hazard Traits, End Points and Other Relevant Data.* Sacramento, CA, August 10, 2010. See http://oehha.ca.gov/multimedia/green/pdf/081110prereghazard.pdf, accessed June 20, 2011.

California Office of Environmental Health Hazard Assessment. *Proposition 65 in Plain English,* See www.oehha.ca.gov/prop65/background/p65plain.html, accessed June 12, 2009.

California Office of Environmental Health Hazarad Assessment. *Proposition 65 List of Chemicals.* See www.oehha.ca.gov/prop65/prop65_list/newlist.html, accessed February 25, 2013.

Campaign for Safe Cosmetics. *Market Shift: The Story of the Safe Cosmetics Compact and the Growing Demand for Safer Products.* See www.safecosmetics.org/downloads/MarketShift_CSC_June15_2012.pdf, accessed March 22, 2013.

Cano-Ruiz, J. A., and G. J. McRae. "Environmentally Conscious Chemical Process Design." *Annual Review of Energy and Environment*, 23, 1998, pp. 499–536.

Carroll, William F., Jr., and Douglas J. Raber. *The Chemistry Enterprise in 2015.* Washington, DC: American Chemicals Society, 2005. See http://www.acs.org/content/dam/acsorg/membership/acs/welcoming/industry/the-chemistry-enterprise-2015.pdf, accessed January 11, 2012.

Carson, Rachel. *Silent Spring.* New York: Ballantine, 1962.

Cavers, David F. "The Food Drug and Cosmetics Act of 1938: Its Legislative History and Its Substantive Provisions." *Law and Contemporary Problems*, 10, 1939, pp. 2–42.

Centi, G., P. Ciambelli, S. Perathoner, and P. Russo. "Environmental Catalysis: Trends and Outlooks." *Catalysis Today*, 1–13, 2002, p. 2691.

Challener, Cynthia. "Sustainable Development at a Crossroads." *Chemical Market Reporter*, 260, July 16, 2001, pp. Fr3–Fr4.

Chan, G., R. Stavins, R. Stowe, and R. Sweeney. "The SO_2 Allowance-Trading System and the Clean Air Act Amendments of 1990: Reflections on 20 years of Policy Innovation." *National Tax Journal*, 2012, 65:419–452.

Chandler, Alfred J., Jr. *Shaping the Industrial Century: The Remarkable Story of the Evolution of the Modern Chemical and Pharmaceutical Industries.* Cambridge, MA: Harvard University Press, 2005.

Charpentier, Jean-Claude. "Four Main Objectives for the Future of Chemical and Process Engineering Mainly Concerned by the Science and Technology of New Material Production." *Chemical Engineering Journal*, 107, 2005, pp. 3–17.

Charter, Martin, and Ursula Tischner. *Sustainable Solutions: Developing Products and Services for the Future.* Sheffield, UK: Greenleaf Publishing, 2001.

CHEMLinked. "China New Chemical Substance Notification—China REACH." See https://chemlinked.com/chempedia/china-reach#sthash.z1ateSCI.dpbs, accessed September 30, 2014.

Chemical Abstract Service. See http://www.cas.org/, accessed September 23, 2012.

Chemical Industry Vision 2020 Technology Partnership. *Annual Report, 2003.* See http://www1.eere.energy.gov/manufacturing/resources/chemicals/pdfs/vision2020_annual_report.pdf, accessed February 19, 2012.

Chemical Inspection and Regulation Service. *Chemicals Regulated by Japan Chemicals Substance Control Law.* See www.cirs-reach.com/Japan_CSCL/Chemicals%20Regulated%20by%20Japan%20Chemical%20Substances%20Control%20Law.pdf, accessed December 28, 2013.

Chemical Scorecard. See http://scorecard.goodguide.com/chemical-profiles/, accessed February 6, 2013.

Christiani, D. C., and K. T. Kelsey, eds. *Chemical Risk Assessment and Occupational Health.* Westport, CT: Auburn House, 1994.

Clapp, Richard, Genevieve Howe, and Molly Jacobs. *Environmental and Occupational Causes of Cancer: A Review of Recent Scientific Literature.* Lowell, MA: Lowell Center for Sustainable Production, University of Massachusetts Lowell, September 2005. See http://www.sustainableproduction.org/downloads/Causes%20of%20Cancer.pdf, accessed January 4, 2014.

Clark, James H., et al. "Green Chemistry and the Biorefinery: A Partnership for a Sustainable Future." *Green Chemistry*, 8, 2006, pp. 853–860.

Clark, James H., and Fabian E. I. Deswarte, eds. *Introduction to Chemicals from Biomass*. New York: John Wiley, 2008.

Clark, James H., and Fabian E. I. Deswarte. "The Biorefinery Concept—An Integrated Approach." In James H. Clark and Fabian E. I. Deswarte, eds., *Introduction to Chemicals from Biomass*, New York: John Wiley, 2008, pp. 1–20.

Clean Production Action. See www.cleanproduction.org, accessed October 10, 2013.

Clean Production Action. *Green Screen for Safer Chemicals*. See http://www.greenscreenchemicals.org, accessed October 1, 2014.

Coglianese, Cary, and Laurie K. Allen. "Building Sector-Based Consensus: A Review of the U.S. EPA's Common Sense Initiative." In Theo de Bruijn and Vicky Noberg-Bohm, eds., *Industrial Transformation: Environmental Policy Innovation in the United States and Europe*. Cambridge: MIT Press, 2005, pp. 65–92.

Colburn, Theo, Diane Dumanoski, and John Peterson Meyers. *Our Stolen Future: Are We Threatening Our Fertility, Intelligence and Survival: A Scientific Detective Story*. New York: Dutton, 1996.

Coleman, James S. *Medical Innovation: A Diffusion Study*. New York: Bobbs-Merrill, 1966.

Colton, C. E., and P. N. Skinner. *The Road to Love Canal: Managing Industrial Waste Before EPA*. Austin: University of Texas Press, 1996.

Constable, D. J. C., A. D. Cuzons, L. M. F. dos Santos, G. R. Green, R. E. Hannah, et al. "Green Chemistry Measures for Process Research and Development." *Green Chemistry*, 3, 2001, pp. 7–9.

Costner, Pat, Beverley Thorpe, and Alexandra McPhearson. *Sick of Dust: Chemicals in Common Products—A Needless Health Risk in Our Homes*. Clean Production Action, unpublished report, March 2005. See http://www.cleanproduction.org/static/ee_images/uploads/resources/Dust_Report.pdf, accessed October 1, 2014.

Council of Chemical Research, et al. *Vision 2020: Chemical Industry of the Future Technology—Roadmap for Materials*, 2000. See http://all-experts.com/assets/roadmaps/518__materials_tech_roadmap.pdf, accessed September 30, 2014.

Cranor, Carl. *Regulating Toxic Substances: A Philosophy of Science and the Law*. New York: Oxford University Press, 1993.

Cranor, Carl F. *Legally Poisoned: How the Law Puts Us at Risk from Toxicants*. Cambridge, MA: Harvard University Press, 2011.

Curzons, Alan, David Constable, David Mortimer, and Virginia Cunningham. "So You Think Your Process Is Green? Using Principles of Sustainability to Determine What Is 'Green': A Corporate Perspective." *Green Chemistry*, 3, 2001, pp. 1–6.

Dahl, J. A., B. L. S. Maddux, and James E. Hutchinson. "Toward Greener Nanosynthesis." *Chemical Reviews*, 107, 2007, pp. 2228–2269.

Davies, J. Clarence. "Nanolessons for Revamping Government Oversight of Technology." *Issues in Science and Technology*, 26(1), Fall 2009, pp. 43–48.

Davies, Kate. *The Rise of the U.S. Environmental Health Movement*. New York: Rowman and Littlefield, 2013.

Deffree, Suzanne. "China RoHS: Ready or Not, It's Here." *Electronics News*, March 12, 2007, p. 2.

deForest Lamb, Ruth. *American Chamber of Horrors: The Truth about Food and Drugs*. New York: Farrar and Reinhart, 1936.

Dennison, Richard A. *Not So Innocent: A Comparative Analysis of Canadian, European Union and United States Policies on Industrial Chemicals*. Unpublished report. Washington, DC: Environmental Defense Fund, April 2007.

Dennison, Richard, *High Hopes, Low Marks, A Final Report Card on the High Production Volume Chemical Challenge*, Washington, DC: Environmental Defense Fund, July, 2007.

Dennison, Richard A. "Ten Essential Elements in TSCA Reform." *Environmental Law Reporter*, 39, 2009, pp. 10020–10028.

DHI Water and Environment. *The Impact of REACH on the Environment and Human Health*. Horsholm, Denmark, 2005 (ENV.C.3/SER/2004/0042r), p. 3. See http://catt.univ-pau.fr/live/digitalAssets/91/91688_Final_report_2005_09.pdf, accessed December 15, 2013.

Doherty, Richard. "A History of the Production and Use of Carbon Tetrachloride, Tetrachloroethylene, Trichloroethylene and 1,1,1-Trichloroethane in the United States: Part 2—Trichloroethylene and 1,1,1-Trichloroethane." *Environmental Forensics*, 1(2), 2000, pp. 83–93.

Douglas, Mary, and Aaron Wildavsky. *Risk and Culture: An Essay on the Selection of Technological and Environmental Dangers*. Berkeley, CA: University of California Press, 1983.

Dukema, G. P. J., J. Grievik, and M. P. C. Weijnen. "Functional Modeling for a Sustainable Petrochemical Industry." *Transaction of the Institution of Chemical Engineers*, 81, 2003, pp. 331–339.

Duncan, Marvin. "Federal Initiatives to Support Biomass Research and Development." *Journal of Industrial Ecology*, 7(3–4), 2004, pp. 193–201.

Ecolabel Index: Who's Deciding What's Green. See www.ecolabelindex.com, accessed January 8, 2013.

Edwards, Sally, Mark Rossi, and Pam Civie. *Alternatives Assessment for Toxics Use Reduction: A Survey of Methods and Tools, Methods and Policy Report No. 23*. Lowell, MA: Massachusetts Toxics Use Reduction Institute, 2005.

Electronic Product Environmental Assessment Tool. See http://www.epeat.net/, accessed December 12, 2011.

Emsley, John. *Natures Building Blocks: An A-Z Guide to the Elements*. New York: Oxford University Press, 2001.

Environment Canada. *Existing Substances Evaluation*. See http://www.ec.gc.ca/ese-ees/, accessed October 1, 2014.

Environmental Defense Fund. *Toxic Ignorance: The Continuing Absence of Basic Health Testing for Top-Selling Chemicals in the United States.* Washington, DC, 1997.

Environmental Working Group. *Skin Deep* website. See http://www.ewg.org/skindeep/, accessed October 1, 2014.

Esposito, Frank. "Dow Chemical to Separate Chlorine Business." *Plastics News,* December 2, 2013. See http://www.plasticsnews.com/article/20131202/NEWS/131209999, accessed October 1, 2014.

European Chemicals Agency. *Candidate List of Substances of Very High Concern for Authorization.* Updated December 12, 2012. See www.echa.europa.eu/candidate-list-table, accessed February 25, 2013.

European Chemicals Agency. *CMR Substances from Annex VI of the CLP Regulation Registered under REACH and/or Notified under CPL: A First Screening-Report 2012.* See www.echa.europa.eu/documents/10162/13562/cmr_report_en.pdf, accessed March 7, 2013.

European Chemicals Agency. "ECHA Receives 3.1 Million Classification and Labeling Notifications." Press Release, Helsinki, January 4, 2011. See http://echa.europa.eu/documents/10162/13585/pr_11_01_clp_deadline_20110104_en.pdf, accessed October 1, 2014.

European Chemicals Agency. *General Approach for Substances of Very High Concern (SVHC) for inclusion in the List of Substances Subject to Authorization.* Helsinki, May 28, 2010. See http://echa.europa.eu/documents/10162/13640/axiv_prioritysetting_general_approach_20100701_en.pdf, accessed March 22, 2013.

European Chemicals Agency, *Information on Chemicals.* See http://echa.europa.eu/information-on-chemicals, accessed October 1, 2014.

European Chemicals Agency. *REACH Statistics.* Helsinki, 2013. See http://echa.europa.eu/information-on-chemicals/registration-statistics, accessed March 22, 2013.

European Commission. *European Union Ecolabel.* See http://ec.europa.eu/environment/ecolabel/, accessed February 12, 2010.

European Chemicals Agency. See www.echa.europa.eu/regulations/clp/classification, accessed January 30, 2013.

European Commission. *Classification, Labeling and Packaging of Chemical Substances and Mixtures.* See http://ec.europa.eu/enterprise/sectors/chemicals/classification/index_en.htm, accessed September 30, 2014.

European Commission. *European Union Strategy for Sustainable Development.* Luxemburg: Office of Official Publications of the European Commission, 2002, p. 35.

European Commission. *European White Paper on a Strategy for Future Chemicals Policy* (COM 88 Final), Brussels, February 2001. See http://www.isopa.org/isopa/uploads/Documents/documents/White%20Paper.pdf, accessed September 30, 2014.

European Commission. *Green Public Procurement.* See http://ec.europa.eu/environment/gpp/pdf/brochure.pdf, accessed February 12, 2013.

European Commission. *High Level Group on the Competitiveness of the European Chemicals Industry: Final Report.* Brussels, 2009. See http://www.cefic.org/Documents/PolicyCentre/HLG-Chemical-Final-report-2009.pdf, accessed February 7, 2012.

European Commission. *Registration, Evaluation, Authorization of Chemicals (REACH),* Regulation (EC) No. 1907/2006, December 18, 2006. See http://ec.europa.eu/enterprise/sectors/chemicals/documents/reach/review-annexes/index_en.htm, accessed February 11, 2012.

European Parliament and Council of the European Union. *Council Directive 202/96/EC on Waste Electrical and Electronic Equipment.* See http://eur-lex.europa.eu/LexUriServ/LexUriServ.do?uri=OJ:L:2003:037:0024:0038:en:PDF, accessed February 11, 2012.

European Parliament and Council of the European Union. *Council Directive 2002/95/EC on the Restriction of Certain Hazardous Substances in Electrical and Electronic Equipment.* See http://eur-lex.europa.eu/legal-content/EN/TXT/PDF/?uri=CELEX:32002L0095&from=EN, accessed September 30, 2014.

European Parliament and Council of the European Union. *Council Regulation (EC) 1223/2009 on Cosmetics Products Regulation,* 2009. See http://eur-lex.europa.eu/LexUriServ/LexUriServ.do?uri=OJ:L:2009:342:0059:0209:en:PDF, accessed February 11, 2012.

European Parliament and Council of the European Union, *REACH Regulation,* Annex I, Regulation (EC) No. 1907/2006. Brussels, December 18, 2006.

European Parliament and Council of the European Union. *Toy Safety Directive* (2009/48/EC), 2009. See http://eur-lex.europa.eu/LexUriServ/LexUriServ.do?uri=OJ:L:2009:170:0001:0037:en:PDF, accessed February 25, 2013.

European Parliament, Scientific Technology Options Analysis. *The Role of Nanotechnology in Substituting Hazardous Chemicals.* Brussels, 2006. See https://www.itas.kit.edu/downloads/etag_fied07a.pdf, accessed September 30, 2014.

European Trade Union Institute. See www.istas.net/risctox/en/, accessed February 6, 2013.

Faber, Daniel. *Eco-Pragmatism: Making Sensible Environmental Decisions in an Uncertain World.* Chicago: University of Chicago Press, 1999.

Faber, Daniel A. *Emissions Trading and Environmental Justice.* Center for Law, Energy and the Environment, University of California, August 2011. See www.law.berkeley.edu/files/Emissions_Trading_and_Social_Justice.pdf, accessed October 22, 2013.

Fava, James, Frank Consoli, Richard Denison, Kenneth Dickson, Tim Mohin, and Bruce Vigon. *Conceptual Framework for Life Cycle Assessment: A Code of Practice.* Pensacola, FL: Society of Environmental Toxicology and Chemistry, 1993.

Felcher, E. Marla. *The Consumer Product Safety Commission and Nanotechnology.* Washington, DC: Woodrow Wilson International Center for Scholars, August

2008. See http://www.nanotechproject.org/publications/archive/pen14/, accessed March 22, 2013.

Finlay, Mark. "Old Efforts at New Uses: A Brief History of Chemurgy and the American Search for Biobased Materials." *Journal of Industrial Ecology*, 7(3–4), 2004, pp. 33–46.

Fise, M. R. "Consumer Product Safety Regulation." In K. J. Meier, E. T. Garman, and L. R. Keiser, eds., *Regulation and Consumer Protection: Politics, Bureaucracy and Economics*. Houston: DAME Publications, 1998.

Foxall, Gordon R. *Understanding Consumer Choice*. New York: Palgrave Macmillan, 2005.

Friedman, Milton. "The Social Responsibility of Business Is to Increase Its Profits." *New York Times Magazine*, September 13, 1970.

Frienkel, Susan. *Plastics: A Toxic Love Story*. New York: Houghton-Mifflin Harcourt, 2011.

Geiser, Kenneth. *Materials Matter: Towards a Sustainable Materials Policy*. Cambridge, MA: MIT Press, 2001.

Goldman, Lynn. "Preventing Pollution? US Toxic Chemicals and Pesticides Policies and Sustainable Development." *Environmental Law Reporter*, 32, 2002, pp. 11018–11041.

Goldman, Lynn, and S. Koduru. "Chemicals in the Environment and Developmental Toxicity in Children: A Public Health and Policy Perspective." *Environmental Health Perspectives*, 108 (supp. 3), 2000, pp. 443–448.

Goleman, Daniel. *Ecological Intelligence: How Knowing the Hidden Impacts of What We Buy can Change Everything*. New York: Broadway Books, 2009.

Goodale, Wing, and Chris DeSorbo. *Chemical Screening of Osprey Eggs in Casco Bay, Maine: 2009 Field Season*, Report 2010–09. Gorham, ME: Biodiversity Research Institute, 2010. See http://www.cascobayestuary.org/wp-content/uploads/2014/08/2010_goodale_osprey_2009_season_report.pdf, accessed October 1, 2014.

GoodGuide. See www.goodguide.com/, accessed April 5, 2012.

Gottberg, Annika, Joe Morris, Simon Pollard, Ceceilia Mark-Herbert, and Matthew Cook. "Producer Responsibility, Waste Minimization and WEEE Directive: Case Studies in Eco-Design from the European Lighting Sector." *Science of the Total Environment*, 359(1–13), April 15, 2006, pp. 38–56.

Gottlieb, Robert. *Forcing the Spring: The Transformation of the American Environmental Movement*. Washington, DC: Island Press, 1993.

Gottlieb, Robert. *Reducing Toxics: A New Approach to Policy and Industrial Decision Making*. Washington, Island Press, 1995.

Graedel, Thomas. *Streamlined Life Cycle Assessment*. Englewood Cliffs, NJ: Prentice-Hall, 1998.

Grandjean, P., and P. J. Landigren. "Developmental Neurotoxicity of Industrial Chemicals—A Silent Pandemic." *Lancet*, 368:9353, December 2006, pp. 2167–2178.

GreenBlue. *CleanGredients*. See www.cleangredients.org, accessed February 26, 2010.

Green Chemistry and Commerce Council. *Meeting Customer Needs for Chemical Data: A Guidance Document for Suppliers*, February 2011. See www.greenchemistryandcommerce.org/downloads/GC3_guidance_final_031011 .pdf, accessed March 15, 2013.

Green Chemistry and Commerce Council and National Pollution Prevention Roundtable. *Growing the Green Economy through Green Chemistry and Design for the Environment*. Lowell, MA: Lowell Center for Sustainable Production, University of Massachusetts Lowell, 2009. See http://www.p2.org/wp-content/uploads/growing-the-green-economy.pdf, accessed December 11, 2012.

Green Seal. See http://www.greenseal.org/, accessed October 2, 2014.

Greenwood, Mark A. "TSCA Reform: Building a Program That Can Work." *Environmental Law Review*, 39, 2009, pp.10034–10041.

Griener, Tim, Mark Rossi, Beverley Thorpe, and Bob Kerr. *Healthy Business Strategies for Transitioning the Toxic Chemical Economy*. Clean Production Action, 2006, pp. 25–29. See http://dgcommunications.com/documents/dg-CPA_Bus_Report.pdf, accessed April 4, 2011.

Grossman, Elizabeth. *Chasing Molecules: Poisonous Products, Human Health and the Promise of Green Chemistry*. Washington, DC: Island Press, 2009.

Gussman, Sam, Konrad von Moltke, Francis Irwin, and Cynthia Whitehead. *Public Policy for Chemicals: National and International Issues*. Washington, DC: The Conservation Foundation, 1980.

Guth, Joseph H., Richard A. Dennison, and Jennifer Sass. "Require Comprehensive Safety Data for All Chemicals." *New Solutions*, 17(3), 2007, pp. 233–258.

Haber, Ludwig F. *The Chemical Industry in the Nineteenth Century*. New York: Oxford University Press, 1958.

Hancock, Kenneth, and Margaret Cavanaugh. "Environmentally Benign Chemical Synthesis and Processing for the Economy and the Environment." In P. T. Anastas and C. A. Farris, eds., *Benign by Design: Alternative Synthetic Pathways for Pollution Prevention*. Washington, DC: American Chemical Society, 1994, pp. 23–30.

Handley, Susannah. *Nylon: The Story of a Fashion Revolution*. Baltimore: Johns Hopkins University Press, 1999.

Hansen, Steffan Foss, Lars Carlsen, and Joel Tickner. "Chemicals Regulation and Precaution: Does REACH Really Incorporate the Precautionary Principle." *Environmental Science and Policy*, 10, 2007, pp. 395–404.

Hassell, Scott, Noreen Clancy, Nicholas Burger, Christopher Nelson, Rena Rudavsky, and Sarah Olmstead. *An Assessment of the U.S. Environmental Protection Agency's National Environmental Performance Track Program*. Santa Monica, CA: RAND Corporation, 2010. See www.rand.org/content/dam/rand/pubs/technical_reports/2010/RAND_TR732.pdf, accessed September 30, 2014.

Hawke, Neil. *Environmental Policy: Implementation and Enforcement*. Hempshire, UK: Ashgate, 2008.

Hawkin, Paul. *Blessed Unrest: How the Largest Movement in the World Came into Being and Why No One Saw it Coming.* New York: Viking, 2007.

Health Canada. *Categorization of Substances on the Domestic Substances List.* See http://www.hc-sc.gc.ca/ewh-semt/contaminants/existsub/categor/index-eng.php, accessed February 4, 2011.

Health Care for America Now. See http://healthcareforamericanow.org/about-us/mission-history/, accessed January 1, 2014.

Health Care without Harm. *The Global Movement for Mercury-Free Health Care.* Arlington, VA, October 15, 2007, p. 24. See https://noharm-global.org/sites/default/files/documents-files/746/Global_Mvmt_Mercury-Free.pdf, accessed October 2, 2014.

Health Care without Harm. *Green Guide for Healthcare,* Version 2.1. See www.gghc.org, accessed March 22, 2013.

Health Product Declaration Collaborative. See http://hpdcollaborative.org/, accessed October 2, 2014.

Healthy Building Network. *About Pharos.* See www.pharosproject.net, accessed February 26, 2010.

Heintz, James, and Robert Pollin. *The Economic Benefits of a Green Chemical Industry in the United States.* Political Economy Research Institute, Amherst: MA: University of Massachusetts Amherst, May, 2011.

Herremoes, Poul, et al., eds. *Late Lessons from Early Warnings: The Precautionary Principle, 1896–2000.* Copenhagen: European Environment Agency, 2001.

Herremoes, Poul, et al. eds., *Late Lessons from Early Warnings: Science, Precaution and Innovation.* Copenhagen: European Environment Agency, 2013. See www.eea.europa.eu/publications/environmental_issue_report_2001_22, and www.eea.europa.eu/publications/late-lessons-2, accessed March 7, 2013.

Hewlett Packard. *HP 2012 Global Citizenship Report,* 2012, p. 41. See http://h20195.www2.hp.com/V2/GetPDF.aspx/c03742928.pdf, accessed January 30, 2014.

Hilts, P. *Protecting America's Health: The FDA, Business and One Hundred Years of Regulation.* New York: Knopf, 2003.

Hogue, Cheryl. "Pushing for a Phthalate." *Chemical and Engineering News,* 89(9), February 28, 2011, pp. 10.

Holder, H. A., P. H. Mazurkiewicz, C. D. Robertson, and C. A. Wray. "Hewlett Packard's Use of the GreenScreen™ for Safer Chemicals." *Issues in Environmental Science and Technology: Chemical Alternatives Assessment,* Royal Society of Chemistry, 2013, pp. 157–175.

Home Depot. *Wood Purchasing Policy.* See https://corporate.homedepot.com/corporateresponsibility/environment/woodpurchasing/Pages/default.aspx/, accessed October 2, 2014.

Horvath, Islvan T., and Paul T. Anastas. "Innovations in Green Chemistry." *Chemical Reviews,* 107(6), 2007, pp. 2169–2173.

Howard, Philip H., and Derek C. G. Muir. "Identifying New Persistent and Bioaccumulatve Organics among Chemicals in Commerce." *Environmental Science and Technology*, 44(7), 2010, pp. 2277–2285.

ICIS. *Top 100 Chemical Companies*. New York, 2013. See http://img.en25.com/Web/ICIS/FC0211_CHEM_201309.pdf, accessed January 10, 2014.

Industrial Union Department, AFL-CIO v American Petroleum Institute, 448 US 607, 1980.

Innes, Robert, and Abdoul Sam. "Voluntary Pollution Reductions and the Enforcement of Environmental Law: An Empirical Study of the 33/50 Program." *Journal of Law and Economics*, 51(20), 2004, pp. 271–296.

Innovest. *Overview of the Chemicals Industry*. Unpublished report, March 2007, p. 28. See http://www.precaution.org/lib/07/innovest.pdf, accessed December 12, 2011.

Institute of Medicine, Board of Population Health and Public Health Practice. *The Future of Drug Safety: Promoting and Protecting the Health of the Public*. Washington, DC: National Academy Press, 2007. See http://www.iom.edu/~/media/Files/Report%20Files/2006/The-Future-of-Drug-Safety/futureofdrugsafety_reportbrief.pdf, accessed February 11, 2013.

International Chemicals Secretariat. *Substitute it Now List*. See http://www.chemsec.org/what-we-do/sin-list, accessed October 1, 2014.

International Council of Chemical Associations. *Global Product Strategy*. See http://www.icca-chem.org/en/Home/Global-Product-Strategy/, accessed December 12, 2013.

International Institute for Sustainable Development. *The ISO 14020 Series*. See https://www.iisd.org/business/markets/eco_label_iso14020.aspx/, accessed October 2, 2014.

International Organization for Standardization. *ISO 14000-Environmental Management*. See http://www.iso.org/iso/home/standards/management-standards/iso14000.htm, accessed March 14, 2011.

Interstate Chemicals Clearinghouse. See www.newmoa.org/prevention/ic2/, accessed February 19, 2012.

Interstate Chemicals Clearinghouse, *U.S. State Chemicals Policy Database*. See www.theic2.org/chemical-policy, accessed September 30, 2014.

Interstate Chemicals Clearinghouse. *Alternatives Assessment Wiki*. See www.ic2saferalternatives.org/, accessed March 10, 2011.

Jackson, Tim. *Prosperity without Growth: Economics for a Finite Planet*. London: Earthscan, 2009.

Jakl, Thomas, and Petra Schwager, eds. *Chemical Leasing Goes Global: Selling Services Instead of Barrels*. New York: Springer Wien, 2008. See www.chemicalstrategies.org, accessed March 14, 2011.

Jasanoff, Sheila. "Beyond Epistemology: Relativism and Engagement in the Politics of Science." *Social Studies of Science*, 26(2), 1997, pp. 393–418.

Johnson, Richard, James Pankow, David Bender, Curtis Price, and John Zogorski. "MTBE: To What Extent Will Past Releases Contaminate Community Water Supplies." *Environmental Science and Technology*, 34(9), May 1, 2000, pp. 210A–217A.

Judson, Richard, Ann Richard, David J. Dix, Keith Houck, Matthew Martin, Robert Kavlock, Vicki Dellarco, Tala Henry, Todd Holderman, Philip Sayre, Shirlee Tan, Thomas Carpenter, and Edwin Smith. "The Toxicity Data Landscape for Environmental Chemicals." *Environmental Health Perspectives*, 117(5), May 2009, pp. 685–695.

Judson, R. S., Houck, K. A., Kavlock, R. J., Knudsen, T. B., Martin, M.T., et al. "*In vitro* Screening of Environmental Chemicals for Targeted Testing Prioritization: The ToxCast Project." *Environmental Health Perspectives*, 118, 2010, pp. 485–492.

Kallett, Isadore, and F. I. Schink. *100,000,000 Guinea Pigs*. New York: Grosset and Dunlop, 1933.

Karabelas, A. J., K. V. Plakas, E.S. Solomou, V. Drossou, and D. A. Sarigiannis. "Impact of European Legislation on Marketed Pesticides—A View from the Standpoint of Health Impact Assessment Studies." *Environment International*, 35, 2009, pp. 1096–1107.

Karha, Juhana, and Eric J. Topal. "The Sad Story of Vioxx, and What We Should Learn from It." *Cleveland Clinic Journal of Medicine*, 74(12), 2005, pp. 933–939.

Kates, R., W. Clark, R. Corell, J. Hall, C. Jaeger, et al. "Sustainability Science." *Science*, 292(5517), 2001, pp. 641–642.

Kerton, Francesca M. "Green Chemical Technologies." In James H. Clark and Fabian E. I. Deswarte, eds., *Introduction to Chemicals from Biomass*. New York: John Wiley, 2008, pp. 47–76.

Kester, Corinna, and Dana Ledyard. *The Sustainable Apparel Coalition: A Case Study of a Successful Industry Collaboration*. Berkeley, CA: Center for Responsible Business, University of California Berkeley, 2012. See http://responsiblebusiness.haas.berkeley.edu/CRB_SustainableApparelCaseStudy_FINAL.pdf, accessed November 6, 2013.

Kletz, Trevor. "Inherently Safer Plants—The Concept, Its Scope and Benefits." *Loss Prevention Bulletin*, 51, June 1983, pp. 1–8.

Korten, David C. *The Great Turning: From Empire to Earth Community*. San Francisco: Berrett-Koehler, 2006.

Krayer von Krauss, Martin, and Poul Harremoes. "MTBE in Petrol as a Substitute for Lead." In Poul Herremoes et al., eds., *Late Lessons from Early Warnings: The Precautionary Principle, 1896–2000*. Copenhagen:, European Environment Agency, 2001, pp. 110–125.

Kuczenski, Brandon, and Roland Geyer. *Chemical Alternatives Analysis: Methods Models and Tools*. Unpublished paper, Bren School of Environmental Science and Management, University of California, Santa Barbara, 2010. See http://

www.dtsc.ca.gov/PollutionPrevention/GreenChemistryInitiative/upload/08-
T3629-AA-Report-Final-Aug-24-2010.pdf, accessed February 22, 2011.

Kuhn, Thomas. *The Structure of Scientific Revolutions*. Chicago: University of
Chicago Press, 1962.

Lancaster, M. *Green Chemistry: An Introductory Text*. Cambridge, UK: Royal
Society of Chemistry, 2002.

Lam, C. W., J. T. James, R. McCluskey, S. Arepalli, and R. L. Hunter. "A Review of
Carbon Nanotube Toxicity and Assessment of Potential Occupational and Environmental Health Risks." *Critical Reviews in Toxicology*, 36, 2006, pp.
189–217.

Landers, D. H., S. J. Simonich, D. A. Jaffe, and L. H. Geiser. *The Fate, Transport
and Ecological Impacts of Airborne Contaminants in Western National Parks*.
Corvallis, OR: U.S. Environmental Protection Agency, Office of Research and
Development, NHEERL, Western Ecology Division, EPA/600/R-07/138. See
http://www.cfr.washington.edu/research.cesu/reports/J9088020046-WACAP-
Report-Volume-I-Main.pdf; accessed October 1, 2014.

Landrigan, Philip, Clyde B. Schechter, Jeffrey M. Lipton, Marianne C. Fahs, and
Joel Schwartz. "Environmental Pollutants and Disease in American Children: Estimates of Morbidity, Mortality, and Costs for Lead Poisoning, Asthma, Cancer
and Developmental Disabilities." *Environmental Health Perspectives*, 110(7), July
2002, pp. 721–728.

Langston, Nancy. *Toxic Bodies: Hormone Disruptors and the Legacy of DES*.
New Haven, CT: Yale University Press, 2010.

Lapkin, Alexei, and David Constable. *Green Chemistry Metrics: Measuring and
Monitoring Sustainable Processes*. New York: Wiley, 2008.

Lavoie, Emma T., Lauren G. Heine, Helen Holder, Mark S. Rossi, Robert E. Lee,
Emily A. Connor, Melanie A. Veabel, David M. Difiore, and Clive L. Davies.
"Chemical Alternatives Assessment: Enabling Substitution to Safer Chemicals."
Environmental Science and Technology, 44, 2010, pp. 9244–9249.

Leith, Sly J., and D. O. Carpenter. "Special Vulnerability of Children to Environmental Exposure." *Reviews on Environmental Health*, 27(4), 2012, pp.
151–157.

Lewis, Anthony (quoting C. P. Snow). "Dear Scoop Jackson," *New York Times*,
March 15, 1971, p. 37.

Li, C. J. "Organic Reactions in Aqueous Media with a Focus on Carbon-Carbon
Bond Formations: A Decade Update." *Chemical Reviews*, 105, 2005, pp.
3095–3165.

Living Building Challenge. See http://living-future.org/lbc, accessed October 10,
2013.

Loewentheil, Nathaniel. *Of Stasis and Movements: Climate Change in the
111ᵗʰ Congress*. New Haven, CT: Institution for Social and Policy Studies,
Yale University, 2012. See http://isps.yale.edu/research/publications/isps12-020-
0#.UsT3E1WA270/, accessed January 1, 2014.

Lofstedt, Ragmar E., and David Vogel. "The Changing Character of Regulation: A Comparison of Europe and the United States." *Risk Analysis*, 21(3), 2001, pp. 399–416.

Lovins, Amory B., et al. *Winning the Oil End Game: Innovation for Profits, Jobs and Security*. Boulder, CO: Rocky Mountain Institute, 2005.

Lowell Center for Sustainable Production and the Green Chemistry and Commerce Council. *An Analysis of Restricted Substance Lists (RSLs) and their Implication for Green Chemistry and Design for Environment*. Unpublished paper, Lowell, MA: University of Massachusetts Lowell, November 2008. See http://www.greenchemistryandcommerce.org/downloads/RSLAnalysisandList_000.pdf, accessed September 30, 2014.

Lowell Center for Sustainable Production and the Green Chemistry and Commerce Council. *Gathering Chemical Information and Advancing Safer Chemistry in Complex Supply Chains*, Lowell, MA: University of Massachusetts Lowell, 2009. See http://www.greenchemistryandcommerce.org/downloads/summaryreport_000.pdf, accessed March 15, 2011.

Lowell Center for Sustainable Production. *Decabromodiphenylether: An Investigation of Non-Halogen Substitutes to Electronic Enclosures and Textile Applications* (prepared by Pure Strategies, Inc.). Lowell, MA: University of Massachusetts Lowell, April 2005.

Lowell Center for Sustainable Production. *Options for State Chemicals Policy Reform: A Resource Guide*. Lowell, MA: University of Massachusetts Lowell, January 2008. See http://www.chemicalspolicy.org/downloads/OptionsExecutiveSummary.pdf, accessed December 11, 2011.

Lowell Center for Sustainable Production. *Phthalates and their Alternatives: Health and Environmental Concerns*. Lowell, MA: University of Massachusetts Lowell, 2011. See http://www.sustainableproduction.org/downloads/PhthalateAlternatives-January2011.pdf, accessed December 11, 2013.

Lowell Center for Sustainable Production. *State Leadership in Formulating and Reforming Chemicals Policy: Actions Taken and Lessons Learned*. Lowell, MA: University of Massachusetts Lowell, July 2009. See http://www.chemicalspolicy.org/downloads/StateLeadership.pdf, accessed December 15, 2012.

Maine Technology Institute. *Bioplastics Cluster Looks to Maine Potatoes*. See www.mainetechnology.org/results/success-stories/bioplastics-cluster-looks-to-maine-potatoes-and-forests, accessed March 12, 2011.

Malloy, T. F., P. J. Sinsheimer, A. Blake and I. Linkov, *Developing Regulatory Alternatives Analysis Methodologies for the California Green Chemistry Initiative*, Sustainable Technology and Policy Program, University of California, Los Angeles, CA, October 20, 2011. See http://www.stpp.ucla.edu/sites/default/files/Final%20AA%20Report.final%20rev.pdf, accessed October 1, 2014.

Manley, Julie B., Paul Anastas, and Berkeley W. Cue. "Frontiers in Green Chemistry: Meeting the Grand Challenges for Sustainability in R&D and Manufacturing." *Journal of Cleaner Production*, 16, 2008, pp. 743–750.

Marowitz, Gerald, and David Rosner. *Deceit and Denial: The Deadly Politics of Industrial Pollution.* Berkeley: University of California Press, 2002.

Massachusetts Department of Environmental Protection. *Toxics Use Reduction Planning and Plan Update Guidance.* Boston, MA, December 2009. See http://www.mass.gov/eea/docs/dep/toxics/laws/planguid.pdf, accessed October 1, 2014.

Massachusetts Toxics Use Reduction Act. See http://www.mass.gov/eea/agencies/massdep/toxics/tur/ accessed September 16, 2014.

Massachusetts Toxics Use Reduction Institute. *Cleaning Laboratory.* See http://www.turi.org/Our_Work/Cleaning_Laboratory, accessed February 16, 2012.

Massachusetts Toxics Use Reduction Institute. *Five Chemicals Alternatives Assessment Study.* Lowell, MA: University of Massachusetts Lowell, June 2006. See http://www.turi.org/TURI_Publications/TURI_Methods_Policy_Reports/Five_Chemicals_Alternatives_Assessment_Study._2006/, accessed October 1, 2014.

Massachusetts Toxics Use Reduction Institute. *Toxics Use Reduction Act Program Assessment.* Lowell, MA: University of Massachusetts Lowell, June 9, 2009.

Massachusetts Toxics Use Reduction Institute. *Training and Education.* See http://www.turi.org/Our_Work/Training_and_Education, accessed September 30, 2014.

Massachusetts Toxics Use Reduction Institute. *Trends in the Use and Release of Carcinogens in Massachusetts.* Lowell, MA: University of Massachusetts Lowell, June 2013.

Massey, Rachel, Janet Hutchinson, Monica Becker, and Joel Tickner. *Toxic Substances in Articles: The Need for Information.* Copenhagen: Nordic Council of Minister, 2008. See http://www.norden.org/en/publications/publikationer/2008-596/at_download/publicationfile, accessed February 20, 2010.

Matson, P. A., W. J. Parton, A. G. Power, and M. J. Smith. "Agricultural Intensification and Ecological Properties." *Science*, 277, July 25, 1997, pp. 504–509.

McKibbin, Bill. *Deep Economy: The Wealth of Communities and the Durable Future.* New York: Henry Holt, 2007.

Melosi, Martin. *Pollution and Reform in American Cities.* Austin: University of Texas Press, 1980.

Mendoza, Martha. "How Government Decided Lunch Box Lead Levels." *Washington Post*, February 18, 2007.

Mendoza, Martha. "Mercury-Emissions Treaty Is Adopted after Years of Negotiations." *New York Times*, January 19, 2013.

Metzger, Jurgen O. "Organic Reactions without Organic Solvents and Oils and Fats as Renewable Raw Materials for the Chemical Industry." *Chemosphere*, 43, 2001, pp. 83–87.

Michaels, David. *Doubt is their Product: How Industry's Assault on Science Threatens Your Health.* New York: Oxford University Press, 2008.

Moore, Thomas H. "Statement Submitted to Senate Committee on Commerce, Science and Transportation's Subcommittee on Consumer Affairs, Insurance, and Automotive Safety." Washington, DC: U.S. Congress, March 21, 2007. See www.cpsc.gov//PageFiles/121174/moore2007.pdf, accessed March 22, 2013.

Mulvey, Jeanette. "Toxic Toys Create Silver Lining for Green Toy Companies." *Business News Daily*, December 19, 2010. See http://www.businessnewsdaily .com/512-toxic-toys-create-silver-lining-for-green-companies.html, accessed January 15, 2014.

Mulvihill, Martin J., Evan S. Beach, Julie B. Zimmerman, and Paul T. Anastas. "Green Chemistry and Green Engineering: A Framework for Sustainable Technology Development." *Annual Review of Environment and Resources,* 36, 2011, pp. 271–293.

National Academy of Sciences. *Science and Decisions: Advancing Risk Assessment.* Washington, DC, 2008.

National Library of Medicine. *Toxicology Data Network.* See http://toxnet .nlm.nih.gov/, accessed March 7, 2013.

National Pollution Prevention Roundtable. *Facility Pollution Prevention Planning Requirements: An Overview of State Program Evaluations.* Washington, DC, 1997.

National Pollution Prevention Roundtable, *Growing the Green Economy through Green Chemistry and Design for the Environment.* Lowell, MA: Lowell Center for Sustainable Production Development, University of Massachusetts Lowell, 2009. See http://www.p2.org/wp-content/uploads/growing-the-green-economy.pdf, accessed December 11, 2012.

National Research Council, Committee on Human Biomonitoring for Environmental Toxicants. *Human Biomonitoring for Environmental Chemicals.* Washington, DC: National Academy Press, 2006.

National Research Council, Committee on Improving Risk Analysis Approaches Used by the U.S. EPA. *Science and Decisions: Advancing Risk Assessment.* Washington, DC: National Academy Press, 2009.

National Research Council, Committee on the Health Risks of Phthalates. *Phthalates and Cumulative Risk Assessment: The Tasks Ahead.* Washington, DC: National Academy Press, 2008.

National Research Council, Committee on Toxicity Testing and Assessment of Environmental Agents. *Toxicity Testing in the 21st Century: A Vision and a Strategy.* Washington, DC: National Academy Press, 2007.

National Research Council. *Pesticides in the Diets of Infants and Children.* Washington, DC: National Academy Press, 1993.

National Research Council. *Review of the Environmental Protection Agency's Draft IRIS Assessment of Formaldehyde.* Washington, DC: National Academy Press, 2011.

National Research Council. *Risk Assessment in the Federal Government: Managing the Process.* Washington, DC: National Academy Press, 1983.

National Research Council. *Sustainability in the Chemicals Industry: Grand Challenges and Research Needs.* Washington, DC: National Academy Press, 2005. See http://digilib.bppt.go.id/sampul/Sustainability_in_the_Chemical_Industry_Grand_Challenges_and_Research_Needs_A_Workshop_Report.pdf, accessed March 7, 2011.

NatureWorks, LLC. See www.natureworksllc.com, accessed March 12, 2011.

Navigant Research. "Green Chemical Industry to Soar to $98.5 by 2020." June 20, 2011. See www.navigantresearch.com/newsroom/green-chemical-industry-to-soar-to-98-5-billion-by-2020, accessed December 12, 2013.

Nelkin, Dorthy. *Controversy: The Politics of Technical Decisions*. London: Sage, 1979.

Northeast Waste Management Association, Interstate Mercury Education and Research Clearinghouse. See www.newmoa.org/prevention/mercury/imerc.cfm, accessed January 14, 2011.

NSF International. *Greener Chemicals and Chemical Process Information Standard*. See http://standards.nsf.org/apps/group_public/document.php?document_id=9409/, accessed March 13, 2013.

Nuchter, M. B., W. Ondruschka, W. Bonrath, and A. Gum. "Microwave-Assisted Synthesis—A Critical Technology Overview." *Green Chemistry*, 6, 2004, pp. 128–141.

Oberdorster, G., E. Oberdorster, and J. Oberdorster. "Nanotoxicology: An Emerging Discipline Evolving from Studies of Ultra Fine Particles." *Environmental Health Perspectives*, 113, 2005, pp. 823–839.

O'Brien, Mary. *Making Better Environmental Decisions: An Alternative to Risk Assessment*. Cambridge, MA: MIT Press, 2000.

O'Brien, Th. F., T. V. Bommaraju, and F. Hine. *Handbook of Chlor-Alkali Technology*. New York: Springer, 2005.

OMB Watch. *Product Safety Regulator Hobbled by Decades of Negligence*. Washington, DC, February 5, 2008. See http://dev.ombwatch.org/node/3599, accessed March 22, 2013.

Organization for Economic Cooperation and Development. *Center for PTR Data*. See www.oecd.org/env_prtr_data/, accessed February 10, 2011.

Organization for Economic Cooperation and Development. *OECD Environmental Outlook for the Chemicals Industry*. Paris, 2001, See http://www.oecd.org/env/ehs/2375538.pdf, accessed February 20, 2014.

Organization for Economic Cooperation and Development, Environment Directorate. *Current Landscape of Alternatives Assessment Practice: A Meta Review*, Paris,2013.Seehttp://www.oecd.org/officialdocuments/publicdisplaydocumentpdf/?cote=ENV/JM/MONO(2013)24&docLanguage=En/, accessed October 1, 2014.

Organization for Economic Cooperation and Development. *The Global Portal to Information on Chemical Substances*. See http://www.echemportal.org/echemportal/index?pageID=0&request_locale=en, accessed October 1, 2014.

OSPAR Convention. *The OSPAR List of Substances of Possible Concern*. See www.ospar.org/content/content.asp?menu=00950304450000_000000_000000, accessed March 7, 2013.

Outdoor Industry Association. *Chemicals Management Framework*, 2013. See http://dev.outdoorindustry.org/responsibility/chemicals/cmpilot.html, accessed November 6, 2013.

Park, Dae Young. *REACHING Asia Continued*. Unpublished paper. Gent, Belgium: Department of Public International Law, Gent University, September 16, 2009. See http://papers.ssrn.com/sol3/papers.cfm?abstract_id=1474504, accessed February 9, 2012.

Pellow, David Naguib. *Resisting Global Toxics: Transnational Movements for Environmental Justice*. Cambridge, MA: MIT Press, 2007.

"Pesticide Reregistration Performance, Measures and Goals." *Federal Register*, 74:91, May 13, 2009, pp. 22541–22547. See www.epa.gov/fedrgstr/EPA-PEST/2009/May/Day-13/p10758.pdf, accessed March 7, 2013.

Petkewich, Rachel. "Probing Hazards of Nanomaterials." *Chemical and Engineering News*, 86(2), October 20, 2008, pp. 53–56.

Pimentel, David, and Rajinder Peshin, eds. *Integrated Pest Management: Pesticide Problems, Volume 3*. New York: Springer, 2013.

Pimentel, David, and Wen Dazhong. "Technology Changes in Energy Use in U.S. Agricultural Production." In Ronals Carroll, John H. Vandermeer, and Peter M. Rosset, eds., *Aroecology*. New York: McGraw-Hill, 1990, pp. 150–155.

Plotkin, Jeffrey S. "Petrochemical Technology Development" In Peter H. Spitz, ed., *The Chemical Industry at the Milennium: Maturity, Restructuring and Globalization*. Philadelphia: Chemical Heritage Press, 2003, pp. 51–84.

Porter, Michael, and Claas van der Linde. "Toward a New Conception of the Environment-Competitiveness Relationship." *Journal of Economic Perspective*, 9(4), 1995, pp. 97–118.

Public Citizen. *Hazardous Waits: CPSC Lets Crucial Time Pass Before Warning Public about Dangerous Products*. Washington, DC, January 2008. See www.citizen.org/documents/HazardousWaits.pdf, accessed March 7, 2013.

Pruss-Ustiun, A., C. Vickers, P. Haeflinger, and R. Bertollini. "Knowns and Unknowns on the Burden of Disease Due to Chemicals: A Systematic Review." *Environmental Health*, 10, 2011. See www.ehjournal.net/content/10/1/9, accessed December 12, 2012.

Raffensperger, Carolyn, and Joel Tickner, eds. *Protecting Public Health and the Environment: Implementing the Precautionary Principle*. Washington, DC: Island Press, 1999.

Ragauskas, A. J., et al. "The Path Forward for Biofuels and Biomaterials." *Science*, 311, January 27, 2006, pp. 484–489.

Raloff, Janet. "Danger on Deck." *Science News*, 165(5), January 21, 2004, p. 74.

Rejeski, David. "The Molecular Economy." *The Environmental Forum*, 27(1), 2010, pp. 36–41.

Responsible Purchasing Network. See http://www.responsiblepurchasing.org/about/index.php, accessed December 12, 2011.

Ritter, Stephen K. "Designing Away Endocrine Disruption." *Chemical and Engineering News*, 90(51), December 17, 2012, pp. 33–34.

Ritter, Stephen K. "Teaching Green: Initiative Encourages Faster Uptake of Green Chemistry and Toxicology in the Undergraduate Curriculum." *Chemical and Engineering News*, 90(40), October 1, 2012, pp. 64–64.

Rizzuto, Pat. "Chemical Manufacturers Made Far Fewer Confidentiality Claims in 2010." *Chemical Regulation Reporter*, December 23, 2013. See http://www.bna.com/chemical-manufacturers-made-n17179880950/, accessed December 27, 2013.

Roadmap to Zero Discharge of Hazardous Chemicals. See http://www.roadmaptozero.com/, accessed November 7, 2013.

Rogers, Everett. *Diffusion of Innovations*, Fourth Edition. New York: Free Press, 1983.

Rosenberg, Beth. "The Story of the Alar Ban: Politics and Unforeseen Consequences." *New Solutions*, 6(2), 1996, pp. 34–50.

Ross, Bejamin, and Steven Amter, *The Polluters: The Making of Our Chemically Altered Environment*. New York: Oxford University Press, 2010.

Rossi, Mark, Joel Tickner, and Ken Geiser. *Alternatives Assessment Framework of the Lowell Center for Sustainable Production*. Lowell, MA: Lowell Center for Sustainable Production, University of Massachusetts Lowell, 2006. See www.sustainableproduction.org/downloads/FinalAltsAssess06_000.pdf, accessed March 7, 2013.

Ryan, Bryce, and Neil C. Gross. "The Diffusion of Hybrid Seed Corn in Two Iowa Communities." *Rural Sociology*, 8, 1943, pp 15–24.

Sachs, Noah. "Planning the Funeral at the Birth: Extended Producer Responsibility in the European Union and the United States." *Harvard Environmental Law Review*, 30, 2006, pp. 51–98.

Sachs, Noah M. "Jumping the Pond: Transnational Law and the Future of Chemical Regulation." *Vanderbilt Law Review*, 62(6), November 2009, pp. 1817–1869.

Safer Chemicals, Healthy Families. *Healthy States: Protecting Families from Toxic Chemicals While Congress Lags Behind*, November 2010. See http://saferchemicals.org/wp-content/uploads/pdf/HealthyStates.pdf?465b45/, accessed September 16, 2014.

Sarewitz, Daniel, and David Kriebel. *The Sustainable Solutions Agenda*. Consortium for Science, Policy and Outcomes, Arizona State University and the Lowell Center for Sustainable Production, University of Massachusetts Lowell, 2010, p. 7. See www.sustainableproduction.org/downloads/SSABooklet.pdf, accessed February 12, 2011.

Sass, Jennifer, and Mae Wu, *Superficial Safeguards: Most Pesticides Are Approved by Flawed EPA Process*. Washington, DC: Natural Resources Defense Council, March 2013.

Schaltegger, Stefan, and Roger Burritt. *Contemporary Environmental Accounting: Issues, Concepts and Practice*. Sheffield, UK: Greenleaf, 2000.

Schapiro, Mark. *Exposed: The Toxic Chemistry of Everyday Products and What's at Stake for American Power*. White River Junction, VT: Chelsea Green, 2007.

Shedroff, Nathan. *Design Is the Problem: The Future of Design Must Be Sustainable*. Brooklyn, NY: Rosenfeld Media, 2009.

Schenck, Rita. *Outlook-for-Type-III-Ecolabels-in-the-USA*. Vashon, WA: Institute for Environmental Research and Education, September 23, 2009. See http://lcacenter.org/pdf/Outlook-for-Type-III-Ecolabels-in-the-USA.pdf, accessed October 2, 2014.

Schettler, Ted, Gina Solomon, Maria Valenti, and Annette Huddle. *Generations at Risk: Reproductive Health and the Environment*. Cambridge, MA: MIT Press, 2000.

Schifano, Jessica N., Ken Geiser, and Joel A. Tickner. "The Importance of Implementation in Rethinking Chemicals Management Policies: The Toxics Substances Control Act." *Environmental Law Reporter*, 41, 2011, pp. 10527–10543.

Schwartz, Teresa M. "The Consumer Product Safety Act: A Flawed Product of the Consumer Decade." *George Washington Law Review*, 51(1), 1982, pp. 43–44.

Scott, Joanne. "From Brussels with Love: The Transatlantic Travels of European Law and the Chemistry of Regulatory Attraction." *American Journal of Comparative Law*, 57, pp. 897–942.

Selin, Henrik, and Stacy VanDeveer. "Raising Global Standards: Hazardous Substances and E-Waste Management in the European Union." *Environment*, 48(10), December 2006, pp. 7–18.

Sennett, Richard. *The Culture of the New Capitalism*. New Haven, CT: Yale University Press, 2006.

Service, Robert F. "A New Wave of Chemical Regulations Just Ahead?" *Science*, 235, August 7, 2009, pp. 692–693.

Shon, Donald, and Martin Rein. *Frame Reflection: Towards a Resolution of Intractable Policy Controversies*. New York: Basic Books, 1994.

Sieburth, Scott. "Isosteric Replacement of Carbon with Silicon in the Design of Safer Chemicals." In Stephen C. DeVito and Roger L. Garrett, eds., *Designing Safer Chemicals: Green Chemistry for Pollution Prevention*. Washington, DC: American Chemical Society, 1996, pp. 74–83.

Siemiatycki, J., L. Richardson, K. Straif, et al. "Listing Occupational Carcinogens." *Environmental Health Perspectives*, 112, 2004, pp 1447–59.

Sissell, Kara. "Retailers and States Take the Lead." *Chemical Week*, April 14–21, 2008, pp. 26–31.

Sjostrom, Jesper. "Green Chemistry in Perspective—Models of GC Activities and GC Policy and Knowledge Areas." *Green Chemistry*, 8, 2006, pp. 130–137.

Skocpol, Theda. *Naming the Problem: What It Will Take to Counter Extremism and Engage Americans in the Fight against Global Warming*. Unpublished paper, Harvard University. Cambridge, MA, February 14, 2013. See http://www.scholarsstrategynetwork.org/sites/default/files/skocpol_captrade_report_january_2013_0.pdf, accessed January 1, 2014.

Social Investment Forum Foundation. *Report on Socially Responsible Investing Trends in the United States*. New York, 2010. See http://www.ussif.org/files/Publications/12_Trends_Exec_Summary.pdf, accessed September 30, 2014.

Socolow, Robert, and Valerie Thomas. "The Industrial Ecology of Lead and Electric Vehicles." *Journal of Industrial Ecology*, 1(1), January 1997, pp. 13–36.

Speth, James Gustave. *The Bridge at the Edge of the World: Capitalism, The Environment, and Crossing from Crisis to Sustainability*. New Haven, CT: Yale University Press, 2008.

Spitz, Peter H. *Petrochemicals: The Rise of an Industry*. New York: Wiley, 1988.

Spitz, Peter H., ed. *The Chemical Industry at the Millennium: Maturity, Restructuring and Globalization*. Philadelphia: Chemical Heritage Press, 2003.

Spitz, Peter H. "The Chemical Industry at the Millennium." In Peter H. Spitz, ed., *The Chemical Industry at the Millennium: Maturity, Restructuring and Globalization*. Philadelphia: Chemical Heritage Press, 2003, pp. 311–341.

Spitz, Peter H. "The Chemical Industry and the Environment." In Peter H. Spitz, ed., *The Chemical Industry at the Millennium: Maturity, Restructuring and Globalization*. Philadelphia: Chemical Heritage Press, 2003, pp. 206–243.

Stankiewicz, A. J. A., and A. Moulijn. "Process Intensification: Transforming Chemical Engineering." *Chemical Engineering Progress*, 96(1), 2000, pp. 22–34.

Stevens, E. S. *Green Plastics: An Introduction to the New Science of Biodegradable Plastics*. Princeton, NJ: Princeton University Press, 2002.

Stokstad, Erik. "Putting Chemicals on a Path to Better Risk Assessment." *Science*, 325, August 7, 2009, pp. 694–695.

Stone, Alex, and Damon Delistraty. "Sources of Toxicity and Exposure Information for Identifying Chemicals of High Concern to Children." *Environmental Impact Assessment Review*, 30, 2010, pp. 380–387.

SubsPort. *Moving Towards Safer Alternatives*. See www.subsport.eu/, accessed March 28, 2011.

Sunstein, Cass R. *The Cost-Benefit State: The Future of Regulatory Protection*. Chicago: American Bar Association, 2002.

Sustainable Materials Collaborative. *Biomaterials: Essential for the Next Generation of Products*. See www.sustainablebiomaterials.org/, accessed March 12, 2011.

Swedish National Chemicals Inspectorate. *Risk Reduction of Chemicals: A Government Commission Report*. Solna, Sweden, 1991.

Swedish National Chemicals Inspectorate. *Observation List: Examples of Substances Requiring Particular Attention*, Revised Edition. Solna, Sweden, 1998. See www.chemicalspolicy.org/downloads/Swedish%20Obs%20List.pdf, accessed February 2, 2011.

Swedish National Chemicals Inspectorate. *The Substitution Principle, A Report from the Swedish Chemicals Agency*. Report NR-8/07. Solna, Sweden, 2007, p. 5. See http://www.kemi.se/Documents/Publikationer/Trycksaker/Rapporter/Report8_07_The_Substitution_Principle.pdf, accessed October 1, 2014. Swedish Parliament, Government Bill 1997/98:145, 1997.

Target. *Introducing the Target Sustainable Product Standard*. See https://corporate.target.com/discover/article/introducing-the-Target-Sustainable-Product-Standard/, accessed January 4, 2014.

Tera Choice. *The Sins of Greenwashing: Home and Family Edition*, 2010. See http://sinsofgreenwashing.org/index35c6.pdf, accessed April 8, 2011.

Thaler, Richard H., and Cass R. Sunstein. *Nudge: Improving Decisions about Health, Wealth and Happiness*, Revised Edition. New York: Penguin, 2009.

Thorgerson, John. "Promoting Green Consumer Behavior with Eco-labels." In Thomas Dietz and Paul Stern, eds., *New Tools for Environmental Protection: Education Information and Voluntary Measures*. Washington, DC: National Academy Press, 2002, pp. 83–104.

Thornton, Joe. *Pandora's Poison: Chlorine, Health and the New Environmental Strategy*. Cambridge, MA: MIT Press, 2001.

Tickner, Joel, and Ken Geiser. *New Directions in European Chemicals Policies: Drivers, Scope and Status*. Lowell, MA: Lowell Center for Sustainable Production, University of Massachusetts Lowell, September 2003.

Tickner, Joel, and Ken Geiser. "The Precautionary Principle Stimulus for Solutions- and Alternatives-based Environmental Policy." *Environmental Impact Assessment Review*, 24, 2004, pp. 801–824.

Tickner, Joel, and Yve Torre. *Presumption of Safety: Limits of Federal Policies on Toxic Substances in Consumer Products*. Lowell, MA: Lowell Center for Sustainable Production, University of Massachusetts Lowell, February 2008. See www.chemicalspolicy.org/downloads/UMassLowellConsumerProductBrief.pdf, accessed February 3, 2011.

Torre, Yve. *Best Practices in Product Chemicals Management in the Retail Industry*. Green Chemistry and Commerce Council. Lowell, MA: Lowell Center for Sustainable Production, University of Massachusetts Lowell, 2009. See http://greenchemistryandcommerce.org/downloads/uml-rptbestprac1209.pdf, accessed April 5, 2011.

Trusande, Leo, Philip J. Landrigan, and Clyde Schechter. "Public Health and Economic Consequences of Methylmercury to the Developing Brain." *Environmental Health Perspectives*, 5, May 2005, pp. 509–596.

Tsutsumi, Rie. "The Nature of Voluntary Agreements in Japan—Functions of Environment and Pollution Control Agreements." *Journal of Cleaner Production*, 9, 2001, pp. 145–153.

Turley, David B. "The Chemical Value of Biomass." In James H. Clark and Fabian E. I. Deswarte, eds., *Introduction to Chemicals from Biomass*. New York: John Wiley, 2008, pp. 21–46.

United Nations Conference on Environment and Development. *Agenda 21*. Rio de Janeiro, 1992. See http://sustainabledevelopment.un.org/content/documents/Agenda21.pdf, accessed March 10, 2013.

United Nations. *Consolidated List of Products whose Consumption and/or Sale have been Banned, Withdrawn, Severely Restricted or not Approved by Governments, Second Issue* (ST/ESA192). New York, 1987.

United Nations Economic and Social Council. *A Guide to the Globally Harmonized System of Classification and Labeling of Chemicals* (GHS). See www.unece.org/fileadmin/DAM/trans/danger/publi/ghs/ghs_rev04/English/ST-SG-AC10-30-Rev4e.pdf, accessed March 7, 2013.

United Nations Economic Commission for Europe. *Globally Harmonized System of the Classification and Labeling of Chemicals (GHS)*, Third Revised Edition. See www.unece.org/trans/danger/publi/ghs/ghs_rev03/03files_e.html, accessed March 7, 2013.

United Nations Environment Program, Chemicals Bureau. *Chemicals in Products Project.* See http://www.unep.org/chemicalsandwaste/UNEPsWork/ChemicalsinProductsproject/tabid/56141/Default.aspx, accessed December 20, 2013.

United National Environment Program, Chemicals Bureau. *Global Chemicals Outlook: Towards Sound Management of Chemicals.* Geneva, 2013. See http://www.unep.org/chemicalsandwaste/Portals/9/Mainstreaming/GCO/The%20Global%20Chemical%20Outlook_Full%20report_15Feb2013.pdf, accessed September 30, 2014.

United National Environment Program, Chemicals Bureau. *Global Chemicals Outlook, Trends and Indicators* (Rachel Massey and Molly Jacobs). Unpublished draft, Geneva, November 2011.

United Nations Environment Program. *Plan of Implementation of the World Summit on Sustainable Development.* Johannesburg, 2002, Section 23. See www.unmillenniumproject.org/documents/131302_wssd_report_reissued.pdf, accessed March 10, 2013.

United Nations Environment Program, Persistent Organic Pollutants Review Committee. *Substitution and Alternatives, Addendium: General Guidance on Consideration Related to Alternatives and Substitutes for Persistent Organic Pollutants.* Geneva, October 12–16, 2009.

United Nations Environment Program, Rotterdam Convention. *Annex III Chemicals.* See www.pic.int/TheConvention/Chemicals/AnnexIIIChemicals/tabid/1132/language/en-US/Default.aspx, accessed March 22, 2013.

United Nations Environment Program. *Stockholm Convention on Persistent Organic Pollutants, As Amended in 2009.* Stockholm, 2009. See http://www.env.go.jp/chemi/pops/treaty/treaty_en2009.pdf, accessed January 14, 2014.

United Nations Environment Program, World Summit on Sustainable Development. *Plan of Implementation of the World Summit on Sustainable Development.* Johannesburg, 2002, Section 23, p. 13. See www.un.org/esa/sustdev/documents/WSSD_POI_PD/English/WSSD_PlanImpl.pdf, accessed October 1, 2014.

United Nations, Food and Agriculture Organization. *International Code of Conduct on the Distribution and Use of Pesticides.* See ftp://ftp.fao.org/docrep/fao/009/a0220e/a0220e00.pdf, accessed March 22, 2013.

United Nations. *Rio Declaration on Environment and Development.* Rio de Janeiro, June 1992, Principle 15. See www.unep.org/Documents.Multilingual/Default.asp?documentid=78&articleid=1163, accessed February 2, 2011.

United Nations, Strategic Approach to International Chemicals Management (SAICM), *Dubai Declaration on International Chemicals Management, Overarching Policy Statement, Dubai Global Plan of Action,* 2006. See http://sustainabledevelopment.un.org/content/documents/SAICM_publication_ENG.pdf, accessed January 12, 2012.

U.S. Census Bureau, Census of Manufacture, *The 2012 Statistical Abstract*. See https://www.census.gov/compendia/statab/cats/manufactures.html, accessed October 1, 2014.

U.S. Centers for Disease Control, National Center for Environmental Health. *Fourth National Report on Human Exposure to Environmental Chemicals*. Atlanta, GA, July 2009. See http://www.cdc.gov/exposurereport/pdf/FourthReport .pdf, accessed January 4, 2014.

U.S. Congress, Office of Technology Assessment. *Managing Industrial Solid Wastes from Manufacturing, Mining, Oil and Gas Production, and Utility Coal Production*. Washington, DC: U.S. Government Printing Office, 1992, p. 9. See http://ota-cdn.fas.org/reports/9225.pdf, accessed October 1, 2014.

U.S. Congress. *Toxic Substances Control Act: Report by the House of Representatives Committee on Interstate and Foreign Commerce*. Report No. 94–1341, July 4, 1976, 94th Congress, 2nd Session.

U.S. Department of Agriculture, Economic Research Service. *Fertilizer Use and Price*. See www.ers.usda.gov/data-products/fertilizer-use-and-price.aspx#26720/, accessed October 1, 2014.

U.S. Department of Commerce, NOAA Inter-organizational Committee on Principles and Guidelines for Social Impact Assessment. *Guidelines and Principles for Social Impact Assessment*, Tech Memo MNFS-F/SPO-16, 1993. Reprinted in *Impact Assessment*, 12(2), 1994, pp. 107–152.

U.S. Department of Defense. *Green Procurement Strategy*. Washington, DC, 2008. See www.wbdg.org/pdfs/dod_gpp_082704.pdf, accessed April 17, 201.

U.S. Department of Energy. *Top Value Added Chemicals from Biomass: Volume 1: Results of Screening for Potential Candidates from Sugars and Syngas*. Washington, DC, August 2004. See http://www1.eere.energy.gov/bioenergy/pdfs/35523 .pdf, accessed January 14, 2014.

U.S. Department of Health and Human Service. *Healthy People, 2020*, 2010. See http://www.cdc.gov/nchs/healthy_people/hp2020.htm, accessed January 20, 2014.

U.S. Department of Health and Human Services, National Library of Medicine. *Household Products Database*. See http://householdproducts.nlm.nih.gov, accessed February 12, 2011.

U.S. Energy Information Administration. *Annual Energy Review, 2011*. Washington, DC. See http://www.eia.gov/totalenergy/data/annual/pdf/aer.pdf, accessed January 10, 2014.

U.S. Environmental Protection Agency, Office of Chemical Safety and Pollution Prevention. *ChemView*. See http://www.epa.gov/chemview/, accessed January 10, 2014.

U.S. Environmental Protection Agency, Office of Chemical Safety and Pollution Prevention. *Existing Chemical Action Plans*. See http://www.epa.gov/oppt/ existingchemicals/pubs/ecactionpln.html#posted/, accessed February 16, 2012.

U.S. Environmental Protection Agency, Office of Chemical Safety and Pollution Prevention. *Harmonized Test Guidelines—Master List*. See www.epa.gov/ocspp/ pdfs/OCSPP-TestGuidelines_MasterList.pdf, accessed March 7, 2013.

U.S. Environmental Protection Agency, Office of Chemical Safety and Pollution Prevention. *Pesticide Industry Sales and Usage, 2006 and 2007 Market Estimates.* Washington, DC, 2011. See www.epa.gov/opp00001/pestsales/07pestsales/market_estimates2007.pdf, accessed March 7, 2013.

U.S. Environmental Protection Agency, Design for Environment Program. *Safer Ingredient List for Use in DfE Labeled Products.* See www.epa.gov/dfe/saferingredients.htm, accessed January 8, 2013.

U.S. Environmental Protection Agency, Design for Environment Program. *Safer Product Labeling Program.* See www.epa.gov/dfe/pubs/projects/formulat/saferproductlabeling.htm, accessed January 8, 2013.

U.S. Environmental Protection Agency, Computational Toxicology Research Program. *ACToR.* See www.epa.gov/ncct/actor/, accessed January 14, 2011.

U.S. Environmental Protection Agency, Office of Inspector General. *EPA Needs to Comply with the Federal Insecticide, Fungicide, and Rodenticide Act and Improve its Oversight of Exported Never-Registered Pesticides,* Report No. 10-P-0026. Washington, DC: November 10, 2009. See http://www.epa.gov/oig/reports/2010/20091110-10-P-0026.pdf, accessed December 12, 2012.

U.S. Environmental Protection Agency, Office of Pollution Prevention and Toxics. *Design for Environment Program Alternatives Assessment Criteria for Hazard Evaluation.* Washington, DC, November 2010. See http://www.epa.gov/dfe/alternatives_assessment_criteria_for_hazard_eval.pdf, accessed March 7, 2011.

U.S. Environmental Protection Agency, Office of Pollution Prevention and Toxics. *Exposure Assessment Tools and Methods.* See www.epa.gov/oppt/exposure/, accessed March 7, 2013.

U.S. Environmental Protection Agency, Office of Pollution Prevention and Toxics. *High Production (HPV) Challenge.* See www.epa.gov/chemrtk/pubs/general/basicinfo.htm, accessed February 3, 2011.

U.S. Environmental Protection Agency, Office of Pollution Prevention and Toxics. *HPV Chemical Hazard Data Availability Study.* Washington, DC, 1998. See http://www.epa.gov/chemrtk/pubs/general/hazchem.htm, accessed February 3, 2011.

U.S. Environmental Protection Agency, Office of Pollution Prevention and Toxics. *Overview: Office of Pollution Prevention and Toxics Programs.* Washington, DC, December 24, 2003. See http://www.chemicalspolicy.org/downloads/TSCA10112-24-03.pdf, accessed March 7, 2013.

U.S. Environmental Protection Agency, Office of Pollution Prevention and Toxics. *Overview: Office of Pollution Prevention and Toxics Programs.* Washington, DC, 2007. See www.epa.gov/oppt/pubs/oppt101c2.pdf, accessed March 7, 2013.

U.S. Environmental Protection Agency, Office of Pollution Prevention and Toxics. *TSCA Work Plan Chemicals: Methods Document.* Washington, DC, February 2012. See www.epa.gov/oppt/existingchemicals/pubs/wpmethods.pdf, accessed March 7, 2013.

U.S. Environmental Protection Agency, Office of Pollution Prevention and Toxics. *2006 Inventory Update Rule Data Summary.* Washington, DC, 2008. See http://www.epa.gov/cdr/pubs/2006_data_summary.pdf, accessed March 7, 2013.

U.S. Environmental Protection Agency, Office of Prevention, Pesticides and Toxics. *Implementing the Food Quality Protection Act: Progress Report.* Washington, DC, August 1999. See http://www.epa.gov/pesticides/regulating/laws/fqpa/fqpareport.pdf, accessed March 7, 2013.

U.S. Environmental Protection Agency, Office of Reinvention. *The Common Sense Initiative: Lessons Learned.* Washington, DC, 1998.

U.S. Environmental Protection Agency, Office of Research and Development. *Total Exposure Methodology Study*, Volumes I–IV. Washington, DC, 1985. See http://exposurescience.org/pub/reports/TEAM_Study_book_1987.pdf, accessed February 11, 2011.

U.S Environmental Protection Agency, Office of Solid Waste. *The National Biennial RCRA Hazardous Waste Report: Based on 2011 Data.* Washington, DC: no date. See www.epa.gov/epawaste/inforesources/data/br11/national11.pdf, accessed March 7, 2013.

U.S. Environmental Protection Agency, Office of Solid Waste and Emergency Response. *Municipal Solid Waste Generation, Recycling, and Disposal in the United States: Facts and Figures for 2010.* Washington, DC, 2011. See www.epa.gov/epawaste/nonhaz/municipal/pubs/msw_2010_rev_factsheet.pdf, accessed February 19, 2012.

U.S. Environmental Protection Agency, Office of Wastewater Management. *Environmentally Acceptable Lubricants.* Washington, DC, November 2011.

U.S. Environmental Protection Agency, Presidential Green Chemistry Challenge. *2002 Designing Greener Chemicals Award.* See http://www2.epa.gov/green-chemistry/2002-designing-greener-chemicals-award/, accessed January 7, 2014.

U.S. Environmental Protection Agency, Presidential Green Chemistry Challenge. *2007 Greener Synthetic Pathways Award.* See http://www2.epa.gov/green-chemistry/2007-greener-synthetic-pathways-award/, accessed January 7, 2014.

U.S. Environmental Protection Agency, Sustainable Futures. *OncoLogic: A Computer System to Evaluate the Carcinogenic Potential of Chemicals*, and *Ecological Structure Activity Relationships (ECOSAR).* See www.epa.gov/oppt/sf/pubs/oncologic.htm, and www.epa.gov/oppt/newchems/tools/21ecosar.htm, accessed March 7, 2013.

U.S. Environmental Protection Agency, Toxics Release Inventory. *Industry Sector Profile: The Chemical Industry.* Washington, DC, 2012. See http://www2.epa.gov/toxics-release-inventory-tri-program/2011-tri-national-analysis-overview, accessed October 1, 2014.

U.S. Environmental Protection Agency. *Analog Identification Methodology (AIM).* See www.epa.gov/oppt/sf/tools/aim.htm, accessed March 7, 2013.

U.S. Environmental Protection Agency. *Cap and Trade Acid Rain Program Results.* See www.epa.gov/capandtrade/documents/ctresults.pdf, accessed February 16, 2012.

U.S. Environmental Protection Agency. *CERCLA Overview.* See http://www.epa.gov/superfund/policy/cercla.htm, accessed December 2, 2013.

U.S. Environmental Protection Agency. *Cleaner Technologies Substitution Assessment*, EPA 774R–95–002. Washington, DC, December 1996.

U.S. Environmental Protection Agency. *Design for the Environment*. See www.epa.gov/dfe/, accessed February 19, 2012.

U.S. Environmental Protection Agency. *Evaluation of U.S. EPA's Pesticide Product Registration Process: Opportunities for Efficiency and Innovation*. Washington, DC, March 2007. See http://www.epa.gov/evaluate/pdf/pesticides/eval-epa-pesticide-product-reregistration-process.pdf, accessed September 16, 2014.

U.S. Environmental Protection Agency. *FY 2013: Budget in Brief*. Washington, DC, 2012. See http://yosemite.epa.gov/sab%5CSABPRODUCT. NSF/2B686066C751F34A852579A4007023C2/$File/FY2013_BIB.pdf, accessed January 14, 2014.

U.S. Environmental Protection Agency. *Green Chemistry*. See www.epa.gov/greenchemistry/, accessed December 12, 2012.

U.S. Environmental Protection Agency. *Ground Water and Drinking Water*. See water.epa.gov/drink/index.cfm/, accessed January 14, 2011.

U.S. Environmental Protection Agency. *Guidelines for Exposure Assessment*. See http://cfpub.epa.gov/ncea/cfm/recordisplay.cfm?deid=15263/, accessed January 12, 2014.

U.S. Environmental Protection Agency. *Information About the U.S. Presidential Green Chemistry Challenge*. See http://www2.epa.gov/green-chemistry/information-about-presidential-green-chemistry-challenge/, accessed December 12, 2013.

U.S. Environmental Protection Agency. *PBT Profiler*. See www.pbtprofiler.net, accessed March, 7, 2013.

U.S. Environmental Protection Agency. *Presidential Green Chemistry Challenge, Award Recipients, 1996–2013*. Washington, DC, 2009. See http://www2.epa.gov/sites/production/files/documents/award_recipients_1996_2012.pdf, accessed March 12, 2012.

U.S. Environmental Protection Agency. *Presidential Green Chemistry Challenge Winners*. See www2.epa.gov/green-chemistry/presidential-green-chemistry-challenge-winners/, accessed December 12, 2013.

U.S. Environmental Protection Agency. *Principles of Green Engineering*. See http://www.epa.gov/oppt/greenengineering/pubs/whats_ge.html, accessed December 12, 2012.

U.S. Environmental Protection Agency. *Reinventing Environmental Protection, 1998 Annual Report*. Washington, DC, 1998.

U.S. Environmental Protection Agency. *Sustainable Futures*. See http://www.epa.gov/oppt/sf/, accessed February 19, 2012.

U.S. Environmental Protection Agency. *Toxicity Forecaster (ToxCast)*. See www.epa.gov/ncct/download_files/factsheets/toxcast_12-13-2010.pdf, accessed February 11, 2011.

U.S. Environmental Protection Agency. *Toxics Release Inventory Program.* See www.epa.gov/tri/, accessed February 10, 2011.

U.S. Environmental Protection Agency, *Toxicological Priority Index (ToxPi).* See http://www.epa.gov/ncct/ToxPi/, accessed October 5, 2014.

U.S. Environmental Protection Agency. *U.S. EPA and Other Organization Tools for Chemicals Assessment and Design.* See http://www.chemicalspolicy.org/downloads/EPAandOtherResources.pdf, accessed September 30, 2014.

U.S. Environmental Protection Agency. *Verified Technologies.* See http://www.epa.gov/etv/verifiedtechnologies.html, accessed December 12, 2012.

U.S. Environmental Protection Agency. "12 Principles of Green Chemistry." See www.epa.gov/sciencematters/june2011/principles.htm, accessed March 13, 2013.

U.S. Environmental Protection Agency. *2006 Inventory Update Reporting: Data Summary,* 2006. See http://www.epa.gov/oppt/cdr/pubs/2006_data_summary.pdf, accessed September 30, 2014.

U.S. Environmental Protection Agency. *2008 Sector Performance Report.* Washington, DC, 2008. See www.epa.gov/sectors/pdf/2008/2008-sector-report-508-full.pdf, accessed February 4, 2011.

U.S. Environmental Protection Agency. *2012 Chemical Data Reporting Results.* See www.epa.gov/cdr/pubs/guidance/cdr_factsheets.html, accessed November 20, 2013.

U.S. Federal Trade Commission. *Guides for the Use of Environmental Marketing Claims.* See http://www.ftc.gov/enforcement/rules/rulemaking-regulatory-reform-proceedings/guides-use-environmental-marketing-claims, accessed October 1, 2014.

U.S. Fifth Circuit Court of Appeals. *Corrosion Proof Fittings v EPA* (947 F.2d 1201), 1991.

U.S. Food and Drug Administration. *FDA's Authority Over Cosmetics.* See http://www.fda.gov/Cosmetics/GuidanceRegulation/LawsRegulations/ucm074162.htm, accessed September 16, 2014.

U.S. Green Building Council. See http://usgbc.org/leed, accessed October 10, 2013.

U.S. Government Accounting Office. *Chemical Regulation: Options Exist to Improve EPA's Ability to Assess Health Risks and Manage Its Chemical Review Program* (GAO-05–458). Washington, DC, June 2005.

U.S. Government Accounting Office. *Chemical Risk Assessment: Selected Federal Agencies' Procedures, Assumptions and Policies* (GAO-01–810). Washington, DC, 2001.

U.S. Government Accounting Office. *Consumer Product Safety Commission: Better Data Needed to Help Identify and Analyze Hazards.* Washington, DC, 1997.

U.S. Government Accounting Office. *EPA's Chemical Testing Program Has Made Little Progress* (T-RCED-90–88). Washington, DC, June 20, 1990.

U.S. Government Accounting Office. *Toxic Substances, EPA's Chemical Testing Program Has Not Resolved Safety Concerns* (T-RCED-91–136). Washington, DC, June 19, 1991.

U.S. Government Accounting Office. *Workplace Safety and Health: Multiple Challenges Lengthen OSHA's Standard Setting.* Washington, DC, April 2012. See www.gao.gov/assets/590/589825.pdf, accessed March 12, 2013.

U.S. Government Services Administration. *Environmental Products Overview.* See http://www.gsa.gov/portal/content/104543?utm_source=FAS&utm_medium=print-radio&utm_term=enviro&utm_campaign=shortcuts/, accessed October 2, 2014.

U.S. House of Representatives, Committee on Government Reform. *The Chemical Industry, the Bush Administration and the European Effort to Regulate Chemicals: A Special Interest Case Study.* Washington, DC, April 1, 2004. See http://oversight-archive.waxman.house.gov/documents/20040817125807-75305.pdf, accessed February 2, 2011.

U.S. Office of the President, Council on Environmental Quality. *Toxic Substances.* Washington, DC, 1971.

U.S. Office of the President, Executive Order 12866. *Federal Register*, 58:190, October 4, 1993. See http://www.archives.gov/federal-register/executive-orders/pdf/12866.pdf, accessed December 14, 2012.

U.S. Office of the President. *Executive Order 13514, Federal Leadership in Environmental, Energy and Economic Performance.* Washington, DC, October 5, 2009.

U.S. President's Cancer Panel. *2008–2009 Annual Report*, Washington, DC, 2010. See http://deainfo.nci.nih.gov/advisory/pcp/annualReports/pcp08-09rpt/PCP_Report_08-09_508.pdf, accessed October 1, 2014

U.S. Senate. *S. 847, Safe Chemicals Act of 2011.* See www.govtrack.us/congress/bills/112/s847/text/, accessed September 30, 2014.

Vidovic, Martina, and Neha Khanna. "Can Voluntary Pollution Prevention Programs Keep Their Promise." *Journal of Environmental Economics and Management*, 53(2), 2007, pp. 180–195.

Vogel, David. "Trading Up and Governing Across: Transnational Governance and Environmental Protection. *Journal of European Public Policy*, 4(4), December 1997, pp. 556–571.

Vogel, Sarah. *Is It Safe: BPA and the Struggle to Define the Safety of Chemicals.* Berkeley: University of California Press, 2013.

Wal-Mart. *Policy on Sustainable Chemistry in Consumables*, 2013. See http://az204679.vo.msecnd.net/media/documents/wmt-chemical-policy_130234693942816792.pdf, accessed November 16, 2013.

Wargo, John. *Our Children's Toxic Legacy: How Science and Law Fail to Protect Us from Pesticides*, Second Edition. New Haven, CT: Yale University Press, 1998.

Warner, John, Amy Cannon, and Kevin Dye. "Green Chemistry." *Environmental Impact Assessment Review*, 24, 2004, pp. 775–799.

Weir, David, and Mark Schapiro. *Circle of Poison: Pesticides and People in a Hungry World*. San Francisco, CA: Institute for Food and Development Policy, 1981.

Wexler, Phillip, Jan van der Kolk, Asish Mohapatra, and Ravi Agarwal. *Chemicals, Environment, Health: A Global Management Perspective*. New York: CRC Press, 2011.

Whorton, James. *Before Silent Spring; Pesticides and Public Health in Pre-DDT America*. Princeton, NJ: Princeton University Press, 1974.

Wilson, Michael, Daniel Chia, and Bryan Ehlers. *Green Chemistry in California: A Framework for Leadership in Chemical Policy and Innovation*. Berkeley: California Policy Research Center, University of California, Berkeley, 2006. See http://coeh.berkeley.edu/FINALgreenchemistryrpt.pdf, accessed January 14, 2014.

Wingspread Statement on the Precautionary Principle. January 25, 1998. See www.gdrc.org/u-gov/precaution-3.html, accessed February 2, 2011.

Wolfe, Katy. *Tert-Butyl Acetate: Safer Alternatives in Cleaning and Thinning Applications*. Report prepared by the Institute for Research and Technical Assistance for U.S. Environmental Protection Agency, March 2007. See http://www.irta.us/reports/TBAC%20Report.pdf.pdf, accessed October 1, 2014.

Wolfe, Raymond M. *Business R&D Performed in the United States Cost $291 Billion in 2008 and $282 Billion in 2009*. Washington, DC: National Science Foundation (NSF 12–309), March 2012. See www.nsf.gov/statistics/infbrief/nsf12309/, accessed February 12, 2012.

Woodhouse, E. J., and S. Breyman. "Green Chemistry as Social Movement?" *Science, Technology and Human Values*, 30, 2005, pp. 199–222.

Woodrow Wilson Institute. *Project on Emerging Nanotechnologies*. See www.nanotechproject.org/, accessed March 13, 2013.

World Commission on Environment and Development. *Our Common Future*. New York: Oxford University Press, 1987.

World Health Organization. *Global Burden of Disease, 2004 Update*. Geneva, 2008. See www.who.int/healthinfo/global_burden_disease/GBD_report_2004update_full.pdf, accessed March 7, 2013.

World Health Organization. *The WHO Recommended Classification of Pesticides by Hazard*. Geneva, 2009. See http://www.who.int/ipcs/publications/pesticides_hazard_2009.pdf, accessed January 4, 2014.

World Trade Organization. *International Trade Statistics, United States Profile*. See http://stat.wto.org/TariffProfile/WSDBTariffPFView.aspx?Language=E&Country=US/, accessed October, 2, 2014.

Wynn, Brian. "Knowledge in Context." *Science, Technology and Human Values*, 16(1), 1991, pp. 11–121.

Zarker, Kenneth, and Robert Kerr. "Pollution Prevention through Performance-Based Initiatives and Regulation in the United States." *Journal of Cleaner Production*, 16, 2008, pp. 673–685.

Index

Urban and Industrial Environments
Series editor: Robert Gottlieb, Henry R. Luce Professor of Urban and Environmental Policy, Occidental College

Steve Lerner, *Diamond: A Struggle for Environmental Justice in Louisiana's Chemical Corridor*

Jason Corburn, *Street Science: Community Knowledge and Environmental Health Justice*

Peggy F. Barlett, ed., *Urban Place: Reconnecting with the Natural World*

David Naguib Pellow and Robert J. Brulle, eds., *Power, Justice, and the Environment: A Critical Appraisal of the Environmental Justice Movement*

Eran Ben-Joseph, *The Code of the City: Standards and the Hidden Language of Place Making*

Nancy J. Myers and Carolyn Raffensperger, eds., *Precautionary Tools for Reshaping Environmental Policy*

Kelly Sims Gallagher, *China Shifts Gears: Automakers, Oil, Pollution, and Development*

Kerry H. Whiteside, *Precautionary Politics: Principle and Practice in Confronting Environmental Risk*

Ronald Sandler and Phaedra C. Pezzullo, eds., *Environmental Justice and Environmentalism: The Social Justice Challenge to the Environmental Movement*

Julie Sze, *Noxious New York: The Racial Politics of Urban Health and Environmental*

Justice Robert D. Bullard, ed., *Growing Smarter: Achieving Livable Communities, Environmental Justice, and Regional Equity*

Ann Rappaport and Sarah Hammond Creighton, *Degrees That Matter: Climate Change and the University*

Michael Egan, *Barry Commoner and the Science of Survival: The Remaking of American Environmentalism*

David J. Hess, *Alternative Pathways in Science and Industry: Activism, Innovation, and the Environment in an Era of Globalization*

Peter F. Cannavò, *The Working Landscape: Founding, Preservation, and the Politics of Place*

Paul Stanton Kibel, ed., *Rivertown: Rethinking Urban Rivers*

Kevin P. Gallagher and Lyuba Zarsky, *The Enclave Economy: Foreign Investment and Sustainable Development in Mexico's Silicon Valley*

David N. Pellow, *Resisting Global Toxics: Transnational Movements for Environmental Justice*

Robert Gottlieb, *Reinventing Los Angeles: Nature and Community in the Global City*

David V. Carruthers, ed., *Environmental Justice in Latin America: Problems, Promise, and Practice*

Tom Angotti, *New York for Sale: Community Planning Confronts Global Real Estate*

Paloma Pavel, ed., *Breakthrough Communities: Sustainability and Justice in the Next American Metropolis*

Kelly Sims Gallagher, *The Global Diffusion of Clean Energy Technology: Lessons from China*

Vinit Mukhija and Anastasia Loukaitou-Sideris, eds., *The Informal City: Settings, Strategies, Responses*

Roxanne Warren, *Rail and the City: Shrinking Our Carbon Footprint While Reimagining Urban Space*

Ken Geiser, *Chemicals without Harm: Policies for a Sustainable World*